Catalytic and Kinetic Waves in Polarography

Catalytic and Kinetic Waves in Polarography

Stal' G. Mairanovskii

Institute of Organic Chemistry
Academy of Sciences of the USSR, Moscow

Translated from Russian by
Bela M. Fabuss
Technical Director, Environmental Pollution Division
Lowell Technological Institute Research Foundation
Lowell, Massachusetts

Translation Editor
Petr Zuman
Heyrovský Institute of Polarography
Czechoslovak Academy of Sciences
Prague, Czechoslovakia

℗ Springer Science+Business Media, LLC 1968

Stal' Grigor'evich Mairanovskii was born in 1926. He studied at the Lomonosov Institute of Fine Chemical Technology and pursued his postgraduate studies at the Institute of Chemical Physics of the Academy of Sciences of the USSR, obtaining his doctorate in 1962. Currently Dr. Mairanovskii is the leader of the polarographic group at the Institute of Organic Chemistry of the Academy of Sciences of the USSR in Moscow. Mairanovskii developed the theory of catalytic hydrogen evolution and has published more than 110 papers dealing with the polarographic behavior of different classes of organic compounds and with the kinetics of electrode processes and chemical reactions occurring at the electrode. While at the Institute of Organic Chemistry, he also took part in the work performed at the Institute of Electrochemistry of the Academy of Sciences of the USSR under the guidance of Academician A. N. Frumkin.

The Russian text, originally published by Nauka Press in Moscow in 1966 for the Institute of Organic Chemistry of the Academy of Sciences of the USSR, was corrected by the author for this edition.

CATALYTIC AND KINETIC WAVES IN POLAROGRAPHY

KATALITICHESKIE I KINETICHESKIE VOLNY V POLYAROGRAFII

КАТАЛИТИЧЕСКИЕ И КИНЕТИЧЕСКИЕ ВОЛНЫ В ПОЛЯРОГРАФИИ
Сталь Григорьевич Майрановский

Library of Congress Catalog Card Number 67-10535

ISBN 978-1-4899-2833-7   ISBN 978-1-4899-2831-3 (eBook)
DOI 10.1007/978-1-4899-2831-3

© 1968 Springer Science+Business Media New York

Originally published by **Plenum Press** in 1968.

Preface

As our knowledge of the mechanism of electrode processes increases, it becomes more and more apparent that the kinetic currents first observed by R. Brdička and by K. Wiesner in the 1940's are very widely encountered. Very many electrode processes contain a chemical stage.* This is true primarily of electrode processes that involve organic compounds. Therefore, to understand the mechanism of electrode processes and, particularly, to correctly interpret the results of polarographic investigations, it is important to know the characteristics and relationships controlling the chemical reactions taking place at the electrode surface. Generally, these reactions are substantially different from ordinary chemical reactions taking place in the bulk of the solution, since the reactions at the electrodes are often affected by the electric field of the electrode and the adsorption of the participating compounds.

The fact that hydrogen ions usually take part in the electrochemical reduction of organic compounds makes possible the use of electrochemical methods, particularly polarography, for the study of protolytic reactions. These reactions play an important role in organic chemistry: the majority of reactions of organic compounds in solutions go through a stage in which a hydrogen ion is removed or added (see, for example, [1, 2]). Therefore, the polarographic study of protolytic reactions can supply much important information to theoretical organic chemistry.

---

* By electrode processes we mean the combination of electrochemical and chemical stages as well as the supply and removal of compounds, while the term "electrochemical reaction" will denote only the process of electron transfer.

v

The present work deals mainly with the polarographic investigation of fast protolytic reactions, i.e., with reactions of hydrogen ion addition to the anions of weak acids and to un-ionized organic bases. In the latter case, the so-called catalytic hydrogen wave is used for the determination of the constant. A separate chapter is devoted to the description and theory of catalytic hydrogen waves.

Because of its complexity, catalytic hydrogen evolution on a dropping mercury electrode represents a separate, extremely interesting area of polarography. This process consists of the protolytic reaction, electron transfer, and the bimolecular interaction of the products of the electrode reaction. Therefore, a chapter dealing with the theory of catalytic hydrogen waves follows chapters dealing with the processes taking place with antecedent protonation and processes in which the electrochemical stage is followed by a fast dimerization of the products that form.

Certain stages of the catalytic reaction cycle are substantially affected by the adsorption of the compounds at the mercury electrode, by the structure of the electric double layer, by stirring under the conditions of the polarographic maximum of the second kind, etc. Therefore, a study of catalytic polarographic hydrogen waves makes it possible to investigate relatively simple processes and phenomena that are hardly (or not at all) accessible by other methods: the rate constants for fast protolytic reactions, adsorption phenomena at extremely low coverage of the electrode surface by adsorbed particles (less than 0.5%), the rate of bimolecular reactions of certain free radicals, and certain other phenomena.

The catalytic hydrogen waves are important from a practical standpoint, too. Their application makes possible the attainment of a sensitivity of 10^{-7} M for analytical purposes using ordinary polarographic equipment. This means that the sensitivity can be increased by several orders of magnitude compared with "classical" polarography where diffusion currents are used. Particularly interesting is the application of catalytic waves in biology and medicine, where several diagnostic methods have been developed. The well-known serological polarographic test for cancer developed by Brdička [3] is an example.

The objectives of this book are: first, to show examples and methods of investigations of electrode processes occurring at the dropping electrode that are complicated by chemical reactions taking place at the electrodes and, second, to demonstrate how the kinetic parameters of fast reactions in solutions can be calculated on the basis of investigation of electrode processes. Finally, we intend to show how the reactivity of organic compounds can be determined from the value of the polarographic half-wave potential $(E_{1/2})$, taking into consideration the effect of the structure of the electric double layer, adsorption, and accompanying chemical reactions on the half-wave potential.

This work does not discuss all published material on kinetic waves; instead, its aim is to demonstrate, with selected characteristic examples, the importance of kinetic currents and their relationships. The book is based mainly on the data from investigations conducted over a period of several years at the N. D. Zelinskii Institute of Organic Chemistry of the Academy of Sciences of the USSR.

In the first chapters of the book some aspects of the theory of polarography and the kinetics of electrode processes are discussed, and a short review of the investigations dealing with kinetic currents is given. General information on the theory of the polarographic method and also of electrochemical kinetics can be found in polarography handbooks [4-6] and also in the well-known monographs of Frumkin, Bagotskii, Iofa, and Kabanov [7].

The author wishes to use this occasion to express his deep gratitude to Academician Aleksandr Naumovich Frumkin, for his interest and helpful criticism, and for his evaluation of experimental data which helped to clarify a number of phenomena of the kinetics of electrode processes that were complicated by chemical reactions and adsorption. The author further acknowledges the help for several years of Doctor of Chemical Sciences Valentina Alekseeva Klimova, who carried out many of the experiments described in this book. The author also acknowledges the assistance of B. I. Khaikin, G. A. Tedoradze, A. B. Ershler, and E. D. Belokolos for valuable suggestions and evaluations of certain results and also the help of S. I. Zhdanov, B. D. Bezuglyi, and L. G. Feoktistov in reading the manuscript. L. K. Gladkova helped in the preparation of the manuscript and proof.

Contents

Chapter I

Basic Principles of the Polarographic Method

Let us consider the electrochemical process of reduction of compound A to compound B

$$A + ne^- \underset{k_b}{\overset{k_f}{\rightleftarrows}} B, \tag{I}$$

where n is the number of electrons taking part in this process, and k_f and k_b are the heterogeneous rate constants of the forward and reverse electrochemical reaction (cm/sec).

The values of k_f and k_b are functions of the potential, and on the basis of the theories of slow discharges [7, 8] they can be expressed by

$$k_f = k_f^0 \exp\left(-\frac{\alpha nFE}{RT}\right), \tag{1}$$

$$k_b = k_b^0 \exp\left(\frac{(1-\alpha)nFE}{RT}\right), \tag{2}$$

where k_f^0 and k_b^0 are the rate constants of the forward and reverse processes at a potential E = 0; α is the transfer coefficient which expresses the fraction of the potential carrying out the cathodic process; F is the Faraday number; R is the universal gas constant; and T is the absolute temperature.

Generally, the current flowing through the electrode (i) is equal to the algebraic sum of the cathodic (i_c) and anodic (i_a) currents, corresponding to the forward and reverse directions of reaction (I). In the absence of concentration polarization these cur-

1

rents are

$$i_c = nFsc_A k_f, \tag{3}$$

$$i_a = nFsc_B k_b, \tag{4}$$

where s is the surface area of the electrode, and c_A and c_B are the concentrations of compounds A and B.

If reaction (I) is reversible at the so-called equilibrium potential, E_e, the sum of the values of cathodic and anodic currents taken with the appropriate signs is equal to zero, and it follows from Eqs. (3) and (4), considering Eqs. (1) and (2), that

$$E_e = \frac{RT}{nF} \ln \frac{k_f^0}{k_b^0} + \frac{RT}{nF} \ln \frac{c_A}{c_B}. \tag{5}$$

Equation (5) is the well-known Nernst equation. The first term of Eq. (5) is very nearly equal to the standard potential (E_0) of the oxidation−reduction system (I) and is expressed in this case by the rate constants of the reversible electrode process [9].

It can be shown that at the standard potential, E_0, the rate constants of the forward and reverse electrode processes are equal [9]: $k_f = k_b = k_s$, and the value of k_s can serve as a measure of the reversibility of the electrode process [10]. This will be discussed further.

With concentration polarization the concentration values at the surface of the electrode, $c_{A,s}$ and $c_{B,s}$, must be used in Eqs. (3) and (4). If the rate of the electrode process is sufficiently high, the concentration of one of the depolarizers decreases practically to zero at the electrode surface, and the current is determined by the rate of transport of the depolarizer to the electrode. For the case of diffusion transport of the depolarizer to the dropping mercury electrode, Ilkovič [11] gave an equation that forms the basis for quantitative polarographic analysis. The limiting diffusion current averaged over the lifetime of the drop, at 25°C, according to Ilkovič's equation is

$$i_D = 607 m^{2/3} t^{1/6} n D^{1/2} c_0, \tag{6}$$

where i_D is the diffusion current (μA); m is the flow velocity of mercury from the drop electrode (mg/sec); t is the drop period (sec); D is the diffusion coefficient (cm^2/sec); and c_0 is the concentration of depolarizer in the bulk of the solution (mM).

For many calculations Ilkovič's equation can be given more conveniently in the form

$$i_{\text{D}} = 1.235c_0 D^{1/2} snF t^{-1/2},\tag{6a}$$

where the average current i_{D} during the drop time is expressed in amperes, c_0 is given in moles/cm^3, and the average surface area of the dropping electrode s is in cm^2.

In developing the Ilkovič equation several effects which influence the value of the diffusion current were neglected. Among them are the screening off of the dropping electrode by the capillary tip [12, 13] or the glass plate [13] (with regulated drop formation), the sphericity of the diffusion field [14-16], the concentration change due to electrolysis on previous drops [17-19] (transfer of concentrating polarization), and the stirring of the solution due to the tangential motion of the mercury surface (maximum of the second kind) [20, 21]. Taking into consideration some of these factors, a number of so-called "corrected" Ilkovič equations have been developed [12-19, 22]. However, the analysis of a great number of studies dealing with the verification of the Ilkovič equation [12-28] leads to the conclusion that if proper conditions for obtaining polarographic curves are maintained, these effects compensate each other to such a degree that in many instances the most exact evaluation can be obtained by the original "uncorrected" Ilkovič equation [6].

The polarographic wave corresponding to a reversible electrochemical process can be described by the Heyrovský–Ilkovič equation [29]:

$$E = E_{1/2} - \frac{RT}{nF} \ln \frac{i}{i_{\text{D}} - i},\tag{7}$$

where E is the potential in a given point of the wave at a current intensity i, and $E_{1/2}$ is the potential at i = $i_{\text{D}/2}$.

The value of the half-wave potential is related to the normal potential E_0 of the oxidation–reduction system by

$$E_{1/2} = E_0 - \frac{RT}{nF} \ln \left(\frac{f_{\text{B}}}{f_{\text{A}}} \right) \left(\frac{D_{\text{A}}}{D_{\text{B}}} \right)^{1/2},\tag{8}$$

where f_{A} and f_{B} are the activity coefficients of compounds A and B and D_{A} and D_{B} are their diffusion coefficients.

The half-wave potential of reversible waves is independent of the depolarizer concentration and of the characteristics of the dropping electrode.

The degree of reversibility (or irreversibility) of the electrode reaction depends on the ratio of the velocities of two processes: (1) the material transfer process to the electrode and (2) the electron transfer process. In polarography, irreversible electrochemical processes are those in which the material transport to the electrode proceeds at a rate comparable to or greater than that of the electrochemical stage itself [30]. Under the usual conditions applied in polarography, when the drop period is $t \simeq 3$ sec, practically reversible waves can be obtained if $k_S > 2 \cdot 10^{-2}$ cm per sec. Completely irreversible waves can be characterized by a value of $k_S < 3 \cdot 10^{-5}$ cm/sec [10]; the overpotential in this case exceeds 100 mV.* When a rotating dropping mercury electrode is used [31-34] the material transport is substantially higher compared with the normal dropping electrode and the irreversibility of the wave of a given system is much higher [30, 31]. If electrodes with regulated drop time are used the drop period can be about 10 times shorter than with ordinary electrodes. Consequently, the material transport is smaller, and the k_S values characterizing the limit of reversibility and irreversibility of polarographic waves are about 3 times higher than the previously given values for ordinary dropping electrodes [30].

The shape of the polarographic wave corresponding to an irreversible electrochemical reaction, where the depolarizer is introduced and the products formed are removed by diffusion only, can be expressed for the dropping electrode by the following equation based on the theory of slow discharge [8]:

$$E = E_{1/2} - \frac{RT}{\alpha' n_a F} \ln \frac{i}{i_D - i}, \tag{9}$$

where n_a is the number of electrons transferred in the potential-determining stage.

---

*In electrochemistry overpotential is the potential difference which has to be employed at the electrodes to increase the current density at the electrode to a value that would be observed if the same electrochemical reaction took place reversibly on the same electrode.

In Eq. (9) the constant α' is slightly different from the true value of α, the transfer coefficient, namely, $\alpha'/\alpha \approx 1.03$ [35-37]. For simplicity in the remainder of this discussion, we will not differentiate between α and α'. The different forms of irreversible waves and the deviations in α values obtained from the slopes of waves on dropping and stationary electrodes [in the latter case the true α value can be determined from the slope of the waves, according to Eq. (9)] are caused, as pointed out by Frumkin [38], by the changes in the concentration of the irreversibly discharging depolarizer at the surface of the electrode during the life of the drop while the potential is kept constant on the growing surface of the dropping electrode.

The half-wave potential of irreversible waves, in contrast to that of reversible waves, is determined not only by the thermodynamic value E_0 but also by the rate constant of the electrode process and the drop time [7]*

$$E_{1/2} = \frac{RT}{\alpha n_a F} \ln \frac{0.886 k_f^0}{D^{1/2}} + \frac{RT}{2\alpha n_a F} \ln t. \tag{10}$$

The dependence of the half-wave potential on the drop time given in Eq. (10) was checked on irreversible waves of inorganic [10, 39] and organic [40] compounds.

The range of k_S values from about $3 \cdot 10^{-5}$ to about $2 \cdot 10^{-2}$ cm/sec corresponds to the transition from irreversible to reversible processes. Calculations show [10] that if the k_S value increases, the slope of the wave increases from the value of $2.3 RT/\alpha n_a F$ determined by Eq. (9) for completely irreversible waves to a value of $2.3 RT/nF$, as given by Eq. (7) for reversible waves ($\alpha < 1$, $n_a \leq n$). Since the transition from irreversible to reversible waves corresponds to a relatively narrow range of k_S values, most processes on the dropping electrode can be considered as either completely reversible or completely irreversible.

If the electrode process is complicated by adsorption phenomena or if simultaneously with the diffusion-controlled transport of participating materials (or products) material is transferred by

---

* The value 0.886 for Eq. (10) corresponds to the rigorous solution of the depolarization scheme; the approximate method of solution, which is often used in this book for derivation of equations, gives the value 0.81 for this coefficient [7].

a chemical reaction, then the form of the wave must be described
by a more complex expression. For example [41], the shape of
the reversible anodic waves for depolarization by hydroxyl, rhod-
anide, cyanide, and certain other ions corresponding to the formation of salts of bivalent mercury (formed by anodic oxidation of
metallic mercury of the dropping electrode) can be described by
Eq. (11):

$$E = \text{const} + \frac{RT}{2F} \ln \frac{i}{(i_D - i)^2}. \tag{11}$$

Chapter II

Currents Limited by the Rate
of Chemical Reaction

An important stage in the development of polarography was the discovery of kinetic waves by Brdička and Wiesner [42, 43]; in kinetic waves the transport of electrochemically active materials to and from the electrode is determined not only by diffusion but also by the chemical reaction taking place at the electrode surface. With kinetic currents the chemical reactions can take place before the electron transfer, can follow the electron transfer, or may take place parallel with the electrochemical stage. Accordingly, the kinetic waves can be classified into waves corresponding to antecedent, successive, or parallel chemical reactions [44].

In the following sections of this chapter we will discuss the most frequently encountered variety of kinetic currents, namely those with antecedent chemical reactions. Several examples of the other two cases will be discussed in later chapters of the book.

A. THE MEANING OF KINETIC CURRENTS

Kinetic currents (this expression also includes catalytic currents) are in principle different from currents determined by diffusion. To clarify the meaning of kinetic currents and to derive their characteristic properties, let us consider an electrode process preceded by chemical reaction of first kinetic order*

$$C \underset{\rho\sigma}{\overset{\rho}{\rightleftarrows}} A \xrightarrow{el} B. \tag{II}$$

---

*The symbol "el" here and in the following discussion will be used for the electrochemical reaction.

In this case the compound entering into the electrochemical reaction A $\xrightarrow{el}$ B, such as a reduction reaction, is present in the solution in two forms, A and C, which are in equilibrium with each other. These two forms are different in their depolarizing properties (A can be reduced much easier than C). The equilibrium between these forms in the solution is given by the corresponding constant σ = [C]/[A], and the rate of attainment of equilibrium depends on the ratio of rate constants of the forward (ρ) and reverse ($\rho\sigma$) reactions.

Let us assume that the equilibrium of the C $\rightleftharpoons$ A reaction in the solution is shifted toward the electrochemically less active form C (which means that $\sigma \gg 1$). If, in this case, the approach to equilibrium of the C $\rightleftharpoons$ A reaction is very slow, then at the potential of the limiting current of reduction of the A compound in the solution the current intensity will be very small owing to lack of A particles in the solution (their presence could cause a substantial diffusion current) and also owing to the insignificant rate of formation of A from C. If the rate of conversion of C into A increases (which means a higher ρ value), the decrease in concentration of the A compound, which disturbs the C $\rightleftharpoons$ A equilibrium on the electrode due to the electrochemical reaction, will be compensated to a certain degree by formation of A from C. For this reason a wave will appear on the polarogram that will show a limiting current determined by the rate of the conversion C → A. When the current depends in such a way on the rate of reaction (II), it is called a kinetic current.

B. DISCUSSION OF KINETIC CURRENTS ON THE BASIS OF THE REACTION-LAYER CONCEPT

To develop the quantitative relationships controlling kinetic currents, the simple but approximate method of Brdička and Wiesner, which is based on the concept of reaction layer, can be used [42, 43].

If compound A is used up by the electrochemical reaction, the equilibrium C $\rightleftharpoons$ A is shifted most completely at the surface of the electrode, and the degree of shift decreases with the distance from the electrode toward the bulk of the solution. The layer in which the shift of equilibrium takes place is called the reaction layer. It is evident that the kinetic current is determined by the number of electrochemically active A particles formed within the reaction layer in unit time. If we denote the thickness of the reaction layer by μ, then the volume of the layer will be μs, and the limiting kinetic current can be given by

$$i_{\lim} = nF\mu s\rho c_s, \qquad (12)$$

where s is the average surface area of the dropping electrode during the life of the drop, and c_S is the concentration of the electrochemically inactive compound C in the reaction layer.

The concentration c_S depends on the rate of the C → A reaction. If the rate of this reaction increases, the value of c_S decreases and, in the limiting case, becomes zero. Under these conditions the kinetic current will be determined by the diffusion of compound C from the bulk of the solution; this means that it will attain the value of the diffusion current i_D for compound C. When we give the Ilkovič equation in the form

$$i_D = \varkappa c_0 \qquad (13)$$

(where $\varkappa$ is a proportionality constant and c_0 is the concentration of compound B in the bulk of the solution), the concentration c_S can be expressed approximately by

$$c_s = \frac{i_D - i_{\lim}\varkappa}{\varkappa}. \qquad (14)$$

Substituting the value of c_S into Eq. (12), we obtain

$$i_{\lim} = \frac{nsF\mu\rho/\varkappa}{1 + nsF\mu\rho/\varkappa} i_D \varkappa \qquad (15)$$

If compound C is not surface active, which means that the C ⇌ A process proceeds as a volume reaction only in the reaction layer of the solution at the electrode surface, then the effective thickness of this reaction layer can be determined, according to Wiesner [45], by the average free path length that can be traversed by an electrochemically active particle in its average lifetime. The mean lifetime of the particle is inversely proportional to the rate of the reaction leading to the disappearance of the particle or, using the above-introduced formulation, is inversely proportional to the $\rho\sigma$ value [see (II)]. The thickness of the reaction layer μ can be expressed by [46-48]

$$\mu = \sqrt{D/\rho\sigma}\,,\tag{16}$$

which differs from the equation first derived by Wiesner [45] by the absence of the coefficient $\sqrt{1/2}$. Substitution of the μ value into Eq. (15) results in an equation giving the dependence of limiting kinetic current on different factors.

Hanuš [49] took a slightly different approach to the solution of the depolarization scheme of electrode processes accompanied by chemical reactions. His method is also based on the concept of a reaction layer. Hanuš expressed the concentration gradient of the electrochemically active A form at the electrode for a process represented by scheme (II), and for simplicity he assumed that this gradient is constant within the limits of the reaction layer. The gradient is equal to the difference between the concentration at the solution side of the effective reaction layer and that on the surface of the electrode, divided by the effective thickness of the reaction layer

$$\left(\frac{\partial [A]}{\partial x}\right)_{x=0} = \frac{[A]_\mu - [A]_{x=0}}{\mu}\,.\tag{17}$$

According to Hanuš the value of this gradient determines the kinetic current

$$i = nsFD\left(\frac{\partial [A]}{\partial x}\right)_{x=0}\,.\tag{18}$$

It is not hard to prove that Hanuš's solution leads to the same results as does Brdička and Wiesner's method. Thus, according to the definition of the reaction layer, at the outer boundary of the reaction layer the concentration of the electrochemically active form $[A]_\mu$ is equal to its equilibrium value. This means

$$[A]_\mu = [C]_\mu \frac{1}{\sigma}\,.\tag{19}$$

Considering that at the limiting kinetic current $[A]_{x=0} = 0$, we obtain from Eq. (17)

$$\left(\frac{\partial [A]}{\partial x}\right)_{x=0} = \frac{[C]_\mu}{\mu\sigma}\,.\tag{20}$$

Substituting this expression into Eq. (18) and taking the μ value from Eq. (16), we obtain

$$i_{\lim} = nFsD^{1/2}[C]_\mu \rho^{1/2}/\sigma^{1/2}. \qquad (21)$$

If the thickness of the reaction layer is small, $[C]_\mu \approx [C]_s$; by substituting into Eq. (21) the value $[C]_s = c_s$ from Eq. (14) we obtain

$$i_{\lim} = \frac{nFsD^{1/2}\rho^{1/2}/\sigma^{1/2}\kappa}{1 + nFsD^{1/2}\rho^{1/2}/\sigma^{1/2}\kappa} i_D. \qquad (22)$$

This equation is identical to Eq. (15) if the μ value is replaced according to Eq. (16).

The characteristic properties of kinetic currents are expressed by Eqs. (15) and (22). When the reaction rate of the antecedent reaction is not too high and when the limiting kinetic current is substantially smaller than the diffusion current, the second term in the denominator of Eqs. (15) and (22) can be neglected compared with the value of the first term (one), and the equations reduce to Eqs. (12) and (21). Under these conditions ($i_{\lim} \ll i_D$) c_μ (or c_s) = c_0; this means that the concentration of the inactive form will be the same at the surface of the electrode as in the bulk of the solution. The kinetic current in this case is independent of the diffusion of the inactive form and can be called a completely kinetic current.

The basic characteristic of a completely kinetic current follows from Eq. (21). Its value is independent of the height of the mercury column h_{Hg} measured to the tip of the dropping electrode. Since the values of the outflow velocity m and the drop time t characterizing the dropping electrode are used in the calculation of the average surface of the dropping electrode during the life of the drop, according to

$$s = {}^3/_5 0.85 m^{2/3} t^{2/3} \qquad (23)$$

(the numerical constant corresponds to 25°C) and since m is directly proportional to h_{Hg}, their product mt and also the value of the surface s are independent of h_{Hg}.

Unlike the completely kinetic currents, the diffusion currents are proportional to $\sqrt{h_{Hg}}$, as can be easily shown from the Ilkovič equation (6).

The second characteristic property of most of the kinetic currents is their relatively high temperature coefficient. With diffusion currents, when the electrode process is limited by the rate of diffusion, which has an experimental activation energy not greater than 5 kcal/mole, the temperature coefficient is equal to 1.5-2.0%/deg. For limiting diffusion currents, limited by the rate of a chemical reaction the activation energy of which is usually larger than 10 kcal/mole, the temperature coefficient is usually higher than 2.5-3%/deg.

As the value of the second term in the denominator of Eq. (15) increases, corresponding to an increase in the influence of the rate of diffusion on the transport of the electrochemically less active form of the depolarizer to the electrode surface, the limiting current gradually loses its kinetic characteristics. The limiting current begins to depend on the height of the mercury column, and the temperature coefficient decreases. If the rate of the chemical reaction increases significantly, the second term in the denominator of Eq. (15) becomes much larger than the first term. Then, as has already been pointed out, the limiting current practically reaches the value of the diffusion current (governed by the diffusion of the inactive form) and acquires the properties of the diffusion current, and loses the characteristics of the limiting kinetic current.

C. THE RIGOROUS SOLUTION OF THE DEPOLARIZATION SCHEME; COMPARISON OF THE EXACT AND APPROXIMATE METHODS

In the calculation method based on the concept of reaction layer certain constant concentrations averaged over the lifetime of the drop are used. Within the limits of the reaction layer the change of concentration of compounds with the distance from the electrode is considered to be linear. The thickness of the effective reaction layer is also treated somewhat freely. All these assumptions determine the approximate character of the solution to a certain degree. The greatest disadvantage of this method is well understood: the approximate method can be used only when, in a time interval of the order of the drop time, the rate of chemical reaction substantially exceeds the rate of material transport by diffusion and when the deactivation of the electrochemically active form of the depolarizer A (conversion of A into C) takes place in a reaction that is fast enough to permit the determination of the thickness of the reaction layer.

The so-called rigorous method of solving the depolarization scheme for electrode processes with chemical reactions developed by Koutecký [50-63] is nearly free of these deficiencies. Koutecký's method is based on the solution of a system of partial differential equations that describe the transport of material as a result of diffusion and take into consideration the movement of the growing surface of the drop electrode. These equations also consider the formation (or disappearance) of compounds due to chemical reactions in the layer at the electrode. Koutecký's method can be applied to fast as well as to relatively slow processes. The method yields an expression describing the transport of the depolarizer to the electrode and also an expression for the current intensity as a function of time and parameters characterizing the diffusion and chemical reaction.

The method was first used for the solution of the problem with an antecedent reaction at a stationary electrode, and the result obtained was applied with certain approximations to the dropping electrode [50]. Then the rigorous solution of the depolarization scheme was given for a dropping electrode for a process with a very fast chemical reaction preceding the electron transfer [53]. Further, the solution was obtained for processes in which regeneration of the depolarizer occurs by a reversible first-order reaction [52] and also for processes with partial regeneration of the starting compound from the products of the electrochemical reaction [62]. The rigorous method was then extended to different systems with fast first-order chemical reactions and partly to the examples where the electrochemical process proper can have a smaller rate [55] and the diffusion coefficients of reacting compounds can be different [57]. Also discussed were cases of consecutive chemical reactions that are antecedent to the electron transfer [56], the depolarization scheme with regeneration of the depolarizer by disproportionation of the electrode product [63], and the process in which a by-product of the electrode reaction affects (in proportion to its concentration) the rate of the antecedent chemical reaction [59]. In addition, depolarization schemes were solved for the case of regeneration of the original depolarizer by second-order reaction from the product of the electrochemical reaction [61] and for the process of a slow electrochemical stage accompanied by very fast antecedent and consecutive reactions [60].

The rigorous method was applied to the calculations of the limiting current which is limited by a fast reversible antecedent decomposition reaction of an electrochemically inactive compound into an active and simultaneously into an inactive compound. The latter affects the equilibrium of the antecedent reaction [64]. This method provides the solution of the depolarization scheme for one of the frequently observed cases, namely, a process that is limited by the rate of formation of the electrochemically active product from two inactive compounds that are present in the solution in stoichiometric quantities [65]. For the general solution of this depolarization scheme a somewhat modified version of the approximate method was used [65].

In the solution of the depolarization schemes Koutecký modified the starting equations by introducing new, dimensionless, independent parameters. Neiman [66] used a rather similar approach in developing the equations for the irreversible polarographic waves which, as Bagotskii has shown [37], closely represent the experimental results. It must be mentioned that the solution of a depolarization scheme with antecedent chemical reaction is very similar to that of the problem of an irreversible process, i.e., a process with a slow electrode process proper [35, 48].

An equation for kinetic currents was also developed by Henke and Hans [67] by a method different from that used by Koutecký. Nevertheless, the resulting equations, as has been shown [68], can be transformed into the basic equations of Koutecký [52].

Kinetic currents were also discussed in detail by Delahay and his co-workers [69-73]. They used the Laplace transformation for the solution of the system of differential equations. His results are similar to those given earlier by Koutecký, but in several cases Delahay committed an error due partly to the improper formulation of the thickness of the reaction layer. A note by Brdička and Koutecký [74] deals with criticism of Delahay's early work and with ascertaining the priority of the Czechoslovakian scientists in the development of the theory of kinetic currents.

The similarity of results obtained by both methods is not surprising, as was shown by Smutek [75], since the method of dimensionless parameters and the Laplace transformation are interrelated. The general theory of kinetic currents and several examples of their application are summarized in the publication of Koutecký and Koryta [76] and also by Brdička, Hanuš, and Koutecký [77].

The results of the rigorous solution of the depolarization scheme were given by Koutecký and co-workers in the form of tables giving separately the instantaneous and average (over the life of the drop) polarographic currents. In these tables the function of the current given (usually the ratio of limiting kinetic current to the hypothetical diffusion current that would be observed for an infinitely fast chemical reaction) depends on a parameter which includes the unknown rate constant (see, for example, [35, 52]).

Koutecký and Čižek [78, 79] showed that in dealing with kinetic currents corresponding to slow electrochemical reactions the sphericity of the diffusion at the dropping electrode has to be considered in order to determine exactly the rate constant from polarographic data. Koutecký and Čižek calculated the corrections for the spherical diffusion. These corrections (given by the authors mentioned in tables for both instantaneous and average currents) depend on the characteristics of the dropping electrode and on the kinetic parameters of the reaction. The method of using these corrections will be shown in Section C of Chapter IV.

Spherical diffusion does not naturally play a role in processes on rotating disk electrodes [80, 81] for which Koutecký and Levich [81] gave a rigorous solution of the depolarization scheme for a kinetically governed current.

The rigorous solution of the depolarization scheme for polarographic currents governed by the rate of a chemical reaction is a very tedious and complicated process. Although the solution is available for the most frequently encountered types of kinetically limited currents, such a solution for new, more complex systems can become even more tedious and complex. In some cases it cannot be carried out using the available rigorous methods. The only possible method then is to use the approximate calculation technique, based on the concept of reaction layer, which was discussed in the previous section. To prove that the precision of the approximate method for the given scheme is sufficient, the equations developed by this method must be compared with those obtained by the rigorous solution results for the identical depolarization scheme.

Let us discuss first the most frequently encountered type of electrode process, that in which the electrochemical step follows a slow chemical reaction. Brdička and Wiesner [82], using the reaction layer concept, obtained Eq. (22) in the case of the limiting kinetic current. After inserting the following values (at 25°C): $\varkappa = 607 \ m^{2/3}t^{1/6}D^{1/2}n$ (μA/mole) [see Eqs. (6) and (13)], the Faraday value $F = 96,500$ (C/mole), and using s from Eq. (23), one obtains

$$i_{lim} = i_D \frac{0.81 \ \sqrt{\rho t/\sigma}}{1+0.81 \ \sqrt{\rho t/\sigma}}. \tag{24}$$

The rigorous solution of the same depolarization scheme, given in the form of tables, can be well represented, as Weber and Koutecký [35] have shown by the equation

$$i_{lim} = i_D \frac{0.886 \ \sqrt{\rho t/\sigma}}{1 + 0.886 \ \sqrt{\rho t/\sigma}}. \tag{25}$$

A comparison of Eqs. (24) and (25) shows that they are identical in form and differ only in the value of the numerical constant. With completely kinetic currents it can be easily shown that when the difference between the i_{lim} values calculated by Eqs. (24) and (25) is maximum [the denominator of Eqs. (24) and (25) for completely kinetic currents is close to one], the relative difference in the value i_{lim} calculated by Eq. (24) does not exceed 10%. When the rate constant of the chemical reaction is calculated from the experimental values of i_{lim}/i_D the resulting ρ obtained from Eq. (24) is higher by roughly 20% compared with the more exact values determined from the tabulated data [35] or by Eq. (25). For all depolarization schemes to which the approximate method can be applied, expressions were determined that correctly give the experimentally found relationships [49, 64, 82, 83].

The equations derived by Levin [84] are an example of successful application of the approximate method to a case in which the reaction does not affect the limiting current but only the shape and half-wave potential of the wave. These equations were derived for the case of a reversible electron transfer subsequent to a slow protonation of the starting compound in the presence of low activity proton donors in an unbuffered, aprotic medium.

In certain cases, especially in catalytic hydrogen evolution, which will be discussed in Chapter IX of this book, the equations obtained by the approximate method are identical with those obtained by the rigorous method. This occurs, for example, for fast regeneration of the depolarizer from the electrode products when a steady state is attained at the electrode surface very rapidly (compared with the drop period). This steady state corresponds to the balance of the rates of chemical and electrochemical reactions. For the case of a rotating disk electrode the equations of the kinetic current with antecedent chemical reaction developed on the basis of the reaction-layer concept [85] were found to be identical with the equations given by Koutecký and Levich [81].

On the basis of the foregoing discussion, it can be concluded that the approximate method when applied in the range of its validity (which was discussed) yields equations that correctly describe the observed phenomena. The kinetic constants determined by these equations can deviate from their true values at most by only a few tenths of a percent, and the constant values are always of the correct order of magnitude [49, 64, 83].

D.   EXAMPLES OF POLAROGRAPHIC CURRENTS LIMITED BY THE RATE OF CHEMICAL REACTIONS

1.   Catalytic Processes

Kinetically limited currents were first observed with catalytic polarographic waves. For a catalytic wave the product of the electrode reaction interacts chemically with the components of the solution, partly or completely regenerating the initial depolarizer, and the height of the catalytic wave is determined by the rate of this chemical process. The first examples of catalytic currents were the anodic catalytic waves of oxidation of certain hydroquinones [43]. Wiesner observed that if colloidal palladium and gaseous hydrogen are introduced into a solution of the reversible

oxidation–reduction system quinone–hydroquinone, the anodic
wave, corresponding to the oxidation of hydroquinone, increases
substantially. Wiesner [43] explained this observation by assuming that the quinone formed at the electrode due to electro-oxida-
tion is reduced in solution in the neighborhood of the electrode by
hydrogen on the palladium to hydroquinone, which can be again
electro-oxidized to quinone.

Thus the hydroquinone reduced on palladium by hydrogen
causes an additional (catalytic) anodic current which is superimposed on the original diffusion current of the hydroquinone. The
scheme of this process can be given by the following equations:

$$R(OH)_2 \overset{el}{\underset{}{\rightleftarrows}} RO_2 + 2e^- + 2H^+$$

$$RO_2 + H_2(Pd) \overset{k}{\rightarrow} R(OH)_2. \tag{III}$$

This cycle of reactions causes the electrochemical oxidation
of molecular hydrogen to take place

$$H_2 \rightarrow 2H^+ + 2e^-.$$

Hydroquinone remains unchanged and plays the role of a catalyst.
The name "catalytic current" originates from this formulation.

The intensity of the catalytic current is determined by the
rate constant k of the chemical reduction reaction of quinone by
hydrogen on palladium. This follows from the equation developed by
Wiesner [43] which gives the current. In this paper it was first
pointed out that the catalytic current is independent of the height
of the mercury column above the dropping electrode.

At approximately the same time Brdička and Wiesner [42]
observed an additional current in the reduction of hemoglobin or
hematin in the presence of oxygen from air (giving hydrogen peroxide at the electrode). They explained it by catalytic processes.
In 1937 Brdička and Tropp [86] observed that in the polarography
of hydrogen peroxide the addition of hemoglobin displaced the reduction wave of hydrogen peroxide to more positive potentials. At
low hematin concentrations a new wave was observed, preceding
the reduction wave of hydrogen peroxide. The wave height of this
new wave increased at higher hematin concentrations and at lower

pH of the solution. The sum of the wave height of the new wave and the remaining part of the hydrogen peroxide reduction wave remained constant and was equal to the wave height of the original diffusion current of hydrogen peroxide.

This observation was first explained [86] as activation of hydrogen peroxide by hemoglobin or hematin. Later [42, 87] it was shown that the new reduction wave corresponds to the catalytic reduction of hydrogen peroxide in the presence of iron compounds:

$$H_2O_2 + Fe^{2+} \xrightarrow{k} Fe^{3+} + OH^- + OH^{\cdot},$$

$$Fe^{2+} + OH^{\cdot} \rightarrow Fe^{3+} + OH^-,$$

$$Fe^{3+} + e^- \underset{\rightleftarrows}{\overset{el}{\rightleftarrows}} Fe^{2+}. \qquad (IV)$$

The iron, changing its valence, seems to be the transfer agent for electrons from the electrode to hydrogen peroxide. The slowest stage of the process seems to be the first reaction in the reaction series (IV), and therefore the height of the catalytic wave is governed by the rate of this reaction.

On the basis of the reaction layer theory Brdička and Wiesner derived the equation for the limiting catalytic current, which is similar to Eq. (15). They also gave an expression for the form of the wave, considering that the electrochemical stage of the process is reversible and the $[Fe^{3+}]/[Fe^{2+}]$ ratio at the electrode surface is determined by the potential of the electrode.

If the solution is made alkaline the height of the catalytic wave decreases. This phenomenon was explained by the dissociation of H_2O_2 and by the fact that the limiting current is not affected by the reaction between Fe^{2+} and the HO_2^- anion. The equation given by Brdička and Wiesner [42, 87] correctly expresses all experimental facts.

Recently, the catalytic wave caused by myoglobine in solutions of hydrogen peroxide was described [88]. The introduction of cyanide ions into the polarographed solution results in disappearance of the catalytic effect; this is most likely due to a masking of the iron in the catalyst.

The polarographic behavior of the system hydrogen peroxide—iron salts was investigated in detail by Kolthoff and Parry
[89]. They observed the high temperature coefficient of the catalytic wave $-4.6\%/\deg$ (in the 25-30°C temperature interval). Upon
addition of alcohols, acetone, acrylonitrile, and acetic acid to the
solution the limiting catalytic current was substantially reduced.
Kolthoff and Parry explained this observation by the reaction of
the OH˙ radicals formed in the catalytic cycle (IV) with added organic compounds.

Besides iron salts, molybdenum, tungsten, and vanadium
salts [90] are similar catalysts for the electrochemical reduction
of hydrogen peroxide.

Miller [91] formulated the reaction of hydrogen peroxide
with iron, which determines the catalytic wave in a somewhat different way

$$\left.\begin{aligned}2Fe^{2+} + H_2O_2 + 2H^+ \rightarrow 2Fe^{3+} + 2H_2O, \\ Fe^{3+} + e^- \underset{\leftarrow}{\overset{el}{\rightarrow}} Fe^{2+}.\end{aligned}\right\} \qquad \text{(V)}$$

He gave a rigorous solution of this depolarization scheme
for linear diffusion and then modified it for the dropping electrode
by introducing a semiempirical coefficient, taking into consideration the effect of drop growth on the diffusion and kinetic currents
[91]. Although the derivation of Miller is much more precise than
that of Delahay and Stiehl [73], the most exact solution for this depolarization scheme on a dropping electrode must be considered
the solution of Koutecký [52] that fits well the experimental data
[92] measured for this system in dilute sulfuric acid solutions.

The catalytic electrochemical reduction of hydrogen peroxide
also takes place on platinum cathodes at the potential corresponding to the potential range of the limiting current of the reduction
wave of oxygen that is present in the solution [93]. The catalytic
process is indicated by the increased height of the first oxygen
wave on platinum electrodes in the presence of hydrogen peroxide,
and is interpreted as the reaction of the $HO_2^˙$ radical (formed as an
intermediate in the reduction of oxygen) with hydrogen peroxide

$$HO_2^˙ + H_2O_2 \rightarrow HO^˙ + H_2O + O_2. \qquad \text{(VI)}$$

As a result of this reaction the oxygen is regenerated and the easily reducible OH• radical is formed. The catalytic increase of the first oxygen wave is masked by a polarographic maximum [93] on the dropping mercury electrode. A catalytic reduction wave for hydrogen peroxide can also be observed in presence of lead ions [94]. Lead can be determined in concentrations as low as 10^{-10} mole by its catalytic effect in a stationary mercury drop [94].

A catalytic wave for hydrogen peroxide also appears with osmium tetroxide [95]. The wave can be observed in both buffered and unbuffered solutions. In buffered solutions the wave possesses the form of a maximum whose height increases inversely with a decrease in the height of the mercury column above the dropping electrode. The wave can be explained by interaction of the Os(VI) formed on the electrode with hydrogen peroxide; this results in the regeneration of osmium tetroxide [95].

It was shown that the limiting current of the reduction wave of Co(III) in the presence of hydrogen peroxide is much higher than the diffusion current [96]. The increase in height of the Co(III) wave was explained by the catalytic reduction of hydrogen peroxide by the reduction products of Co(III) salts.

Like hydrogen peroxide, other oxidants can play the role of the electron donor, i.e., of the compound being reduced in catalytic processes at the mercury electrode. For example, the limiting reduction current of W(VI) → W(V) increases sharply (nearly 100-fold) if perchloric acid is added to the solution with a trace of hydrochloric acid [97]. The appearance of catalytic current in this case can be explained by the regeneration of W(VI) from the product of the electrode reaction W(V) by oxidation in the solution by perchloric acid. In the presence of dissolved oxygen the catalytic activity of W(VI) sharply decreases [98].

Holtje and Geyer [99] observed the appearance of a catalytic current component in the reduction of molybdate in perchloric acid solutions. This catalytic current was studied by Haight [100-102], who also made use of it for the determination of molybdenum in steel [100]. The catalytic wave caused by molybdate in perchlorate solutions was also studied by Sinyakova and Glinkina [103] and by Rechnitz and Laitinen [104].

In acidic medium molybdenum salts can catalyze not only the reduction of perchlorates but also that of other oxidants such as nitrates [105].

The catalytic reduction currents of chlorates, nitrates, and perchlorates in the presence of molybdenum and tungsten salts were studied in detail by Kolthoff and Hodara [106-109]. A substantial catalytic current can be observed that is due to the reaction of Mo(V) formed at the electrode surface in the reduction of Mo(VI) with the above-mentioned anions. This current can be used for the determination of Mo(VI) at concentrations as low as 10^{-5} M. Especially interesting is the reduction of chloric acid in the presence of Mo(VI). In this case two catalytic waves appear. The first corresponds to the reduction of Mo(VI) to Mo(V) and the second to the reduction of Mo(III). In chloric acid solutions both waves show the shape of peaks, and the height of the maximum of the second wave is strictly proportional to the concentration of molybdenum salt at very low values ($<10^{-4}$ M). The height of the second catalytic wave of molybdenum in solutions of chloric acid is significantly higher than that in nitric acid and much higher than that in perchloric acid [107]. The calculation of reaction rates for Mo(VI) with acids in 1 M H_2SO_4 medium gave the following values: $k_{ClO_3^-} = 1.3 \cdot 10^3$; $k_{NO_3^-} = 2.3 \cdot 10^2$, and $k_{ClO_4^-} = 30$ liter/mole $\cdot$ sec.

Addition of phosphoric acid somewhat reduces the catalytic wave in chloric acid, and significantly decreases it in a nitric acid medium [107].

Addition of W(VI) to Mo(VI) salts in nitric and chloric acid solutions [108] reduces the catalytic effect of Mo(VI). This can be explained by formation of Mo(VI) and W(VI) heteropolyacids. These heteropolyacids do not possess catalytic activity in the region of Mo(VI) reduction potentials. An addition of W(VI) to Mo(VI) solutions in chloric acid strongly increases the catalytic wave, and very small W(VI) concentrations can be determined by this increase [108].

A catalytic reduction of chlorates also takes place in the presence of Cr^{3+} ions which are reduced at the electrode to Cr^{2+} and then react with the chlorates [110]. The uranyl ion which is reduced at the electrode into U^{3+} causes a catalytic current in nitrate solutions [111].

Even oxalic acid can be an oxidant, giving a catalytic wave in the presence of uranyl ions, as was shown by Grabowski and Grabowska [112]. This wave, appearing at a potential of -1.3 V (S.C.E.) at low concentrations of oxalic acid, increases linearly with the acid concentration, has a diffusion character, and corresponds to the transfer of two electrons. In an excess of oxalic acid the wave becomes kinetic. A complex in which two molecules of oxalic acid are bound on a uranium ion takes part in the electrode reaction. As a result of the reaction the reduction products of oxalic acid are formed and the uranyl ion is regenerated.

On the basis of the linear relationship between the height of the catalytic wave and the oxalic acid concentration (at not too high concentrations) a calcium determination method was proposed [113]. The calcium ions react with oxalic acid, and the reduction of oxalic acid concentration can be determined from the decrease of the limiting catalytic current [113].

In acid solutions of hydroxylamine, a catalytic current can be observed on the Ti(IV) → Ti(III) reduction wave which is caused by the chemical oxidation of Ti(III) into Ti(IV) by hydroxylamine [114]. The height of this catalytic wave was used [114] for further confirmation of Koutecký's results [52] obtained using a dropping electrode.

An interesting type of catalytic current was observed by Bower and Kolthoff [115]. In the electrochemical reduction of silver dicyanide, $Ag(CN)_2^-$, two CN^- ions are formed. If colloidal AgBr is added to the solution to be polarographed, a catalytic current increased the reduction current of $Ag(CN)_2^-$, owing to the reduction of additional silver dicyanide that is formed: $2CN^-$ + AgBr → $Ag(CN)_2^-$ + Br^-.

Catalytic waves also can be caused by the disproportionation of electrode products if, as a result of this reaction, the original depolarizer is partly regenerated. As an example the electrochemical reduction process of U(VI) in acid medium studied by Orlemann and Kern [116] can be cited. The U(V) formed as a result of the electrode reaction at the dropping electrode, at potentials of -0.2 to -0.9 V, is unstable and disproportionates into U(IV) and U(VI) according to the following kinetic equation: $d[U(V)]/dt = -k[H^+][U(V)]^2$. The regenerated U(VI) gives an additional reduction current. The equation for the current, determined by diffusion of the depolarizer and its partial regeneration by bi-molecular reaction from the electrode product, was also given by Groden [117].

A very interesting type of catalytic current was described by Kastening [118]. The anion radicals, formed in alkaline medium in the presence of camphor by reversible, single-electron reduction of aromatic nitro compounds, can be oxidized in the solution chemically. In the presence of oxidants [copper (II) − tartarate, IO_4^-], they are oxidized to the original nitro compounds which can be again reduced on the electrode giving an additional kinetically limited current. The anion radicals also are able to take part in disproportionation reactions, which lead to formation of the corresponding hydroxylamine derivatives, and the original nitro compounds are regenerated [119]. The reduction of the latter at the dropping electrode leads to the appearance of an additional kinetic current superimposed on the height of the wave corresponding to a single electron transfer. Thus the resulting wave can even reach the level corresponding to a 4- or 6-electron transfer which corresponds to the complete reduction of the nitro compound.

Catalytic current waves were also observed in the electrolyses of molten salts (without solvent) such as chlorides of lead, cadmium, and tin on platinum electrodes [120]. Chronopotentiometric measurements showed that the current is of catalytic nature. The authors [120] assumed that the catalytic character of the process is caused by discharge of the subcompound MCl which is later regenerated in the reaction of the electrode product (metal) with the electrolyte: $M + MCl_2 \rightarrow 2MCl$.

The theoretically interesting catalytic hydrogen evolution wave observed in the polarography of tellurium (IV) compounds [121-125] takes the form of a sharp maximum on the reduction wave of TeO_3^{2-}. The mechanism of formation of this catalytic current was not explained. It is assumed that it may be caused by discharge of hydrogen ions on elemental tellurium.

Zhdanov [126] investigated the character of processes in the polarography of lanthanum chloride in an unbuffered media. In the presence of small amounts of hydrochloric acid the polarization curve increases sharply after the hydrogen ion discharge wave, indicating an autocatalytic rate increase of the electrode process. When the direction of polarization was changed, i.e., when the curve was recorded from negative to positive potentials, the new curve was observed at less negative potentials thus forming a hysteresis when compared with the old curve; this indicates the presence of autocatalysis [126]. The process involves hydrogen evolution catalyzed by La^{3+} ions [126]; the mechanism of this process is not understood.

Autocatalytic processes also occur in the reduction of nitrate, nitrite, bromate, and iodate anions, and also with hydroxylamine in an unbuffered media in the presence of polyvalent cations [127-134]. In these cases the reaction products, which increase the rate of the electrode process, seem to be the hydroxyl ions, which react with the polyvalent cations and increase their catalytic activity. The mechanism of this process is unknown.

Zhdanov [135], using a method of inserting a large ohmic resistance in series with the cell, showed that the polarization curves have sections with negative slopes. His [135, 136] relationships describe semiquantitatively the form of observed polarographic curves for autocatalytic processes; in several cases these curves show a hysteresis.

The most frequently encountered and, from a theoretical viewpoint, the most interesting are the catalytic waves of hydrogen in solutions of organic compounds. These are also the most important type of catalytic processes. They will be discussed in Chapter IX.

2. Kinetic Currents Limited by the Rate of Recombination of Acid Anions

The most widespread variety of kinetic currents with antecedent reactions are those that are limited by the rate of recombination of acid anions with hydrogen ions.

It is more difficult to reduce anions than undissociated molecules of the same acids. This is caused by the difficulty of electron transfer to negatively charged particles and by the reduced concentration of anions at the surface of the cathode due to electrostatic repulsion.

The ratio of anion and molecule concentrations in the solution depends on the pH of the medium and on the dissociation constant of the acid (K_A)

$$\frac{[A^-]}{[HA]} = \sigma = \frac{K_A}{[H^+]}. \tag{26}$$

In sufficiently dilute solutions, weak acids usually give a single reduction wave on polarographic curves. The height of the wave is governed by the diffusion, i_D, and is proportional to the analytical (overall) acid concentration in the solution

$$c_{acid} = [A^-] + [HA]. \tag{27}$$

In a certain pH range in acidic solutions the wave height is independent of the pH. If the pH of the solution is increased beyond this range, the wave height gradually begins to decrease and at the same time a new wave becomes apparent at more negative potentials. This latter wave increases with increasing pH. The sum of the wave heights of both waves usually remains constant and is independent of the pH. The first wave, which decreases at increasing pH values, corresponds to the reduction of undissociated molecules; the second corresponds to the reduction of acid anions. The sum of both waves is equal to the diffusion current of reduction of undissociated molecules and anions. The change of the limiting current of the first wave with the pH of the solution is similar to the change of the $[HA]/c_{acid}$ ratio (dissociation curve), and it could be supposed at first glance that the first wave corresponds to the diffusion current of undissociated molecules contained in the solution. But a comparison of the so-called polarographic dissociation curve (the curve showing the dependence of the ratio of the limiting current of the first wave on the total diffusion current i_{lim}/i_D on the pH) with the true dissociation curve, which gives the change of the $[HA]/c_{acid}$ ratio as a function of pH

$$[HA]/c_{acid} = \frac{[H^+]}{K_A + [H^+]}, \tag{28}$$

shows that although both curves often have a similar shape the polarographic (apparent) dissociation curve is shifted compared to the true curve by several pH units toward higher pH values [46, 137]. Furthermore, the characteristics of the limiting current of the first wave (at pH values for which $i_{lim} < i_D$) indicates its kinetic nature.

If we assume that the establishment of the protolytic equilibrium in the system

$$A^- + H^+ \overset{\rho}{\underset{\rho\sigma}{\rightleftarrows}} HA \qquad \text{(VII)}$$

is sufficiently fast, the observed phenomena can be easily explained by appearance of an additional kinetic reduction current of the molecules that are formed in the electrode layer due to recombination of anions according to Eq. (VII). In buffered solution at pH values that are higher by one or two units than the pK_A of the acid, the concentration of undissociated molecules in the solution is much smaller than the overall (analytical) concentration of the acid. Therefore the fraction of the diffusion current caused by molecules can be disregarded compared with the value of the kinetic current. Expressed in other words, at these pH values the current is almost entirely determined by the rate of anion recombination [forward direction of reaction (VII)].

The extent of the shift of the polarographic (apparent) dissociation curve in comparison with the true curve depends on the rate of anion recombination. If the recombination did not take place (which means that the rate of recombination would be zero) both curves would be identical. If the rate of recombination increases, the apparent dissociation curve shifts toward much higher pH values compared to the true curve. If we assume that the protonation of anions takes place only in the volume of solution close to the electrode surface and only under the influence of hydroxonium ions, then, on the basis of the shift between the apparent and true dissociation curve, the rate constant for the protonation of the acid anion can be determined [46, 137]. If the apparent dissociation constant is pK_A' (which means the pH at which the height of the first kinetic wave corresponds to half the total diffusion current, $i_{lim}/i_D = 0.5$), it holds that [46]

$$\log \rho = pK_A' - pK_A + 0.105 - \log t, \qquad (29)$$

and in this case $\rho = k_{H^+}[H^+]$, where k_{H^+} is the protonation rate constant of acid anions with hydroxonium ions and t is the drop period of the electrode [the numerical coefficient is calculated according to Eq. (25)].

In a buffered solution, the height of the kinetic wave, limited by the rate of interaction of anions with hydrogen ions can be correctly expressed by Eq. (25). This equation can be used if the diffusion current of undissociated molecules does not form a considerable portion of the total limiting current, which means that $\sigma \gg 1$. The general form of the equation is

$$\rho = k_H \ [H^+] + k_{H_2O} \ [H_2O] + k_{HA_1} \ [HA]_1 + k_{HA_2} \ [HA]_2 + \cdots . \qquad (30)$$

The ρ value seems to be an overall rate constant for interaction of anions of the acid investigated with all proton donors present in the solution. Therefore, the value of ρ generally can depend both on the nature of the buffer components in the solution and on their concentration. In contrast, the σ value is independent of the composition of the buffer solution and is determined only by the pH and the dissociation constant of the acid investigated as given by Eq. (26).

Brdička and Wiesner [82] first described kinetic currents limited by the rate of recombination of anions of weak acids. They also gave the first quantitative explanation of their observation. The first examples of this type of kinetic waves were the currents observed during the polarography of pyruvic acid and phenylglyoxalic acid [138]. Brdička first mentioned in this work that in calculating the rate constant for protonation of the anions one must consider not only the hydroxonium ions but all other proton donors present in the solution.

Later, Clair and Wiesner [139] investigated polarographically a series of substituted pyruvic acids. They determined the height of the kinetic wave as a function of the pH of the solution (in buffer mixtures of Britton and Robinson) and calculated the effective (overall) rate constant of protonation based on the difference between the apparent polarographic and true dissociation constant of the acids [according to an equation like (29)]. The linear relationship between the logarithm of the effective rate constant of protonation and pK_A in a series of acids of similar structure was first given in this paper [139]. The same method was applied in the study of the protonation of anions with deuterium ions [140]. Rate constants for recombination of pyridinecarboxylic acid [141, 142] and quinolinecarboxylic acid [143] derivatives were calculated from polarographic kinetic currents. Furthermore, the mechanism

of reduction and the kinetics of protonation were studied for di-
basic acids, maleic acid, citraconic acid, fumaric acid [137],
phthalic acid [144-146], and for certain of their esters [145], an-
hydrides, and monomethylamides [147].

Ryvolová and Hanuš [145] observed three waves on the polarographic curves of phthalic acid. They assigned the first wave,
which can be observed in strongly acidic solutions, to the reduction of protonated complexes of the undissociated acid with hydrogen ions: $C_6H_4(COOH)_2 + H^+ \rightarrow [C_6H_4(COOH)_2H]^+$. The second
wave was assigned to the reduction of undissociated molecules of
phthalic acid and the third wave to its monoanions. In appropriate
pH ranges all three waves show kinetic character caused by addition of hydrogen ions to the bases: $C_6H_4(COO^-)_2$ for the third
wave, $C_6H_4(COOH)COO^-$ for the second, and to the undissociated
acid for the first wave. The reduction of the protonated (cationic)
form of the phthalic acid at potentials of the first wave was recently investigated by Buck [146], who showed spectrophoto-
metrically (in H_2SO_4 solutions) that equal concentrations of
$C_6H_4(COOH)_2$ and $[C_6H_4(COOH)_2H]^+$ are attained in 24.4 N sulfuric
acid.

The presence of the second (and third) dissociation step of
an acid causes an additional reduction of the concentration of undissociated acid at increasing pH values as compared with monobasic acids. Therefore, the dissociation curves (true and polarographic) of multibasic acids are more complex than those of monobasic acids. According to Hanuš and Brdička [137], the dissociation curve for the first stage of a dibasic acid is asymmetrical,
and its form can be expressed by

$$\log \frac{i_{\lim}}{i_D - i_{\lim}} = pK_1' - 2pH - \log([H^+] + K_2), \qquad (31)$$

where K_1' is the apparent dissociation constant of the first stage
and K_2 is the true dissociation constant of the second stage.

It follows from Eq. (31) that the steepness of the dissociation
curve increases at higher K_2 values and tends toward the twofold
value as compared with the steepness of monobasic acids when
both dissociation constants are equal. If the K_2 value in Eq. (31)
is very small the equation reduces into the expression for the

symmetrical dissociation curves of monobasic acids

$$\log \frac{i_{\lim}}{i_D - i_{\lim}} = pK' - pH. \tag{32}$$

The dissociation curve of the first stage is even more curved for tribasic acids than for dibasic acids [145], and therefore the polarographic dissociation curve (the curves showing the dependence of limiting currents of the individual waves on pH) of phthalic acid show different steepness. The curve showing the dependence of $i_{\lim}$ on pH for the first wave (i.e., the reduction of the protonated complex, which can be considered as the undissociated form of a tribasic acid) is the steepest. The curve of the second wave, corresponding to the dibasic acid, shows an intermediate steepness. (Under the conditions of appearance of the second wave its character is not affected by the formation of the cationic protonated form.) Finally, the curve corresponding to the third, most negative wave, caused by the process $C_6H_4(COO^-)_2 + H^+ \rightarrow C_6H_4(COOH)COO^-$, shows a form characteristic of monobasic acids.

It must be mentioned that the dependence of $i_{\lim}$ on pH is only approximately given by these simple equations, since the following effects were neglected in their derivation: (1) the limiting kinetic current depends not only on the pH of the solution but also on the nature and concentration of buffer components [see, for example, Eq. (30)]; (2) the height of preceding waves also depends on the rate of the processes that determine the subsequent waves on the polarograms. For example, the wave limited by the rate of the $A^- + H^+$ reaction is affected to a certain degree by the rate of formation of A^- in the $A^{2-} + H^+$ reaction; (3) with the pH of the solution increasing the wave is usually shifted toward more negative potentials; this changes the character of the influence of the structure of the double layer and adsorption of reactants on the kinetic currents (this will be discussed in detail in Chapters V and VI).

Kůta and Krejčí [148] observed three waves corresponding to the different forms of the depolarizer at different pH values of the solution, and they also found three apparent dissociation curves in a polarographic study of trans-urocanic acid.

Two waves caused by the reduction of undissociated acid molecules and anions and the kinetic character of the first wave were observed in the polarographic study of isomeric nitrobenzoic acids [149] and also with the glyoxalic [150] and terephthalic acids [151].

Recently, Tur'yan and co-workers [152-154] reported kinetic currents in the polarography of organic acids. They investigated the behavior of nitrophthalic and nitroterephthalic acids [153] and trimelitic acid [154] in different buffer solutions and calculated the overall rates of anion protonation from the values of limiting currents, assuming that the recombination is caused mainly by hydroxonium ions. These authors assumed that the processes investigated are entirely volume processes, but owing to the rather large size of the molecules of these organic acids and the relatively positive potentials at which their reduction is observed (−1.3 V for the trimelitic acid and −0.54 and −0.76 V for the nitrophthalic acids) the waves most probably have a substantial surface component (see section B of Chapter IV). For this reason the rate constant values determined by these authors are less dependable.

In several other cases the reduction waves of the anions and molecules are not clearly separated on the polarograms. This occurs in reduction of α-bromoalkanoic [155], chloroacetic [156], and iodomethanesulfonic acids [157]. However, the change in the half-wave potential of the waves with pH in these cases indicates a separate reduction of anions and molecules.

The kinetic waves resulting from the reduction of phenylglyoxalic acid were investigated on a rotating, copper-treated, platinum electrode [158]. The observed currents at different electrode rotation speed and pH values agreed very well with the data calculated by an equation given earlier by the same author [47]. The recombination rate constant, calculated on the basis of the i_{lim} and pH relationship, obtained at the rotating electrode is practically identical with the data obtained at the dropping electrode [158].

Similarly to carboxylic acids, several organic nitrogen-containing compounds forming onium structures (Brönsted acids) also gave kinetic currents limited by the rate of proton addition. Thus Kolthoff and Liberti [159] observed two waves on the polarograms of nitrosophenylhydroxylamine (cupferron) solutions at pH 7-9. The first of these waves, corresponding to the reduction of the protonated compound, decreases with increasing pH and has a kinetic character. A decrease of wave height at increasing pH, which is caused by the decrease of the rate of protonation, can also be observed with oximes [160], hydrazines, semicarbazones, and other

derivatives with an azomethine group [161], as well as for p-dimethylaminobenzaldehyde [162]. 2-Ethyl-4-thiocarbamoyl pyridine can be protonated at two places, at the nitrogen of the pyridine
ring and on the amino group. Therefore, on the graphs of the i_{lim}
and pH relationship two sections can be distinguished (in acidic
and basic solutions), corresponding to the apparent dissociation
curves [163].

The effect of the nature and concentration of the buffer solution on the rate of antecedent protonation was investigated in
several papers. On the basis of experimental results determined
in solutions of different buffer capacity, the partial protonation
constants corresponding to the different proton donors were determined [162, 164-166].

It was shown that the partial rate constants of protonation of
a given compound by different proton donors are related to the dissociation constant of these donors by the Brönsted (see [167]) relationship.

The rate constants of protonation of weak acids and those of
the reverse reaction, deprotonation, can also be determined on the
basis of hydrogen ion discharge waves which are formed in unbuffered solutions at the electrode as a result of dissociation of the
molecules of weak acids. This possibility was first mentioned by
the author of this book and Neiman [168]. In this paper an attempt
was made to use the discharge waves of hydrogen ions, partly
limited by the rate of dissociation in solutions of weak acids, for
the determination of their dissociation rate constants. Since in
the equations used in the work cited [168] the physical meaning of
the coefficient corresponding to the thickness of the reaction layer
in the Brdička—Wiesner theory was not properly interpreted and
also since acids that were too strong were selected for the experimental verification [169], it was not possible [168] to evaluate
numerical values for the dissociation rate constants. Kůta [170]
observed kinetic hydrogen waves which were limited by the rate
of dissociation of boric acid in unbuffered solutions. He applied
[171] the equations given by Hanuš [49] including the different diffusion coefficients of hydrogen ions, anions, and undissociated
molecules and obtained the value of the rate constant of dissocia-
tion of boric acid into ions: $1.3 \cdot 10^3 \text{ sec}^{-1}$ (at 18°C in 0.02 M LiCl
solution).

Levin [172] observed an interesting type of kinetic current with antecedent protonation in the polarography of naphthalene sulfo acids. The electrode reaction of reduction of aromatic sulfo acids consists of a transfer of two electrons with cleavage of the C-S bond, formation of sulfite ion, and exchange of the SO_3H group in the aromatic ring by hydrogen [173]. If the molecule contains several sulfo groups their reduction occurs stepwise. The height of the first wave, corresponding to the reduction of the first sulfo group, is limited by the rate of diffusion with transfer of two electrons, and the subsequent waves have a kinetic character. The heights of the subsequent waves are much lower than those of the preceding waves so that for the naphthalene tetrasulfoacid the wave corresponding to the splitting of the fourth sulfo group cannot be observed at all. With increasing pH of the solution the height of subsequent waves is reduced. These observations were explained by Levin by the reduced protonation of the aromatic anion radical which is formed in the transfer of two electrons and splitting of SO_3^{2-} from the anion of the original sulfo acid. The proton transfer accordingly occurs before the electrochemical cleavage of the second sulfo group

$$R(SO_3H)\,SO_a^- + 2e^- \rightarrow R(SO_3H)^- + SO_3^{2-}, \qquad \text{(VIII)}$$

$$R(SO_3H)^- + H^+ \xrightarrow{\substack{\text{(limiting stage of} \\ \text{the second wave)}}} RH\,(SO_3H). \qquad \text{(IX)}$$

In concluding the discussion of this problem, the indirect polarographic method of investigation of the kinetics of protolytic reactions must be mentioned. In this case neither the investigated acid nor hydrogen ions take part directly in the electrochemical process; instead, a compound added to the solution is reactive. This compound abstracts a proton from the acid and then in the protonated form enters into the electrochemical reaction. Basically, such a compound, a proton transfer agent from the donor (which is the acid) to the electrode, can be any organic or inorganic compound which can exist in the solution in two forms: a protonated or acid form and an unprotonated basic form. Under the specified conditions only the protonated form can take part in the electrode reaction. The value of the limiting current under these conditions is determined by the rate of protonation of this compound introduced into the solution by interaction with the acids that are present in the solution.

In one variant of this method catalytic hydrogen waves are utilized; in this case the catalyst introduced into the solution, the proton transfer agent, is regenerated from the products of the electrode reaction. A detailed analysis of the mechanism of these processes will be given in Chapter IX.

Another variant of the method was proposed by Rüetschi [174]. Rüetschi observed two reduction waves on the polarographic curves of ill-buffered azobenzene solution. The first wave corresponded to the discharge of the protonated form; the second corresponded to the discharge of the unprotonated form. The total current of both waves was equal to the diffusion current of azobenzene which can be observed in well-buffered solutions. Rüetschi showed [174, 175] that on the basis of the height of the first wave, which is governed both by diffusion and the rate of the chemical reaction, the rate constants of the protolytic reactions in the solution can be determined.

In the calculation of the rate constant Rüetschi [175] used the dependence of the height of the kinetic wave on the drop time of the dropping electrode. To investigate the dependence of i_{lim} on t, he transformed the basic Eq. (25) of the limiting kinetic current into

$$\frac{i_{lim}}{\varkappa \sqrt{t}} = 0.886\, c_{acid}\, \sqrt{\rho/\sigma} - 0.886\, \frac{i_{lim}}{\varkappa} \sqrt{\frac{\rho}{\sigma}}, \qquad (33)$$

which can be further given in the following form, by changing the height of the mercury column and using the same capillary ($\varkappa = km^{2/3}t^{1/6} = k't^{-1/2}$)

$$i_{lim} \sqrt{t} = k'' - \frac{i_{lim}}{0.886\, \sqrt{\rho/\sigma}}. \qquad (34)$$

Rüetschi determined the rate constant of the chemical process preceding the corresponding electrochemical stage from the slope of the plot of $(i_{lim}\sqrt{t})$ vs. i_{lim} constructed from the experimental i_{lim} values measured at different t values for the kinetic wave of azobenzene in an acetate buffer solution in 50 % ethyl alcohol.

Both variants of the indirect method given by Rüetschi were used by Nürnberg and co-workers [176-178] for the determination of the dissociation rate constants which they erroneously attributed to the dissociation of weak acids. In reality, these authors dealt with the overall rate of transfer of protons from acids present in the solutions to the molecules of the catalyst; this transfer resulted in catalytic hydrogen formation or production of p-nitro-aniline, which they used as proton-transfer agent rather than azobenzene.

Wolf [179, 180] used the dependence of the height of the kinetic wave on the drop period for the determination of the rate constant of an antecedent reaction. The drop period in these experiments was regulated by a periodic tapping on the end of the capillary; this kept the rate of mercury flow practically constant. The rate constant values were determined from the slope of a plot constructed on the basis of experimentally observed currents in $(i_{lim}/t^{1/16})$ vs. $(i_{lim}/t^{2/3})$ coordinates according to Eq. (33) in which, for the specified conditions of the experiments, $m = $ const and $\varkappa = kt^{1/6}$. The value determined for the rate constant of protonation of the anions of pyruvic acid in a citrate buffer solution was close to the value found by Strehlow and Becker [181, 182].

Rüetschi's method was also used for the determination of the rate constant of protolytic reactions by the chronopotentiometric method; this involves observing the curves of the electrode potential as a function of time during electrolysis with a current of constant intensity applied [183] (see section E of this chapter).

3. Kinetic Currents Limited by the Rate of Dehydration of Carbonyl Groups

Many organic compounds containing a carbonyl group are present in aqueous solutions in their hydrated form (see, for example, [184]). In the hydration of the carbonyl group methyleneglycol derivatives are formed

$$RR_1C = O + H_2O \underset{k_2}{\overset{k_1}{\rightleftarrows}} RR_1C \begin{matrix} OH \\ \diagup \\ \diagdown \\ OH \end{matrix}, \qquad (X)$$

which are usually electrochemically inactive compared with compounds containing a free carbonyl group. In the polarography of solutions of carbonyl compounds when the equilibrium (X) is shifted strongly toward the right a kinetic current is observed corresponding to the reduction of the aldehyde or ketone that is limited by the rate of dehydration of the carbonyl group.

The effect of hydration on the polarographic reduction current of carbonyl compounds was first observed on the waves of formaldehyde. Neiman and Gerber [185] observed that the polarographic reduction current of formaldehyde increases sharply at higher temperatures and explained this phenomenon by the shift of the equilibrium (X) toward the left at higher temperatures. Bieber and Trümpler [186] explained the polarographic behavior of formaldehyde in solution by a strong shift of equilibrium (X) to the right and by the effect of temperature and pH on the equilibrium. The value of the limiting current of the formaldehyde wave was first related to the rate of dehydration by Brdička and co-workers [187]. At increasing pH of the solution the kinetic current of formaldehyde reduction increases; this was explained by the catalytic effect of OH^- ions and other bases on the rate of dehydration of formaldehyde [187]. At pH > 13 the formaldehyde wave decreases with increasing pH due to the acidic dissociation of formaldehyde and formation of its nonreducible anionic form [187]. Loshkarev and Chernikov [188] showed that the hydroxyl ions formed during the reduction catalytically increased the rate of dehydration of formaldehyde, making the process autocatalytic.

The mechanism of the reduction of formaldehyde was studied in greatest detail by Brdička, who investigated this problem several times [189-191]. Brdička showed that in the polarography of formaldehyde in unbuffered solutions with increasing formaldehyde concentrations the height of the kinetic wave increases faster than the formaldehyde concentration. Assuming the catalytic action of the hydroxyl ions formed in the reduction of formaldehyde, Brdička developed on the basis of the reaction-layer concept an equation for the relationship of i_{lim} with the formaldehyde concentration [189, 190]. This equation was in good agreement with the observations. The rigorous solution of this depolarization scheme was given by Koutecký [58]. Complete agreement between the theory and the measured relationship of i_{lim} with formaldehyde concentrations could be attained only [191] when a dropping electrode with a vertical orifice was used (after Smoler [192, 193]). This electrode assured nearly complete regeneration of the solution at the dropping electrode, so that the OH^- ions formed in the electrolysis on the former drops did not take part in the reaction at the electrode on successive drops. This effect is most pronounced on the i vs. t curves taken during the life of a single drop [191]. The transfer of concentration polarization demands that on an electrode of the usual type with a vertical capillary the theoretical relationship for the dependence of the instantaneous current on time corresponding to this autocatalytic process ($i = kt^{7/6}$) can be observed only on the very first drop after the beginning of the polarization. The i vs. t curves measured in the same solution with the help of the dropping electrode with a vertical orifice are nearly identical on the first and succeeding drops [191].

Kinetic currents limited by the rate of dehydration of the keto group were observed in the reduction of dehydroascorbic acid (A) [194, 195]. The i_{lim} value of the polarographic wave of reduction of this acid is smaller by about three orders of magnitude, as can be predicted by the Ilković equation, and shows a very high temperature coefficient. The latter corresponds to the activation energy of dehydration of the keto group of the acid, which is about 13,000 Cal/mole. Similar kinetic waves limited in height by the rate of dehydration of the carbonyl group were observed in the polarography of solutions of mesoxalic aldehyde (B), dehydroreductic acid (C), and alloxan (D), which all contain in their structure a group similar to dehydroascorbic acid

A   B

$$
\begin{array}{cc}
\text{C structure} & \text{D structure}
\end{array}
$$

For keto acids [197-199] such as 2-ketogulonic acid limitations are caused not only by the kinetic limitations resulting from
the rate of dehydration of the undissociated acid and its anion but
also by the reduced rate of protonation of its anion. Both hydrated
forms (the anion and undissociated acid) are polarographically inactive; the dehydrated acid and its anion are reduced on the dropping electrode, giving two separate waves. Similar behavior can
be observed for mesoxalic acid, dioxytartaric acid [200], and a
series of other keto acids. The protolytic equilibrium and the kinetics of recombination of the anion, together with the acid—base
catalysis of dehydration result in a very complex relationship between the height of the reduction wave of carbonyl-containing acids
and the pH of the solution [197-202]. A characteristic example is
the polarographic behavior of glyoxalic acid, which was investigated in detail by Kůta [201].

In acid solutions, at pH ≤ 5, one wave can be observed that
is limited by the rate of dehydration of glyoxalic acid; this wave
increases when the solution is made more acidic (by increasing
the sulfuric acid concentration) and reaches a maximum in 10 M
H_2SO_4. At pH > 5 two waves appear. The first stage to appear
corresponds to the discharge of the undissociated acid; the second
corresponds to the reduction of its anion. At pH > 7.0 only one
wave of anion reduction remains that is limited by the rate of dehydration. At increasing pH the height of this wave increases and
reaches a maximum at pH 12.32; then it begins to decrease. The
increase of the wave height in strongly acidic and basic solutions
is caused by the acid—base catalysis of the dehydration reaction
of the undissociated acid and its anions, respectively. The reduction of the wave height in a strongly alkaline medium is caused by
the acidic dissociation of the aldehyde group [201]. Using the de-

pendence of the wave height on the buffer capacity at constant pH values and ionic strength of a borate buffer, Küta [201] determined the value of the rate constant of reactions catalyzed by borate anions and hydroxyl anions. It is interesting to note that if unbuffered solutions are used the wave height of glyoxalate anions increases because of the formation of hydroxyl ions in the reduction of the carbonyl group of glyoxalic acid [201]. The hydroxyl ions are effective dehydration catalysts, as was already mentioned.

Ono, Takagi, and Wasa [200, 202] explained in a somewhat different way the effect of pH on the rate-limiting role of dehydration on the α-keto acid waves. According to their view, the basic factor determining the dependence of the wave height on pH is not the acid–base catalysis of dehydration but the difference in rate of dehydration of hydrated acid molecules and their anions. They assumed that the anions dehydrate faster than acid molecules. For this reason, a change in pH, which affects both the ratio of the hydrated acid and its anion concentrations in the solution and also affects the rate of the protolytic reactions, causes a change of wave height. The increase in the wave height in strongly acidic solutions was explained by these authors by the enolization of the keto group. It must be mentioned here that Müller and Baumberger [203], who first observed the change of wave height of such compounds with pH, explained it by the keto–enol tautomerism. For certain acids, for example, phenylpyruvic acid, the wave height slowly increases with time in strong acids. It was shown spectroscopically that this increase is caused by the slow enolization of such acids, and the equilibrium is attained in about two hours [202]. In our opinion, the dehydration of the carbonyl group follows basically the scheme proposed by the Czechoslovakian investigators [187, 189, 190, 201] that takes into consideration the acid–base catalysis, although keto–enol tautomerism may have a substantial effect on the wave height [202].

The Japanese investigators [200] mentioned above were able to determine separately the effect of hydration and keto–enol tautomerism on the wave height in the reduction of phenylpyruvic acid and oxalylacetic acid. It was not possible to observe a substantial effect of enolization with pyruvic acid and α-ketoglutaric acid [200].

Because of the kinetic character of the waves of α-keto acids and other compounds which contain a hydrated carbonyl group in aqueous solutions, the polarographic method cannot be used for their exact determination; however, the analysis of such compounds on the basis of the wave height of their solutions becomes possible if amines such as o-phenylenediamine are present in the solution [204].

In the investigation of recombination kinetics of the anions of α-keto acids the hydration of their carbonyl group must be considered, as was shown by Becker and Strehlow [165, 166].

Kinetic currents can be observed in the polarography of aldehydic pyridine derivatives. Volke [205] showed that in the reduction of isomeric formylpyridines the limiting currents are determined by the dehydration rate of the aldehyde group. This rate is catalytically affected by acids and bases; this results in a complicated relationship between the wave height and the pH of the solutions and their buffer capacity. In the polarography of aminoaldehydes and ketones both the effect of hydration and the effect of protonation of their molecules on the nitrogen atom can be observed.

The formation of positively charged onium compounds, which promote the reduction, plays the same role as the recombination of acid anions: the protonated form discharges at less negative potentials than the unprotonated form (see, for example, [206]).

The kinetics of dehydration — and in basic medium, the kinetics of protonation — determine the wave height in the reduction of two isomers: 1,3,5- and 1,2,6-trimethyl-γ-piperidones [206]. On the polarograms of these compounds two waves can be observed, each of them corresponding to the transfer of one electron (as in the case of reduction of aromatic aldehydes and ketones in acid medium). In this respect the γ-piperidone derivatives differ from the behavior of previously discussed keto acids, since in their reduction only a single wave can be observed, corresponding to the transfer of two electrons.

If the solution is changed to a neutral solution and further to a weakly basic solution the wave height of γ-piperidone derivatives increases and reaches a maximum at a pH 9. Then, if the pH is further increased the wave height decreases. The increase in wave height during the transition from acidic to neutral solutions is caused by the catalytic rate increase of the dehydration due to increased concentration of hydroxyl ions and other bases (the hydration of the keto group was proven spectroscopically with one of the isomers of trimethylpiperidone [206]). Only the dehydrated (at the C = O group) and protonated (at the nitrogen) molecules of γ-piperidone derivatives can be reduced. Therefore, if the solution is made alkaline the wave height, corresponding to the transfer of the first electron, decreases since these compounds are present already in their unprotonated form (pK$_A$'s of these two investigated isomers are 7.2 and 8.2, respectively). Their limiting currents in this case are determined mainly by the rate of protonation of the nitrogen of the γ-piperidone ring. If the positive charge on the nitrogen atom is preserved even in alkaline medium (as, for example, in case of the tetrasubstituted cation N,N-dimethyl-γ-piperidonium), the height of the first wave is independent of pH in alkaline solutions [206]. The change of wave height with the pH of the solution observed in the polarography of α-aminoaldehydes [207] is similar in nature to that observed for the γ-piperidone derivatives. Here, the wave height is also determined by the dehydration processes of the carbonyl group and protonation (more exactly, deprotonation) of the amine group. For trisubstituted α-aminoaldehyde salts the current either remains constant in alkaline medium or (for other derivatives) the current increases more slowly than with the original α-aminoaldehydes [207].

The hydration of the carbonyl group reduces its electron-acceptor properties and consequently reduces its effect on the reduction of other groups in the molecule of the organic compound. For example, in solutions of aliphatic α-chloro- and α-bromoaldehydes the first waves, corresponding to the electrochemical cleavage of the carbon–halogen bond, are substantially lower than the value corresponding to the diffusion current. The limiting currents of these waves are characterized by a very high temperature coefficient (from 16 to 26%/deg!), and their value is independent of the mercury column height. This shows that these waves have a definite kinetic character [208]. The limiting currents in this case are determined by the dehydration rate of the aldehyde group, since the presence of a free electron acceptor C = O group facilitates the nucleophilic electrochemical splitting of the neighboring carbon–halogen bond. The formation of acetals from α-halogen-aldehydes makes the cleavage of the carbon–halogen bond more difficult: reduction waves of α-bromoacetaldehyde acetal were observed [208] only in the presence of tetramethylammonium salt as the supporting electrolyte ($E_{1/2}$ = -1.95 V in 50% dioxane), whereas the $E_{1/2}$ of α-bromoacetaldehyde is about -0.5 V (in aqueous 0.1 N LiCl solution).

The limiting current of the wave, corresponding to the hydrogenolytic cleavage of the carbon—halogen bond of α-halogen aldehydes, increases with the pH of the solution; this is caused by the catalytic effect of hydroxyl ions. In a neutral, unbuffered medium the wave height increases with the concentration of the haloaldehyde [208] by more than would correspond to a linear relation due to the autocatalytic increase in the dehydration rate by hydroxyl ions (formed in the electrochemical reaction). This was observed also in the polarography of the formaldehyde. If dioxane is added to the solution the limiting current increases because the equilibrium is shifted toward the formation of the free (nonhydrated) aldehyde group [208].

The hydration of the aldehyde group increases substantially in α-dihaloaldehydes when compared with α-monohalogen derivatives of aldehydes. In α-dibromoaldehydes the aldehyde group shows barely any effect on the cleavage of the carbon—halogen bond, and the halogen is reduced in the hydrated molecule ($E_{1/2}$ of the wave is about -1.2 V). The wave is diffusion controlled [209]. Dichloroacetaldehyde gives only a small kinetic wave, whereas chloral gives only one diffusion two-electron wave. The trichloroacetal gives approximately the same wave as chloral [209]. Hence, with chloral the degree of hydration of the carbonyl is so great that the kinetic current, limited by the rate of dehydration of the carbonyl, cannot be observed.

A kinetic current caused by the dehydration of the carbonyl group can be observed in the polarography of 2-chlorocyclohexanone. The wave corresponds to the cleavage of the carbon—chlorine bond [210]. The limiting current of this wave shows a temperature coefficient of 5-8%/deg over the 0-25°C temperature range. The wave height increases with the pH of the solution (catalytic increase of the dehydration rate by hydroxyl ions). If dioxane is introduced into the solution, the wave loses its kinetic character. The limiting current of the waves of cis- and especially of trans-α,α-dichlorocyclohexanones is much higher than that of the monochloro-derivatives. The temperature coefficient of the limiting current is 20%/deg. Similarly to aliphatic haloaldehydes, dichlorocyclohexanone is more completely hydrated than the monochloro-derivatives [210]. Two waves can be observed on the polarographic curves of gem-dichlorocyclohexanone corresponding to the gradual cleavage of both carbon—chlorine bonds. It was shown spectroscopically that the degree of hydration of the carbonyl group of mono- and di- α-chlorocyclohexanones decreases in the following order: trans > cis > gem > monochloro-derivatives.

Very similar to the waves limited by the rate of dehydration of carbonyl group are the waves observed in the polarography of certain sugars [211-213]. The aldoses and certain ketoses are present in aqueous solutions in cyclic, polarographically inactive, acetal forms. The rate of conversion of these compounds at the electrode into noncyclic products with a free carbonyl group determines the heights of kinetic waves of sugars.

The wave height on the polarographic curves of erythrose solutions is limited by the rate of dehydration of the molecules [214]. It increases with increasing pH attaining a maximum height (at pH 10-12) that nearly corresponds to the diffusion current with $n = 2$. At pH > 12, the wave again decreases and shows a kinetic character. In the presence of an ammonia buffer solution the erythrose gives an additional wave (at less negative potentials) corresponding to the reduction of the appropriate aldimine [214].

4. Kinetic Currents in the Polarography of Solutions of Complexes

Koryta and Kössler [215] first observed a kinetic current limitation of the discharge of complex metal ions. They investigated the reduction of cadmium, tin(II), zinc, copper(II), and indium (II) on the dropping electrode in the presence of nitrilotriacetic acid (complexone I).

The discharge wave for free metal ions in the presence of complexone shows in part a kinetic character, due to an increase in the current that is caused by the dissociation of complexone. The kinetic character of the current in this case was determined by the authors (1) on the basis of the change of the limiting discharge current of free Cd^{2+} ions with the height of the mercury column above the electrode, (2) on the basis of the high temperature coefficient of the limiting current of this wave, and (3) by the unusual relationship of its half-wave potential to the concentration of the added complex-forming reagent in the solution. If the wave were determined only by the diffusion of cadmium ions, its half-wave potential should be independent of the complexone concentration; if the equilibrium between the complex and uncomplexed ions is attained very fast and is very mobile, the half-wave potential should change in proportion to the logarithm of the added complex-forming reagent concentration [216]. The experimentally observed displacement of the half-wave potential with increasing complexone concentration toward negative potentials [215] cannot be described by the equation which is valid for mobile equilibria. Hydrogen ions take part in the complex-formation equilibrium which was studied by Koryta and Kössler [215]

$$MX^- + H^+ \underset{k_2}{\overset{k_1}{\rightleftarrows}} M^{2+} + HX^{2-}, \qquad (XI)$$

(where X^{3-} is the anion of the nitrilotriacetic acid); therefore, the pH of the solution shows a substantial effect on the height of the kinetic wave.

In the investigation of this system on the streaming mercury electrode, the kinetic component of the current fell practically to zero. Therefore it was possible to determine the stability constants of cadmium, zinc, and tin complexes formed with nitrilotriacetic acid.

In a later investigation [217] Koryta used the equations derived by Koutecký and Brdička [46] and estimated the rate constant (k_1) of the dissociation of the cadmium complex formed with nitrilotriacetic acid. He showed, on the basis of the value determined, that in the discharge of cadmium ions in an excess of nitrilotriacetic acid on a streaming electrode (the contact time with the solution is $8 \cdot 10^{-4}$ sec) the kinetic component of the current should not play any role [217]. The physical meaning of this observation is that if the time of electrolysis is reduced the effective thickness of the diffusion layer decreases and becomes of the same order of magnitude or smaller than the reaction layer. The current becomes independent of the reaction rate and is determined only by the diffusion of electrochemically active compounds to the electrode.

This was the case [218] in Koryta's experiment on the streaming electrode.

It was already mentioned that if the current intensity is limited by the rate of a chemical reaction, the dependence of the half-wave potential (for a reversible electrode process) of the kinetic wave or the added complex-forming reagent is different from that obtained with diffusion currents. The dissociation constants of complexes can be determined from the reversible diffusion waves on the basis of the dependence of the half-wave potential on the concentration of the added reagent. Koryta [219] developed the equations for the dependence of the half-wave potential on the concentration of the added ligand for the case of kinetically governed reversible waves. On the basis of these equations he calculated the dissociation constants of cadmium complexes with nitrilotriacetic acid in acetate and ammonia buffer solutions [219]. The analysis of the second wave in these solutions, corresponding to the reduction of the complex of cadmium with nitrilotriacetic acid, enabled Koryta [220] to prove that the complex is reduced only after addition of one or two protons.

Dandoy and Gierst [221] showed that kinetic limitations appear in the deposition on nickel ions from solutions not containing complex-forming compounds. The limitations in this case are caused by the slow dehydration of the nickel aquo complex, which precedes the electrochemical reaction.

Stromberg [222] investigated the effect of slow dissociation of complexes on the polarographic anodic—cathodic waves with a reversible electrochemical step on a dropping amalgam electrode.

Tur'yan and Serova [223, 224] studied kinetic currents limited by the rate of complex formation. They found that among the three waves appearing on the polarograms of thiocyanide containing nickel solutions the first two waves are limited by the rates of the antecedent chemical reactions; they were assigned to the reduction of $Ni(CNS)_2$ and $Ni(CNS)^+$, respectively. The third wave corresponds to the reduction of the nickel aquo complex.

The same authors observed [225] double waves for the reduction of nickel and cobalt (II) ions in the polarography of complexes of these metals with pyridine; in a much later study Tur'yan [226], using the Brdička—Wiesner method, calculated the rate constants of formation of these complexes. It was assumed that they are formed only in the volume reaction.

Mark and Reilley [227, 228] also observed more positive waves on the polarograms of nickel salts upon addition of pyridine to the solution. They found that at low pyridine concentrations this wave is much higher than the height of the diffusion current of pyridine. They showed that the wave has a catalytic nature and that its formation is caused by catalytic "conversion" of nickel ions by pyridine during the electrochemical reduction:

$$Ni\,(Py)_x(H_2O)_y^{2+} + 2e^- \xrightarrow{\text{el}} Ni^0 + x\,(Py) + y\,H_2O, \qquad \text{(XII)}$$

$$x\,(Py) + Ni(H_2O)_6^{2+} \rightarrow Ni\,(Py)_x\,(H_2O)_y^{2+} + (6 - y)\,H_2O.$$

The same authors investigated in detail the effect of Ni^{2+} concentration and the pH on this wave and showed that the limiting current of this wave can be used for the analytical determination of pyridine [228]. A well-defined catalytic Ni^{2+} wave was also observed by Mark [229] in the presence of small quantities of o-phenylenediamine; this wave is affected by the pH of the solution.

Kemula and Rosolowski [230] showed the kinetic character of the reduction current of molybdenum silicic, molybdenum phosphoric, and molybdenum arsenic heteropolyacids in buffer solutions, caused by antecedent protonation reactions; for example,

$$[Si\,(Mo_3O_{10})_4]^{4-} + 4H^+ + 4e^- \rightarrow [Si\,(Mo_3O_9)_4\,(OH)_4]^{4-}. \qquad \text{(XIII)}$$

The effect of antecedent chemical reaction kinetics on the polarographic curves of quadri- and trivalent titanium thiocyanate complexes was investigated [231].

It was shown that in the study of the kinetic discharge current of acidic cadmium complexes with nitrilotriacetic acid not only must their interaction with hydrogen be considered but also their interaction with the acetic acid of the buffer solution [232] (compare with scheme XI).

Koryta [233] analyzed theoretically the different cases of complex ion reduction, the reversible and irreversible processes, limited by diffusion or rate of dissociation, tabulated the equations for the waves, and gave the dependence of the half-wave potential on the drop period and concentration of the complexing agent for all these cases. Koryta [234] analyzed the general problem of the effect of complex formation on polarographic waves and derived a method for the determination of the dissociation constant of complexes and rate constants of dissociation from polarographic data.

Koryta's method can be successfully used not only in the investigation of inorganic complexes but also in the study of complex formation between organic compounds. Thus, for example, Peover [235] on the basis of the shift of the half-wave potential of reversible reduction waves of tetracyanoethylene, tetracyanoquinodimethane, tetracyanobenzene, tetrachlorophthalanhydride, dichlorodicyanoquinone, and chloranil showed the complex formation with hexamethylbenzene and pyrene in methylene chloride or chloroform solutions. According to Peover's data, the rate of formation of these complexes is too high to allow the determination of the effect of the kinetics of their formation on the value of i_{lim} or $E_{1/2}$ [235].

E. METHODS OF INVESTIGATION

OF KINETIC CURRENTS

The kinetics of reactions preceding the electrons transfer can be investigated by the classical polarographic method if the rate of these reactions is between certain limits. As the rate constant (ρ) increases, the kinetic current approaches the diffusion current value. The maximum value of ρ that can be determined with sufficient precision on the basis of the kinetic wave height can be judged by Eq. (25). For the determination of ρ (considering the errors in the measurement of i_{lim} and i_D) it is required that $i_{lim}/i_D \leq 0.93$; this inequality determines the upper limit of the value of ρ

$$\rho \leqslant 200 \frac{\sigma}{t}. \tag{35}$$

It follows from Eq. (35) that to increase the upper accessible limit of the rate constants of antecedent reactions (first-order reactions) in the above solution (at σ = const) the value of t must be reduced. This can be done to a certain degree by employing regulated drop formation at the dropping electrode by using a streaming mercury electrode or a rotating disk electrode.

The time of electrolysis can also be substantially reduced by galvanostatic (for example, chronopotentiometric) and potentiostatic methods (see appropriate chapters in [9]).

For example, when the current changes with time are measured and a linearly increasing potential is applied at the electrode (chronoamperometry with linear potential change), very small t values can be attained at extremely fast potential changes. Therefore, the upper limit for the values of ρ, the measurable rate constant, can be significantly increased by this method. If the rate of potential change is, for example, 500 V/sec, the time required for the change of the electrode potential by one volt is only $2 \cdot 10^{-3}$ sec. Still, the possibilities of the simple chronoamperometric method are greatly limited, since at increasing potentials the charging current of the electrical double layer at the electrode becomes much higher than the faradaic current, caused by the electrode reaction. Special instrumentation must be used to reduce the effect of capacity current.

In chronoamperometry it is interesting that with increasing potential and with a linear potential change, the rate of the antecedent chemical reaction ceases to affect the shape of the curves i vs. t (or i vs. E). The current is determined only by the transport of the electrochemically active form of the depolarizer by diffusion. In this case, similarly to the investigation mentioned earlier made by Koryta on the streaming electrode [217], the equilibrium concentration of this active form in the solution can be determined, as can the value of ρ [232, 236]. The i vs. E curves, which can be observed for a reversible electrode process with regeneration of the depolarizer by a first-order chemical reaction by the chronoamperometric method with linear potential change, were described by equations developed by Semerano and Vianello [237]. Saveant and Vianello [238] gave the mathematical formulation of this problem and, using computers, obtained the solution for the curves corresponding to the case of a fast first-order chemical reaction preceding the reversible electrode process. In deriving the equations the authors [238] applied the conditions of semi-infinite linear diffusion. In this work it was proved that for a reversible electrochemical step with antecedent monomolecular reaction, the ρ value determined by the chronoamperometric method with linear potential change is larger by four orders of magnitude than the value determined by the methods of classical polarography.

The chronopotentiometric (galvanostatic) method that has already been mentioned several times in this chapter finds widespread application in the investigation of kinetic currents (and also adsorption phenomena). By this method the potential is recorded as a function of the time of electrolysis at a given current intensity.

Characteristic of the chronopotentiograms are so-called transition times, at which the electrode potential decreases more or less sharply. This is caused by the depletion of the depolarizer at the electrode surface. The potential decreases to a new level at which an electrochemical reaction then proceeds with participation of another depolarizer. For diffusion limited processes the transition time (τ) is determined only by the current intensity (i_0) and the depolarizer concentration (c_0) in the bulk solution

$$\tau^{1/2} = \frac{\pi^{1/2} n F c_0 D^{1/2}}{2 i_0}. \tag{36}$$

If two compounds are present in the solution and if the reduction of these compounds takes place at sufficiently different potentials, two steps can be observed on the chronopotentiograms, and each step can be characterized by its transition time. If the transition time of the process taking place at the more positive potential is determined using Eq. (36), then the transition time of the second process is affected by the concentration of the depolarizer that takes part in the first electrode reaction.

The variation in the chronopotentiometric technique in which the direction of polarization is changed when the transition time is achieved became significant in the investigation of electrode processes. By this method, developed by Berzins and Delahay [239], both cathodic and anodic processes can be investigated. If the polarization is reversed without changing the current intensity, then the transition time of the reverse process is equal to one third the transition time of the direct process.

In the chronopotentiometric investigation of electrode processes with chemical reaction taking place at the electrodes, the transition time (τ_k) is determined by the rates of the forward (ρ) and reverse ($\rho\sigma$) chemical reactions. For antecedent first-order

(or pseudo-first-order) reactions the value of the $i_0 \tau_k^{\frac{1}{2}}$ product is
a linear function of the current intensity. The slope of the plot of
this relationship is determined by the equilibrium constant (σ) between the electrochemically active and inactive form and the sum
of rate constants of the forward (ρ) and reverse $(\rho\sigma)$ reactions. If
the σ value is known, then by this graph, which is very easy to
construct from the experimental data, the ρ value can be determined. If the current intensity is increased (which means that the
transition times are reduced), the conditions of the electrolysis
can be changed sufficiently so that only the electrochemically
active form of the depolarizer present in the solution and transported by diffusion to the electrode will be reduced. The formation of the active form from the inactive form will not show any
effect in practice. Under these conditions the σ value can be determined chronopotentiometrically (basically, the same effect can
be achieved by reducing the electrolysis time if the dropping electrode is replaced by a streaming electrode [217] or by using the
chronoamperometric method with fast potential change [232, 236]).

The application of the chronopotentiometric method for the
investigation of chemical reactions antecedent to the electron
transfer makes possible the determination of reaction rates nearly
two orders of magnitude larger than those that can be determined
by classical polarographic methods. An example of application of
the chronopotentiometric method for the investigation of chemical
reactions preceding the electron transfer is the study of the rate
of protonation of maleic acid in acidic media [146].

Chronopotentiometry with reversal of current flow (cyclic
chronopotentiometry) can be used for the investigation of chemical
reactions consecutive to the electrochemical stage (and for the investigation of the kinetics of reactions in which the products formed
on the electrode take part). Equations were derived for determining the value of the transition time of the forward and reverse polarization in the first and subsequent cycles [240, 241] for electrode processes with subsequent first-order (or pseudo-first-
order) reaction. This method was used for the investigation of the
anodic oxidation of p-aminophenol on a platinum electrode

$$HO-C_6H_4-NH_2 - 2e^- \rightleftarrows O = C_6H_4 = NH + 2H^+. \qquad (XIV)$$

The rate constant of the hydrolysis of the imine formed on the electrode to quinone was determined [241, 242]

$$O{=}C_6H_4{=}NH \ + \ H_2O \rightarrow O{=}C_6H_4{=}O \ + \ NH_3. \qquad (XV)$$

The imine-reduction transition time decreases in reverse polarization, due to the partial conversion of imine to quinone. A new wave, corresponding to the reduction of quinone, appears on the chronopotentiogram because of the reduction of the imine in another step at a more negative potential.

Equations were also derived for calculating the transition time in cyclic chronopotentiometry for the first [243, 244] and subsequent cycles for catalytic processes. The electrode reaction product can be regenerated by a first-order (or pseudo-first-order) reaction into the starting compound, for example, in the reduction of Ti(IV) in acidic solution in the presence of an excess of hydroxylamine [114, 242].

The effect of partial regeneration of the depolarizer from the electrode products on the transition time in polycyclic chronopotentiometry was also treated [242]. The form of the chronopotentiograms obtained in direct and reverse polarization, with disproportionation of the electrode product, was also handled with aid of computers [245]. A comparison of the calculated and experimental curves for the reduction of U(VI) into U(V) and subsequent disproportionation of U(V) gave the value of the disproportionation rate constant. The determined value, $0.95 \cdot 10^3$ liters/mole·sec, is very close to the value determined by Orlemann and Kern from the height of the catalytic wave in classical polarography [116].

Dračka [246] analyzed theoretically the effect of the rate of a first-order chemical reaction in which the intermediate product at the electrode is converted into an electro-inactive species (the depolarizer of the second electrochemical reaction is deactivated) on the ratio of the transition times of the first and second steps, of a two-step electrode process. It is assumed that the electrochemical reactions corresponding to these steps take place at sufficiently different potentials. A chemical reaction in which the intermediate product participates decreases the concentration of the depolarizer of the second step at the electrode and therefore reduces the τ_2 value. The ratio of the transition times of the first and second steps, τ_2 and τ_1, depends on the rate of the forward and reverse chemical reaction in which the product of the first electrode reaction takes part. If the chemical reaction is reversible and is very fast in both directions, or if the chemical reaction is extremely slow, then the τ_2/τ_1 ratio approaches the value characteristic of two-step electrode processes, which are not complicated by chemical reactions [246]. The latter can be determined by the theory of Berzins and Delahay [239].

In the chronopotentiometric investigation of anode processes at the mercury electrode, in which slightly soluble mercury compounds are formed, an abrupt potential change cannot be observed. The salt formation takes place at a certain concentration of mercury ions which is determined by the solubility product of the mercury compound. Therefore, if very high current densities at which a "blocking" effect of the formed layer can be observed are avoided, the chronopotentiograms cannot be used for the determination of the time required for the saturation of the surface. An optical method [247] was developed for the determination of the time required for covering the electrode with a layer of insoluble salt (during polarization with a constant current). This method is based on the change of the elliptic polarization of the light reflected by the mirror surface of the electrode due to formation of a surface layer. (The reflected light passes an analyzer and falls on a photomultiplier; a sudden change of the current of the photomultiplier shows the time required for coverage of the electrode.) The time necessary for covering the electrode depends on the current intensity, the solubility product of the mercury compound, and the rate of the chemical reaction between mercury ions and the investigated reagent. This method, called "chronoellipsometry," was used for the investigation of the mechanism of calomel formation [247].

The potentiostatic method can be used for the investigation of the kinetics of chemical reactions accompanying the electron transfer. In this method the current intensity is measured as a function of time at a given potential (usually on a dropping mercury electrode, during the life of a single drop; the same curves can also be measured on a stable "hanging" mercury drop, and also on solid electrodes). The chemical reaction interposed between two electrochemical reactions was studied using this method. This is the case in p-nitrosophenol reduction. In the first stage of reduction, p-hydroxyaminophenol is formed, and its further reduction (which takes place at the same potential) is preceded by its dehydration [248]. Right at the beginning of the process (at very small t values) the measured current intensity on the i vs. t curves corresponds to the reduction of the nitro to the hydroxylamino group. With further accumulation of the hydroxylamino derivative at the electrode, an additional current can be observed, corresponding to the reduction of the dehydrated hydroxylaminophenol. The magnitude of this additional current depends on the rate constant of dehydration and on the reversibility of the hydration—dehydration process. If the dehydration is sufficiently fast, this stage will not be limiting, and the observed current will correspond to the diffusion-limited process with both electrochemical steps taking place simultaneously. In the work [248] referred to, equations are given for the i vs. t curves, for planar and spherical electrodes, for reversible and irreversible chemical reaction taking place between two electrochemical steps.

The kinetics of an irreversible chemical reaction following a reversible electrochemical step can be studied by oscillopolarography. In this method a voltage of varying frequency, which has the form of a regular step, is applied on the electrode. Using this method, the rate constant of the second-order interaction of the products of the electrode reaction was determined in the reduction of cyclo-octatetraene in 96% dioxane from the change of anodic and cathodic peak areas on the oscillopolarograms with frequency. The reaction products, dianions, are formed by addition of two electrons to the cyclo-octatetraene [249].

The effect of chemical reactions taking place at the electrodes can also be observed in ac (alternating voltage) polarography for a reversible electrochemical step. Thus, using Breyer's method (applying at the electrode, together with a dc potential, a superimposed small sinusoidal potential, and making the ac component independent of the linearly increasing dc electrode potential [250-252]) the electrode process of U(VI) reduction to U(V) was investigated, with disproportionation of U(V) to U(VI) and U(IV) [253]. The current measured by Breyer's method seems to be the sum of the active and capacitance component and therefore cannot be quantitatively evaluated; the results obtained by this method are only

qualitative. A valuable quantitative characteristic of reversible
electrode processes with chemical reactions can be obtained by
methods that separate the capacitance and active components of
the alternating current (with the help of a bridge or a vector polarograph). Many publications deal with the theoretical discussion
of their application to various electrode processes. Thus, for
example, Gerischer [254] discussed the faradaic impedance for
electrode processes with antecedent chemical reaction. Levich,
Khaikin, and Kir'yanov [255] developed equations for the polarization resistance and reaction pseudocapacity of reversible electrode
processes with antecedent pseudo-first-order reaction and second-
order regeneration of the electrochemically inactive form of the
depolarizer from the product of the electrode reaction. The work
of Smith and his co-workers [256, 257], who studied the effect of
several factors (primarily the effect of the frequency of the alternating current) on the phase-shift angle between the current and
potential for different types of electrode processes with chemical
steps, must also be mentioned.

The high-level faradaic-rectification method developed by
Barker [258] offers great possibilities for the investigation of
electrochemical and fast chemical processes accompanying an irreversible electrochemical step at the electrode surface. At an
exactly determined moment in the lifetime of the drop, in addition
to slowly increasing the constant potential, he applied a series of
rectangular impulses. The duration of each impulse is one millisecond, and the duration of the entire series is about 40 milliseconds. Because of the irreversible nature of the electrode process, the faradaic current caused by superimposition of the rectangular impulse passes in only one direction (faradaic rectifica-
tion). The average faradaic current is measured in the second
half of the series of impulses (during the second 20 milliseconds
of the series). This current is proportional to the depolarizer concentration at the electrode and does not contain a capacitance component: the latter decreases practically to zero during the first
half of the period of the series. Since a slowly increasing potential
is also applied to the electrode, the process takes place on each
drop at a chosen constant potential. The results obtained for a
large number of successive drops show a plot of the dependence of
the faradaic current component from the electrode potential which
is similar in shape to a regular polarogram.

This method was successfully applied [258] for the determination of the rate constants of dissociation (and recombination) of a series of weak acids, and the current corresponding to reduction of hydrogen ions, formed in the dissociation of acid molecules, was measured. Application of extremely short impulses, about 10^{-6} sec long, makes possible the use of this method for the investigation of very fast processes [see Eq. (35), in which t in this case means the duration of the impulse]. This method requires the use of very complicated instrumentation.

In concluding this section it must be mentioned that classical polarography is best suited for the investigation of antecedent chemical reactions and for the study of the effect of different factors. The majority of antecedent reaction (in part protonation reactions) are second-order reactions. Therefore, by properly choosing the experimental conditions (for protolytic reactions, for example, the pH of the solution), the conditions prescribed by Eq. (35) can easily be met.

The advantages of the classical method are its universality, its experimental simplicity, and the easy interpretation of measured data. Unlike other methods, the evaluation of data is independent of the precision of the measurement. The method makes the evaluation of factors affecting the kinetics of reactions taking place at the electrodes, such as the adsorption of reacting components and the structure of the electrical double layer, relatively simple. The other methods discussed here briefly are valuable extensions of classical polarography.

Chapter III

Adsorption on the Electrode
and the Electrode Processes

A. BASIC RELATIONSHIPS OF ADSORPTION

The adsorption on a phase boundary (for example, mercury—solution) for simple compounds can be described by the well-known Langmuir equation [259] (which is also called the Langmuir adsorption isotherm)

$$\Gamma = \Gamma_\infty \frac{\beta c}{1 + \beta c}, \tag{37}$$

where Γ is the number of moles of adsorbed compound on 1 cm^2 of mercury—solution boundary surface; Γ_∞ is the quantity of adsorbed compound if the surface is completely covered by a monomolecular layer; c is the concentration of the adsorbed compound in the bulk solution (M); and β is the adsorption coefficient (M^{-1}).

Langmuir [260] gave the following relationship for the β value:

$$1000\beta = \delta e^{\frac{W}{RT}}, \tag{38}$$

where δ is the thickness of the surface layer (adsorbed particles); W is the adsorption work that must be done to remove one mole of compound from the interface of mercury and solution; R is the universal gas constant; and T is the absolute temperature.

The surface fraction θ covered by adsorbed particles can be given, on the basis of Langmuir's equation, as

$$\theta = \frac{\Gamma}{\Gamma_\infty} = \frac{\beta c}{1 + \beta c}. \tag{39}$$

Langmuir's equation was derived without considering the interaction of adsorbed molecules. This condition, as was shown by
Frumkin [261], is very often not fulfilled, mainly with large organic molecules containing polar groups. In this latter case, after a
certain degree of coverage of the electrode surface by adsorbed
molecules, their interaction promotes further adsorption, and the
surface coverage increases faster than would be expected according to the Langmuir equation (39). Frumkin [261] first considered
the interaction of adsorbed molecules in deriving an equation relating the surface tension (which, as is well known, depends on the
quantity of surface-active compounds adsorbed on the phase boundary) to the concentration of the compound in the solution. He
used, by analogy with the van der Waals equation, a so-called attraction factor. The equation derived by Frumkin for the adsorption is known as the Frumkin adsorption isotherm; it can be stated
as

$$\beta c = \frac{\theta}{1 - \theta} e^{-2\gamma\theta},\tag{40}$$

where γ is the attraction factor.

The Frumkin adsorption isotherm expresses the $\theta = f(c)$
relationship. The isotherm has an S-shape; the rising part increases for larger values of the attraction factor. It can be seen
from Eq. (40) that at low degrees of coverage (low θ values) the
$e^{-2\gamma\theta}$ term is close to unity, and the Frumkin isotherm reduces
to the Langmuir equation or, more precisely, to Henry's equation.
Therefore at low degrees of surface coverage the Langmuir adsorption isotherm (37) or (39) or Henry's linear law can be used
instead of the adsorption isotherm of Frumkin (40).

$$\theta = \beta c.\tag{41}$$

The range of surface coverage (θ) over which the Langmuir
or Henry isotherm can be used with sufficient precision is determined by the value of the attraction factor (γ). The value of the
attraction factor depends primarily on the structure of the adsorbed
compound. In a homologous series, with increasing hydrocarbon
chain length in the molecule, the γ value increases. According to
Frumkin's data [261] the γ value increases from 0.8 to 1.2 for
caprylic acid (at 5.5°C) and lauric acid (at 9.5°C), respectively.
The attraction factor decreases with increasing temperature. At

25°C the γ value of caprylic acid is 0.6 [261]. If organic solvents are added to the aqueous solution, the attraction constant decreases (see, for example, [262]). At increasing negative electrode potentials an increase of the γ value is observed in many cases [263-266]. In other cases, as, for example, in the adsorption of aliphatic alcohols and amines from aqueous sodium sulfate solution, the γ value decreases at increasingly negative potentials [266, 267]. From the work of Damaskin [263-267] it follows that the γ value increases approximately linearly with the electrode potential.

It must be mentioned that the Frumkin adsorption isotherm is not only valid for attraction of adsorbed particles, but also for their repulsion. In the latter case, the γ factor in the Frumkin adsorption isotherm equations has a negative value.

Several authors [268-270] have derived different isotherm equations based on different assumptions. A short analysis of these equations was given by Parsons [271]. Seemingly, the best agreement with experimental data is obtained by the Frumkin isotherm; primarily the change of the γ value with potential is given best [265-267, 272].

An S-shaped adsorption isotherm is characteristic for the adsorption of many compounds at the mercury−solution interface. Besides the cases already mentioned, Frumkin and Damaskin [273] found an S-shaped isotherm in the adsorption of the tetra-butylammonium cation. The same shape was observed by Lorenz for tetramyl and n-hexyl alcohols, but he did not find this shape for the adsorption of isopropyl alcohol [275]. S-shaped isotherms are observed in the adsorption of 2,6-lutidine [262, 276] and also for gelatin and triton X-100 in potassium nitrate solutions, which are used to suppress polarographic maxima [277].

Brdička, observing a much faster coverage of the surface than the corresponding increase of bulk concentration for certain dyes called this phenomenon (referring to a publication of Stackelberg), "two-dimensional crystallization" [278].

The relationship between adsorbed, uncharged molecules at the mercury−solution interface and the potential of mercury which was first discovered by Frumkin [279] seems to be extremely important.

Gouy [280] found that the decrease of the surface tension of mercury due to the adsorption of organic molecules on the surface, compared with the surface tension of solution not containing this organic compound, takes place only in certain potential range. This range usually includes the potential of zero charge of the electrode in the supporting electrolyte (which is mostly an inorganic electrolyte).

Maximum adsorption usually can be observed at potentials that lie close to the potential of zero charge of the mercury, and, if the potential of the electrode is shifted in one or another direction, the adsorption of organic molecules on the electrode decreases. Thus the electrocapillary curves for solutions containing organic compounds and those without them overlap at certain potentials. The observed change of adsorption with potential can be explained, in accord with Frumkin [279], in the following way.

The adsorption of organic molecules, which usually have larger dimensions than the water molecules and usually a much lower dielectric constant, causes a reduction of the capacity of the electric double layer. The double layer can be considered to be a condenser, one plate of which is formed by the charged mercury surface, the other by counter ions of opposite charge attracted from the solution by electrostatic forces. If the potential difference increases between the condenser plates (in the electrical double layer) the exchange of adsorbed molecules by water molecules becomes energetically favored and the capacity of the double layer increases. The quantity of organic material remaining in the adsorbed state at a given potential is thus determined by the energy balance, the energy required for the desorption of a part of the organic molecules, and the energy gained due to the increased capacity of the double layer. The competition between the adsorption and the effect of the electrical field, as was shown by Frumkin [279], determines the dependence of the W value in Eq. (38) on the electrode potential.

Frumkin also considered the additional potential difference caused by the adsorption of dipolar molecules (caused by the difference between the potential of zero charge and the potential of maximum adsorption, E_M), and his analysis resulted in an equation that gives the dependence of adsorption constant from the electrode potential:

$$\beta = \beta^0 e^{\frac{1/2(C-C')}{RT\Gamma_\infty}\varphi^2} = \beta^0 e^{-a\varphi^2} .\tag{42}$$

In this equation β^0 corresponds to the maximum value of β found at the potential of maximum adsorption E_M; C and C' are the integral capacities of the double layer in absence of the surface-active compound and for full coverage (monolayer) of the electrode surface, respectively; and φ is the electrode potential related to the potential of maximum adsorption E_M.

On the basis of Eq. (42) the following conclusions can be drawn: if the salt concentration increases in the solution, the value of Eq. (42) increases, since the capacity of the double layer (C) increases with the concentration of supporting electrolytes in the solution and C' remains practically constant. Thus an addition of supporting electrolyte to the solution usually causes (in the absence of other effects, such as salting out from the solution) a reduction of the adsorptivity of organic compounds, mainly at potentials far different from E_M.

The effect of potential on adsorption is expressed not only by the change of the β value in the adsorption isotherm Eqs. (37) or (40), but also by the change of the γ value, as was already shown. This was first found by Frumkin [279].

In the adsorption of charged particles (ions) the potential of maximum adsorption can be substantially different from the potential of zero charge in the solution not containing the adsorbing ions. When the adsorption of anions takes place, the E_M value shifts, as expected, toward positive potentials from the zero-charge electrode potential in the original solution (not containing adsorbing anions); when cations are adsorbed, E_M shifts to negative potentials.

An expression giving the change of surface attraction, causing adsorption of organic material, as a function of potential was published by Butler [281]. The equation is identical with Frumkin's equation (42). Butler also gave the E_M and a values for the adsorption of charged and uncharged compounds on mercury. For organic compounds, the a value increases with increasing chain length; for branched-chain isomers the a value is larger than for normal isomers. Usually the a value is between 1 and 5 V^{-2} [281].

The E_M value also depends on the structure of the adsorbed organic molecule. For aliphatic alcohols, acids, and hydrocarbons, the E_M value usually ranges from -0.8 to -0.95 V relative to the mercury–sulfate electrode in 0.5 M Na_2SO_4 solution (the potential of this electrode is approximately 0.4 V more positive than the saturated calomel electrode). For aldehydes, nitriles, and amides, E_M is somewhat more negative, usually -0.95 to -1.1 V. Introduction of hydroxyl ions into an aromatic compound shifts the E_M value to positive potentials. Thus, for phenol and its homologs $E_M = -0.78$ V, for dihydrooxybenzenes this value is in the -0.6 to -0.65 V range, and for trihydroxybenzenes it is about -0.5 V. Introduction of a carboxyl also shifts E_M to more positive potentials, but introduction of an aldehyde group shifts it toward negative values. The small differences in E_M values for the isomeric o-, m-, and p-dihydroxybenzenes, which have different dipole moments, indicates that these compounds are adsorbed by the plane of the benzene ring and that the OH groups are at right angles to this plane [281].

Up to now we have been discussing equilibrium adsorption, by which is meant a state in which the compound in the bulk of the solution and that adsorbed on the surface is in equilibrium. Many investigations have shown that a long time is required for the attainment of adsorption equilibrium – in many cases tens of minutes (see, for example, [261]). To determine the rate of adsorption Melik-Gaikazyan [282] measured the dependence of so-called desorption-peak heights on the frequency of an applied alternating potential. The desorption peaks were first observed by Proskurnin and Frumkin [283] on the curves of the dependence of differential capacity of the electrical double layer on the electrode potential. These peaks occur on those parts of the curves where the $d\theta/dE$ relationship passes through a maximum, i.e., where the surface coverage shows a maximum change with electrode potential. Frumkin and Melik-Gaikazyan showed [284] that the limiting step of the adsorption process is the diffusion-controlled transport of particles to the surface and not the adsorption process itself. In much later work it was found that the time required for the adsorption of organic compounds of different structures is less than 10^{-5} sec, but the exact value of the rate of adsorption was not determined [286].

Assuming a diffusion-limited rate of adsorption, Koryta [287] calculated the change of coverage of the dropping electrode surface by adsorbed material as a function of time. For the calculation, Koryta used the Ilković equation and assumed that all material transported to the electrode by diffusion remains in the adsorbed state. Delahay and Trachtenberg [288, 289] examined the process of diffusion to planar, spherical, streaming, and dropping electrodes. They also considered that a certain time is required to attain the adsorption equilibrium. They proved that at concentrations of the compound in the solution which undergoes adsorption, which correspond to the linear section of the Langmuir adsorption isotherm (which indicates the possibility of application of Henry's law), adsorption equilibrium is attained only in tens of minutes.

A complete solution of the diffusion process for the adsorption at a dropping electrode at any concentration of the adsorbed compound with consideration of the attainment of adsorption equilibrium can be determined only by computers [290]. The solution obtained by Delahay and Fike [290] (in form of graphs) permits the determination of the degree of approach to equilibrium as a function of the adsorption time, the concentration of adsorbed material in the solution, and the adsorption coefficient β (which is the constant of the Langmuir equation). It is interesting to note that the degree of approach to the adsorption equilibrium $y = \Gamma_t / \Gamma_e$ (where Γ_t and Γ_e are the adsorbed amounts in moles/cm^2, in time t and in the equilibrium state, respectively) at small adsorption times increases nearly linearly with the square root of the time. The rate of approach to equilibrium increases with increasing concentration (c) of the adsorbed compound in the solution. It also increases with the adsorption coefficient β and with the value of D/Γ_∞ (where D is the diffusion coefficient of adsorbed material, and Γ_∞ is the limiting quantity adsorbed in a monolayer on the electrode surface). Delahay and Fike [290] showed a graph of the $y = \Gamma_t / \Gamma_e$ vs. t relationship for the spherical electrode and planar electrode for different values of the βc product with D/Γ_∞ constant and also curves of y vs. t relationship for different values of D/Γ_∞ with βc constant.

Reinmuth showed [291] that in the work of Delahay and Fike the variables determining the y value were not properly selected; therefore, although the general character of the $y = \Gamma_t / \Gamma_e$ vs. t

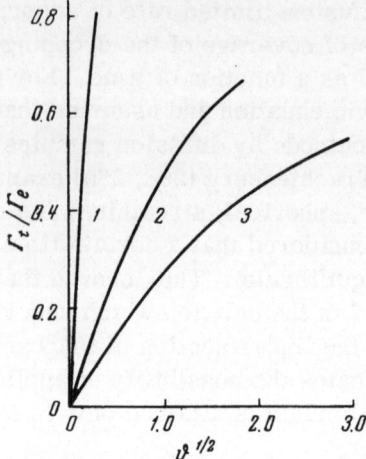

Fig. 1. The approach to equilibrium for the diffusion-limited rate of adsorption as a function of
the square root of the dimensionless parameter
$\vartheta = 4\pi Dt/\beta^2 \Gamma_\infty^2$ (according to Reinmuth [291]).
The values of βc are: (1) 10; (2) 1.0; (3) 0.1.

relationship in their work seems correct, the effect of a single
variable, like D, on the y value is not correct. Reinmuth calculated [291] the dependence of y from a dimensionless parameter
$\vartheta^{1/2}$ ($\vartheta = 4\pi Dt / \beta^2 \Gamma_\infty^2$) for a planar electrode for three βc values
(0.1, 1.0, and 10). Figure 1 shows the plot taken from Reinmuth's
work [291]. From Reinmuth's data, as from the work of Delahay
and Fike, it follows that y increases proportionally with $t^{1/2}$ at low
values, and the rate of approach to equilibrium increases with increasing concentration of the adsorbed compound in the solution.
V. G. Levich and E. D. Belokolos used a dimensionless parameter

$$\tau = \frac{7\pi}{12} \, tDc^2/\Gamma_\infty^2$$

for the calculation of approach to equilibrium at a spherical electrode. With the help of a computer they calculated the y vs. $\tau^{1/2}$
relationship for compounds that are adsorbed according to both the
Langmuir and Frumkin equations.

At larger β values, when all particles on the dropping mercury electrode (in absence of current), transported by diffusion
are adsorbed, the Ilkovič equation, as was already mentioned, can

be used. From this it follows for $\Gamma_t < \Gamma_\infty$ [287]:

$$\Gamma_t = 7.36 \cdot 10^{-4} c D^{1/2} t^{1/2}. \tag{43}$$

The numerical coefficient given in Eq. (43) was calculated for concentrations of the adsorbed compound (c) in the solution in moles/ liter, and D in cm^2/sec. Complete coverage of the electrode surface ($\Gamma_t = \Gamma_\infty$) in this case is attained, according to Koryta [287], at the moment t_1 after the drop formation

$$t_1 = \frac{1.85 \cdot 10^6 \Gamma_\infty^2}{D c^2}, \tag{44}$$

where t_1 is expressed in sec, Γ_∞ in moles/cm^2, D in cm^2/sec, and c in moles.

B. METHODS FOR THE INVESTIGATION OF ADSORPTION ON THE ELECTRODES

The adsorption characteristics of compounds on mercury electrodes can be determined by the change of surface tension on the mercury–solution interface, by the decrease of the capacity of the electrical double layer, and, finally, by the electrochemical effects caused by the adsorbed compounds.

The calculation of the amount of the adsorbed compound on the basis of the decrease in surface tension (σ) at the mercury–solution interface can be carried out by using the Gibbs equation, which, for a single adsorbed compound, can be given in the form

$$\Gamma_e = -\frac{d\sigma}{RT \, d \ln fc}, \tag{45}$$

where f is the activity coefficient of the adsorbed compound.

So-called electrocapillary curves (σ vs. E curves or the dependence of surface tension at the mercury–solution interface on the mercury potential) are measured to determine the adsorption characteristics. These curves are measured in the solution of a supporting electrolyte and in the presence of the surface-active compound. The electrocapillary curves are usually measured by the Lippman capillary electrometer (the principle and operation of which are described in many handbooks, such as in [7]). On the basis of electrocapillary curves measured at several concentrations of the surface-active compound, the quantities of adsorbed

compound are calculated. Thus the complete adsorption behavior
of the compound is determined on the liquid electrode in a solution
of given composition [261, 279]. This method was already in use
for the study of adsorption on the mercury electrode at the end of
the last century and it is still significant. A great number of electrocapillary curves for solutions of organic compounds of different
classes were obtained by Gouy [280, 292], Frumkin [293], Butler
[281], and other investigators. Among these, the work of Gierst
and his co-workers [294, 295], Conway and Barradas [296, 297],
Klyukina and Damaskin [265], Blomgren and Bockris [298], dealing
with the adsorption of pyridine and its derivatives, and the work of
Gerovich [299, 300], dealing with aromatic and unsaturated amines,
must be mentioned. In other studies the change of orientation of
adsorbed molecules with potential was investigated in part, and an
increased adsorptivity was found for certain cations, at positive
potentials, caused by interaction of their π-electrons with the electrons of mercury. The adsorptivity of the cations of aniline salts
increases with the introduction of alkyl substituents (at the nitrogen); thus the adsorptivity increases in the series aniline, methyl-,
dimethyl-, ethyl-, and diethylaniline [301]. Several studies were
made by Frumkin's co-workers on the adsorption of different organic compounds. Besides the works already mentioned [299, 300]
the study of adsorption of thiophene and other sulfur-containing
compounds must be mentioned [302] as must the work of Devanathan
and co-workers (for example, [303]). The latter investigated the
adsorption of tetrasubstituted ammonium iodides and showed that
their adsorptivity increases with increasing size of alkyl substituents. The work of Korchinskii [304] is also of interest. He investigated the adsorption of certain unsaturated organic acids from
absolute ethanol solution, in 0.1 N ammonium nitrate as the supporting electrolyte. He found that the acids of cis-configuration
(for example, maleic acid) are more strongly adsorbed than their
trans-isomers (as fumaric acid).

The electrocapillary curves can be measured not only on the
Lippman electrometer, but also with the help of a dropping mercury electrode. It is well known that the drop period of the dropping mercury electrode at constant flow rate of mercury is proportional to the surface tension. Therefore the curve showing the
dependence of the drop period on the electrode potential is similar

in shape to the electrocapillary curve. This was observed by
Kučera [305], who was the teacher of Jaroslav Heyrovský. The
measurement of the dependence of the drop time on potential (t vs.
E) does not require any special equipment and can be measured
much faster than the electrocapillary curve on a capillary electrometer. Recently, a simple modification was proposed for the auto-
matic registration of t vs. E curves [306]. It must be mentioned
that, for several reasons, the data obtained on the basis of t vs. E
curves are substantially less exact than those obtained from classical electrocapillary curves obtained from the Lippman electrometer. These reasons include variable flow rate, penetration of
the solution between the capillary wall and mercury at negative potentials, incomplete attainment of adsorption equilibrium on the
dropping electrode, and unequal coverage of the surface by the adsorbed compound, and the uplift of the dropping electrode by the
end of the capillary and tangential movement of mercury surface
on some other electrodes. Under strictly controlled conditions a
relatively high precision can be obtained in the measurements of
the t vs. E curves, as was the case, for example, in the previously
mentioned work of Gierst and his co-workers [294].

Even the classical method of measurement of electrocapillary curves does not permit a precision and sensitivity in the investigation of adsorption phenomena such as that obtainable by the
determination of the differential capacity of the electrode. This
method permits the investigation of adsorption effects not only on
liquid electrodes but also on solid electrodes. Differential capacity curves were first determined in 1935 by Proskurnin and
Frumkin [283]. These curves show the following characteristics
for solutions of surface-active materials. In the potential region
of adsorption the differential capacity is nearly independent of the
potential, and its value is substantially lower than that of the supporting electrolyte solution in the absence of the surface-active
material. On both sides of the adsorption region more or less
sharp peaks occur on the differential capacity curves. These are
caused by the desorption of the surface-active compound. Following these desorption regions (i.e., following the "peaks of desorption pseudocapacity") the capacity curves of the solution containing the adsorbable compound and of the pure supporting electrolyte
solution usually coincide.

The appearance of the pseudocapacity can be explained in the following way. The differential capacity curves are usually measured by placing a smaller sinusoidal potential on the polarized electrode. At the half period of the sinusoidal potential where the potential of the electrode is far from the potential of maximum adsorption, the coverage of the surface by the adsorbed material decreases; this results in an increase of the true capacity of the double layer. Therefore, to maintain the potential of the electrode corresponding to this half period, a certain electric quantity must be transferred to the electrode. At the half period of reverse sign of the sinusoidal potential, the electrode potential approaches that of maximum adsorption, and the coverage by the adsorbed compound increases. Thus, the capacity of the double layer reduces and, to maintain the electrode potential, a certain electric quantity must be removed from the electrode. Thus, owing to the change in adsorption and consequently the change of the capacity of the electrode double layer, if a sinusoidal voltage is placed on the electrode, an additional alternating current is generated that coincides in phase with the regular capacitance current caused by the charge–discharge of the double layer. The flow of this additional capacitance current corresponds to the appearance of the pseudocapacity on the differential capacity curves.

At very high frequencies of the applied alternating voltage at which the adsorption–desorption process cannot continue to follow the changes of potential, it is theoretically possible to obtain capacity curves without peaks. On these curves the reduced value of the capacity in the adsorption region will continuously change into the normal values of the capacity of the pure background solution, if the potential is changed. The potential of the peak maximum corresponds to the point where the coverage of the electrode surface changes most with potential [284]. The peak height and its shape (width) under exactly maintained conditions increases with the attraction factor, γ, of the Frumkin adsorption isotherm [263]. It was shown by Damaskin and Tedoradze [307] that at $\gamma > 1.5$ the surface coverage of the electrode at the potential of the desorption peak is half of the maximum, i.e., $\theta = 0.5$. As the concentration of the adsorbed compound increases in the solution, the desorption peak potentials move further from the point of zero charge. On the basis of the dependence of peak potential on the concentration of adsorbed compound in the solution and with the help of other characteristics of the curves of differential capacity, all adsorption properties of the compound investigated can be determined. The dependence of γ on the potential can also be calculated [263, 264, 307]. With increasing concentration of the adsorbed organic compound, the capacity of the double layer between the desorption peaks decreases. This is caused, as mentioned earlier, by the "moving apart" of the condenser plates (of the double layer) and by the reduction of the dielectric constant of the medium between the plates (exchange of water for organic molecules). By considering the capacity decrease ("depression") as a function of the concentration of the adsorbed compound, the adsorption isotherm of the given compound can be determined with quite high accuracy. According to Frumkin [261, 273, 279] the capacity of the double layer (far away from the pseudocapacity peaks, which are caused by the adsorption–desorption process) is a linear function of the electrode surface coverage by the adsorbed compounds

$$C = C_0(1 - \theta) + C'\theta, \tag{46}$$

where C_0 is the capacity of the double layer at the given potential in absence of adsorbed material (at $\theta = 0$); C' is the capacity at full monolayer coverage of the electrode surface ($\theta = 1$).

The adsorption isotherms of the tetrabutyl ammonium cation were determined by a measurement of the double-layer capacity at several potentials [273]. Furthermore, the adsorption characteristics of different organic compounds were determined: alcohols, esters, amines, nitrogen containing heterocyclic compounds, ketones, and several others in 0.1 N solutions of sodium perchlorate [308]; of n-valeric acid in acidic perchlorate solutions [309, 310]; and n-amyl [311] and n-hexyl alcohols [288].

Qualitative data on the adsorption of compounds and also on the potentials of desorption can be determined as was mentioned in the previous section on the so-called Breyer polarograms [252]. These curves give the dependence of the variable component of the current on the potential of the dropping electrode if a smaller sinusoidal voltage is applied at the electrode simultaneously with the constant voltage. At the potentials of desorption of the compound, usually on both sides of the electrocapillary zero, so-called tensiometric waves, which appear as maxima (peaks) on the curves, can be observed on the Breyer polarograms. The height

of these peaks depends on the concentration of the adsorbed compound and on the frequency of the sinusoidal voltage employed. On
the basis of the effect of different factors on the potentials of
tensiometric waves the dependence of the adsorption of the compounds, causing the tensiometric waves, can be determined. By
this method effects of supporting electrolytes on the tensiometric
curves of ethyl, n-amyl, and cyclohexyl alcohols were investigated
[313]. The deficiencies of Breyer's tensiometric waves were mentioned in the previous chapter (see page 53). A critical evaluation
of the method was given by Frumkin and Damaskin [314].

Qualitative information about the adsorption can be obtained
from the "indentations" on oscillographic curves of the Heyrovský–
Forejt type [315]. These are curves of the dE/dt vs. t relationship
(where E is the potential of the electrode with a sinusoidal voltage
applied on the cell). The influence of the adsorption of different
aliphatic alcohols on the dE/dt vs. E curves was investigated by
Matyáš [316]. Kalvoda [317] investigated the effect of the solubility and concentration of many organic compounds, which represented different classes of compounds, on the form and depths of
adsorption–desorption "indentations" on the oscillographic dE/dt
vs. E curves. He was particularly interested in the minimum concentration of compounds at which the "indentations" on the oscillo-
grams could be still observed.

The adsorption of compounds on the electrode surface is indicated by steps (delays in the potential change) on the oscillographic E vs. t curves obtained by the method of Heyrovský and
his co-workers [318]. On the oscillographic i vs. E curves measured by the Ševčik–Randles method [319, 320], the change of the
double-layer capacity corresponding to desorption appears in the
form of additional steps. Kalvoda [321] discussed the effect of adsorption on the form of different types of curves recorded in
oscillographic polarography. The presence of adsorption can be
identified also on the curves recorded with rectangular voltage polarization with the help of Kalousek's commutator [322].

With the methods just discussed, adsorption can be investigated only in that potential range in which electrochemical reactions do not take place, which means that the faradaic current is
practically absent. Dzhaparidze and Tedoradze [323] developed a
method for the measurement of the differential capacity under con-

ditions of faradaic current flow, in the case of catalytic hydrogen evolution. They used as an electrode a mercury drop suspended on a gold-coated platinum wire. In the opinion of Dzhaparidze and Tedoradze [323] only this construction permits sufficiently precise determination of the differential capacity at the mercury electrode under conditions when electrochemical reactions are simultaneously occurring.

The effect of the adsorption of the depolarizer on the rate of electrochemical processes is used in the third group of methods for the investigation of adsorption. Principal among these methods is chronopotentiometry (measurement of the potential as a function of time during electrolysis with a given current intensity), which was discussed in the previous chapter. Lorenz [324] first used chronopotentiometric measurements for the determination of the depolarizer quantity adsorbed on the electrode. Recently a number of papers have been published dealing both with the theory and with the practical application of this method [325-330].

Adsorption of depolarizer on the electrode increases its concentration at the electrode. This causes an increase in the transition time τ (mainly at higher current densities) and changes its dependence on the i_0 value of the applied current. Thus, if the τ value is normally proportional to $(1/i_0)^2$ for an electrode process that is limited by diffusion, then for discharge of adsorbed depolarizer on the electrode the transition time is proportional to $1/i_0$. Depending on the reversibility of the electrochemical stage, and several other factors, the discharge of the adsorbed particles can take place at more positive (easier), more negative (more difficult), or at the same potential as for particles transported by diffusion. Therefore, in certain cases two steps appear on the chronopotentiograms corresponding to the processes taking place with the adsorbed and with the diffusion transported material, respectively. But very often only a single step appears. It is not easy to determine if the adsorbed material enters the electrochemical reaction earlier or later from the shape of chronopotentiograms measured at constant (i_0) current intensity. But this is of importance for the determination of the quantity of adsorbed material and particularly important for the understanding of the mechanism of electrode processes. This problem can be solved much more simply if the chronopotentiograms are measured not at a constant current but at a current that changes in intensity according to a known relationship [331]. For example, it is well known in chronopotentiometry that with a current which changes in intensity proportionally with the square root of time [332], the transition time for simple diffusion-limited processes is proportional to the depolarizer concentration in the solution. (The theory of this method has been developed for cylindrical and spherical electrodes [333].) This is a great advantage for this method from an analytical viewpoint. The application of a linearly changing current ($i_0 = \beta t$) in chronopotentiometry or stepwise changes of current intensity (for a time current i_{01} is applied, then this value is changed to i_{02}) shows whether the adsorption of the depolarizer promotes or retards the electrochemical process [331].

Chronopotentiometry with reversal of current direction (at time τ_1) permits the determination of the adsorption of the products of a reversible electrochemical reaction on the electrode. The transition time of the reverse process (τ_2) increases with the quantity of adsorbed product and tends toward the transition time of the forward process (τ_1) [334, 335]. This method was used for the investigation of iodine adsorption in the oxidation of iodide on a platinum electrode [334, 335] and in the investigation of the adsorption of the leuko-form in the reduction of riboflavin on a mercury electrode [335]. In the latter case, on reversal of the current, the nonadsorbed leuko-form is initially oxidized, and then the particles adsorbed on the electrode are oxidized [335].

The adsorption of the depolarizer on the electrode can be studied by the so-called integral method with potential jump [336]. This method is based on the determination, in the form of a time function, of the quantity of electricity passing the electrode after stepwise changes of the electrode potential up to values at which the electrochemical reaction takes place. Under controlled conditions, the portion of the total electricity passed because of the electrochemical process and not connected with diffusional transport of the material can be calculated. Thus the quantity of depolarizer adsorbed on the electrode can also be calculated. The quantity of adsorbed depolarizer can also be determined from the dependence of the quantity of electricity (Q) and from the rate of electrode potential scanning (V) during an oscillographic investigation of processes with linearly changing potential [337]. For a reversible process with a single electrochemical stage and diffusional transport of the depolarizer, the Q value is proportional to $V^{-1/2}$. If the depolarizer is adsorbed on the electrode, an additional component (Q_{ads}) appears which is proportional to the quantity of adsorbed depolarizer. This component can be determined by the intersection with the ordinate by extrapolation of the linear Q vs. $V^{-1/2}$ graph [337].

The coulostatic method [338] also can be used for the investigation of adsorption phenomena, partly for the study of adsorption kinetics on the electrode. This method is based on analysis of the curves of potential change with time (in open circuit) after the electrode has been charged for a very short time (of the order of one millisecond) with an exactly determined quantity of electricity.

The adsorption of the depolarizer on the electrode also affects the surface kinetics and catalytic currents; these will be discussed in detail in the following sections of this book.

C. THE USE OF SPECIFIC PROPERTIES OF THE DROPPING MERCURY ELECTRODE FOR THE DETERMINATION OF ADSORPTION OF ORGANIC COMPOUNDS

The specific properties of the dropping mercury electrode — the continuous growth of its surface and the appearance of tangential motion under conditions of polarographic maxima of the first and second kind — can be used for the qualitative determination of adsorptivity of organic compounds, and also for the estimation of more or less accurate data on the adsorption. The data obtained with the dropping mercury electrode are particularly interesting in polarographic studies, since very often (without intention or without anticipating that the adsorption equilibrium is established) they can be used in the evaluation of electrochemical effects caused by adsorption.

The methods utilizing suppression of polarographic maxima are very sensitive. But these methods are only capable of evaluating the absorptivity of organic compounds qualitatively.

Polarographic maxima are caused by the tangential movement of the mercury surface, due either to nonuniform polarization of the dropping electrode (maximum of the first kind), or due to the fast outflow of mercury from the capillary (maximum of the second kind); see the appendix in [5]. The adsorption of surface-active compounds has a dampening action on the tangential movement and therefore reduces the polarographic maxima. Under equal conditions, the more the compound is adsorbed at the mercury electrode, the more the polarographic maxima are suppressed. For the determination of surface-active properties of compounds, usually their ability to suppress maxima is followed using the maximum of the first kind observed on the oxygen reduction wave in dilute potassium chloride solutions or other electrolytes saturated with air. Thus, for example, the adsorptivity of a great number of hydrocarbons was determined on the basis of their ability to suppress the oxygen maximum [339]. On the basis of their ability to suppress this maximum the molecular weights of high-molecular-weight compounds were estimated. Examples include the hydrolysis products of starch [340], basic dye stuffs, and phthalic esters of diacetylcellulose [341]. Under identical conditions, the effect of high-molecular-weight compounds of similar structure on suppressing the maximum increases with increasing molecular weight [340, 341].

The relative surface activities of a series of thiobarbituric acid [342] and barbituric acid derivatives [343] were determined on the basis of their suppression of the oxygen maximum. The adsorptivity of barbiturates was also determined [343] by the curves, which showed the dependence of the drop time of the dropping electrode on its potential measured in presence and absence of surface-active compounds.

Besides oxygen reduction waves, the suppression of the maxima of first kind of other depolarizers was also used for the study of surface-active compounds (see, for example, [344]). If alcohol is added to the solution, the maximum suppressing effect of organic compounds is reduced (due to smaller adsorption) [344].

Frumkin and his co-workers developed almost entirely the theory of the formation of maxima and their suppression by surface-active compounds. They also proved experimentally the equations derived. Among the co-workers of A. N. Frumkin, T. A. Kryukova deserves special mention because she devoted much effort to the experimental investigation of maxima.

Kryukova pointed out the importance of a high level of purity of investigated solutions in particular in the study of maxima of the second kind. This enabled her to observe many effects, such as the stirring of the solution resulting from the tangential motion of the electrode surface, that were not observed previously. Kryukova determined quantitatively the dependence between the tangential motion of the surface and the characteristics of the dropping electrode which forms the basis for the selection of optimum conditions to avoid maxima of the second kind in the work with the dropping electrode.

Among the investigations of Kryukova dealing with the ability of organic compounds to suppress the tangential motion, a detailed investigation of the effects of aliphatic alcohols on the motion of the mercury surface under conditions of appearance of a maximum of the second kind must be mentioned. This maximum was studied on the mercury ion reduction wave obtained in mercuric chloride solutions in the presence of large amounts of potassium chloride [345]. The suppressing effect increased with increasing length of the hydrocarbon chain of the alcohol. At low coverage of the surface, the decelerating effect is approximately proportional to the square of the value of that of the alcohol adsorption coefficient. The deceleration of tangential motion under conditions of appearance of maxima of the second kind was also used to determine the molecular weight of polyvinyl alcohol samples [346].

The ability of surface-active compounds to suppress polarographic maxima appears only in the potential region of their adsorption. This permits the determination of this potential region. At low coverages the deceleration of tangential motion depends on the rate of transfer of the surface-active compounds to the electrode. These and other phenomena were investigated in detail by Kryukova [5], together with development of the theory of the origin of maxima.

Valuable information about adsorption at the dropping mercury electrode can be obtained by investigation of charging current [262], which under certain conditions forms a substantial part of the polarographic background current. The simplicity of this method, since it requires no equipment except the polarograph, makes it particularly attractive for semiquantitative evaluation. We will deal with this method in some detail.

In the polarography of solutions from which the oxygen has been removed and that contain only the supporting electrolyte, the current passing through the cell, the so-called residual current, consists of two components. The first is a small faradaic current caused by the presence of traces of electrochemically active compounds (including oxygen) in the solution, and the second is the charging current.

The charging current of the dropping mercury electrode is caused by consumption of electricity for charging the electrical double layer of the newly formed mercury–solution interface. Thus, the charging current depends on the rate of formation of the surface of the dropping mercury electrode and the charge density (ε) at a given electrode potential. The average value of the charging current (in μA) during the drop period t is

$$i_c = 8.5 \cdot 10^{-3} m^{2/3} t^{-1/3} \varepsilon. \tag{47}$$

From the value of the charging current the charge density (ε) can be determined by Eq. (47).

The faradaic component of the residual current usually can be expressed by the Ilkovič equation

$$i_D = 607 m^{2/3} t^{1/6} \Sigma n_i D_i^{1/2} c_i, \tag{48}$$

where $\Sigma n_i D_i^{1/2} c_i$ is the sum of the products of the concentration of individual impurities, the square root of their diffusion coefficient, and the number of electrons required in their reduction.

A comparison of Eqs. (47) and (48) shows that a decrease of the drop period t of the electrode, under otherwise identical conditions, results in an increase of the charging current and in a reduction of the faradaic current, as given by $i_c / i_D = kt^{-1/2}$. Therefore, electrodes with short drop period must be used to determine the capacity component of the residual current. In Figs. 2 and 3 the curves for the residual current at different concentrations of 2,6-lutidine, obtained in oxygen-free solution of 0.1 M aqueous potassium hydroxide and a 16 volume percent ethyl alcohol solution,

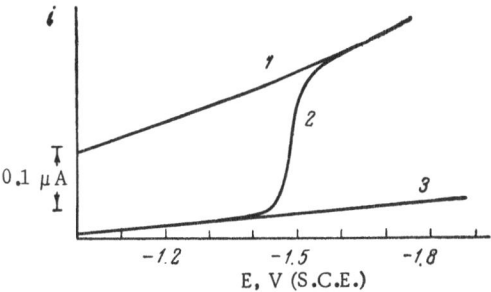

Fig. 2. Residual current curves. (1) 0.1 M aqueous KOH
solution; (2) same solution with addition of 51.5 mM
2,6-lutidine; (3) linear extrapolation of curve (2).

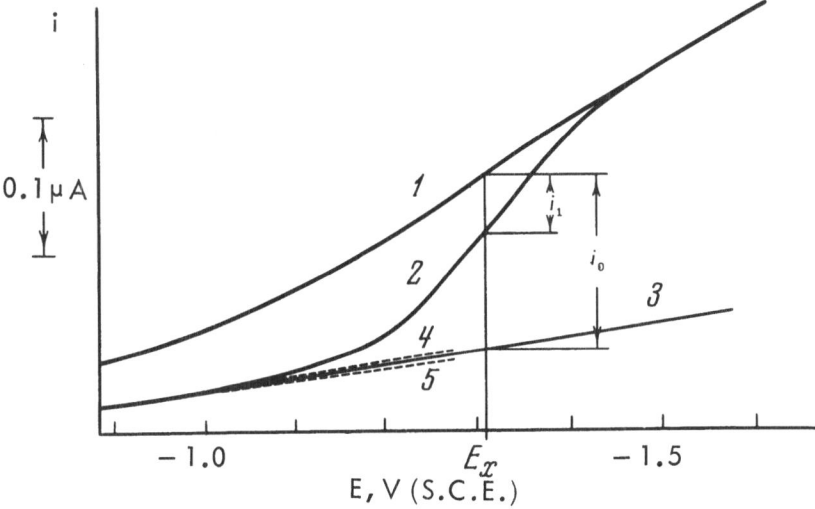

Fig. 3. Residual current curves. (1) 0.1 M potassium hydroxide solution in 16% (by
volume) ethyl alcohol; (2) same with addition of 17.5 mM 2,6-lutidine. For further
information see the text.

respectively, are given. These curves were obtained with an electrode equipped with a glass plate for the regulated detachment of
drops [347] and had the following characteristics: m = 1.07 mg
per sec, t = 0.26 sec. The use of an alkaline solution substantially
removed the catalytic hydrogen current caused by the lutidine and
shifted its wave to sufficiently negative potentials. Besides this,

in an alkaline solution the reduction of many compounds proceeds much more slowly. Therefore, in such a supporting electrolyte the faradaic component of the current is much lower than in neutral or acidic media.

Heyrovský, Šorm, and Forejt [318] first observed the sharp changes of residual current caused by the adsorption of pyridine in alkaline solution on mercury and by its desorption at certain potentials. The residual current curves shown in Figs. 2 and 3 generally have the same shape as the curves in the mentioned work [318].

It follows from Figs. 2 and 3 that addition of lutidine to the solution significantly decreases the residual current at potentials more positive than −1.5 V. This decrease is caused by the adsorption of lutidine that results in a decrease of the double-layer capacity and consequently a reduction in the charging current [see Eq. (47)].

According to Frumkin [261, 273, 279] the electrode surface charge (ε) is a linear function of the coverage (θ) by the adsorbed compound [compare with Eq. (46)].

$$\varepsilon = \varepsilon_0(1 - \theta) + \varepsilon_1\theta, \tag{49}$$

where ε_0 is the charge of the surface free of adsorbed material; ε_1 is the surface charge when the surface is completely covered by the adsorbed material.

From Eq. (49) it follows that

$$\theta = \frac{\varepsilon_0 - \varepsilon}{\varepsilon_0 - \varepsilon_1}. \tag{50}$$

If it is assumed that in a certain potential interval the orientation of adsorbed molecules does not change at the electrode surface, the charge density and consequently also the capacity current can be given as a linear function of the potential. The values of the capacity current, which could be measured in the absence of desorption at more negative potentials, can be determined by a linear extrapolation of the capacity current from the region of more positive potentials, where the electrode coverage at sufficiently high concentrations of the adsorbed compound is practically complete. In other words, if the faradaic current represents a small fraction of the measured current (or if its value does not change with potential) such an extrapolation (compare with the line 3 in Figs. 2 and 3) gives the value of the residual current that would be observed at complete coverage of the electrode surface at sufficiently negative potentials.

By applying Eqs. (47) and (50) and using the difference in the residual current in the absence and in the presence of surface-active compounds, the degree of surface coverage by adsorbed compound can be calculated at a given potential (given in Fig. 3 as E_X) by

$$\theta = \frac{i_1}{i_0} . \tag{51}$$

This equation can be used only if the faradaic component of the residual current is much smaller than the capacity component or if its value is the same for the residual current in presence or absence of surface-active compounds.

In Eq. (51), i_1 is the difference between the residual current in the absence and in the presence of surface-active compounds. This is given by the distance between curves 1 and 2 (see Fig. 3). i_0 is the difference between the residual current in the absence of adsorbed material and its value measured at complete coverage of the surface by adsorbed material is i_0. This difference corresponds to the distance between curve 1 and line 3 (see Fig. 3), at the potential for which θ is being determined.

From the Frumkin isotherm equations (40) and (42) we obtain, after changing to logarithms

$$0.43a\varphi^2 + \log \frac{\theta}{1-\theta} = 0.43 \cdot 2\gamma\theta + \log \beta^0 c. \tag{52}$$

From Eq. (52) it follows that if the expression

$$0.43a\varphi^2 + \log \frac{\theta}{1-\theta}$$

for a given concentration (c) of the adsorbed compound in the solution is plotted as a function of θ, a straight line results. From this line the γ and β^0 values can be calculated. In Figs.. 4 and 5 are given plots in these coordinates for the adsorption of 2,6-lutidine at several concentrations in 0.1 M KOH in

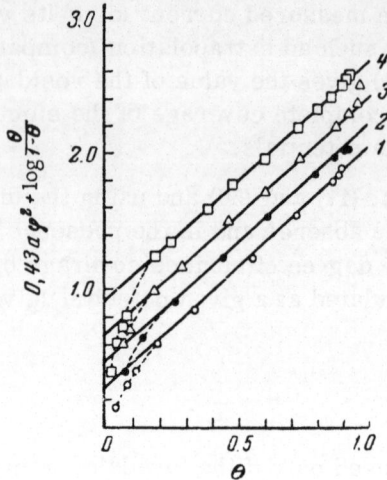

Fig. 4. Dependence of $0.43a\varphi^2 + \log[\theta/(1-\theta)]$ on θ
for the adsorption of 2,6-lutidine on the drop electrode
from 0.1 M aqueous KOH solution. Lutidine concentration in mM: (1) 17.2; (2) 23.7; (3) 34.3; (4) 51.5.

water (Fig. 4) and in 16% ethyl alcohol solution (Fig. 5), respectively. The θ values at different potentials were calculated using
Eq. (51) from the residual current curves. Examples of these
curves are given in Figs. 2 and 3. For the calculation of the a
value, $\Gamma_\infty = 5 \cdot 10^{-10}$ moles/cm^2 was used (which is close to the
values obtained for other pyridine derivatives [295]); the C values
were determined from the slope (average) of the corresponding
residual current curves in the -1.1 to -1.4 V potential region
(versus S.C.E.):

$$C = \frac{1000}{8.5} m^{-2/3} t^{1/3} \frac{\Delta i_c}{\Delta E}.\tag{53}$$

Finally, the C' values were determined from the slope of lines 3
(see Figs. 2 and 3). The values determined for aqueous solutions
are: $C = 21.5\ \mu F/cm^2$ and $C' = 4.3\ \mu F/cm^2$; for solutions in 16%
ethyl alcohol: $C = 25\ \mu F/cm^2$ and $C' = 7.7\ \mu F/cm^2$. Thus, the a
values in these solutions, in the given potential range, are practically identical: $a = 6.8\ V^{-2}$. It must be mentioned that the C
value for the 16% ethyl alcohol solution is higher due to superim-

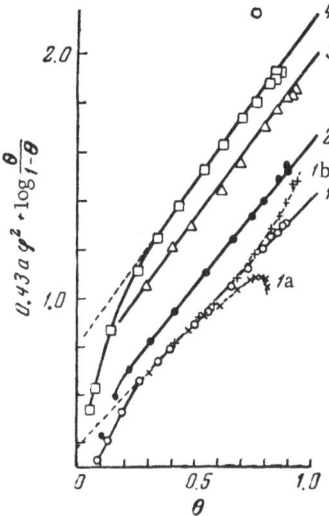

Fig. 5. Dependence of $0.43a\varphi^2 + \log[\theta/(1-\theta)]$ on θ for the adsorption of 2,6-lutidine on the dropping electrode from 0.1 M KOH solution in 16% (volume) ethyl alcohol. Lutidine concentration in mM: (1) 17.6; (2) 24.4; (3) 40.6; (4) 52.7.

position of a pseudocapacitance caused by the desorption of alcohol. This desorption takes place in the same potential range as the desorption of lutidine. Therefore, the given calculation of the adsorption characteristics of 2,6-lutidine in 16% ethyl alcohol solution is only approximate. To obtain more exact values for the adsorption parameters of surface-active compounds in water—alcohol mixtures, only that part of the charging current can be applied for alcoholic solutions that is outside of the potential of practically complete alcohol desorption [276].

The potential of maximum adsorption (E_M) depends generally on the concentration of the compound in the solution; to simplify the calculations, it was assumed that in aqueous and alcoholic aqueous solutions $E_M = -0.70$ V (S.C.E.) and is independent of the lutidine concentration. The resulting error will be discussed later. The $E_M = -0.70$ V value corresponds to the potential of the intersection of the residual current curves in the absence of lutidine and is for a concentration of about 17 mM.

It can be seen from Figs. 4 and 5 that the calculated values of θ from the experimental data can be represented by straight lines in the range 0.2 to 0.85 for aqueous solutions and from 0.3 to 0.85 for the 16% alcohol-containing solution. These straight lines are not parallel with the abscissa, which shows that the adsorption of lutidine in the solutions investigated can be described by the Frumkin adsorption isotherm.

The calculated values do not fall on the straight lines at high surface coverages ($\theta > 0.85$). This can be explained by the increased effect of deviations caused by inexact extrapolation of lines 3 in Figs. 2 and 3. In Fig. 3 the dotted lines 4 and 5 show the result of a slightly different extrapolation of the residual current for $\theta = 1$ in the region of negative potentials. In Fig. 5 the points on curves 1a and 1b correspond to the values calculated from the lines 4 and 5 of Fig. 3, respectively. It follows from Fig. 5 that a small error in constructing lines 3 (see Figs. 2 and 3) causes a substantial deviation from the straight line on Fig. 5 only at very high θ values. Consequently, errors committed in the construction of the straight lines used for extrapolation on the graphs, of the type given in Figs. 2 and 3, do not affect the graph given in Figs. 4 and 5 if only the central portions of the curves are used for the determination of the adsorption isotherm. The systematic deviation of the points from the lines in Figs. 4 and 5, observed at low coverages ($\theta < 0.2$-0.3), is caused by the fact that under these conditions the adsorption equilibrium is not attained on the electrode surface. In adsorption following the Langmuir isotherm ($\gamma = 0$) the rate of approach to adsorption equilibrium increases at higher values of β. The β value, as follows from Eq. (42), increases at more positive potentials. If the adsorption follows the Frumkin isotherm ($\gamma > 0$), at higher θ values the effective value of β increases substantially, and consequently the approach to adsorption equilibrium increases. It is very hard to apply a correction for an incomplete attainment of adsorption equilibrium if the adsorption follows the Frumkin isotherm.

It can be seen from Fig. 4 that the slope of the lines for adsorption from aqueous solutions is independent of the lutidine concentration. The value of the slope corresponds to an attraction constant $\gamma = 1.95$-2.00 (at a potential of about -1.4 V). According to Eq. (52), at increasing lutidine concentrations (c), the lines in

Fig. 4 should shift upward proportionally with $\log c$, but the shift of the lines determined experimentally is much larger than that. The reason for this is that at higher lutidine concentrations the E_M values shift toward more negative potentials, and this change was disregarded in the construction of the lines. Therefore the real potential φ' values are somewhat smaller in absolute value than the potential φ values used for the construction of Fig. 4. Assuming that the a and γ values do not change with the potential, one can calculate the change of the E_M value with potential for increasing lutidine concentrations from the degree of excess shift of the lines. Thus, if at $c = 17.2$ mM, $E_M = -0.70$ V, then at lutidine concentrations of 23.7, 34.3, and 51.5 mM the E_M values are -0.712, -0.717, and -0.733 V, respectively. The β_0 value in these solutions is 115 M^{-1} (0.115 mM^{-1}). The smaller deviations ($\pm 10 - 15$ mV) in this calculation (for the E_M value selected) do not in this case affect the slope of the central portion of the graph given in Figs. 4 and 5. This means that they do not affect the γ value, determined from the slope of the lines. The reason for this is that the φ value has an order of magnitude of 600-700 mV, and the range of potential change for θ values from 0.3-0.7 is much smaller, only 100 mV. The error in determining the E_M value becomes apparent only in β_0 calculated by Eq. (52). However, if Eq. (52) is used for the determination of the equilibrium θ values outside of the potential range of desorption [extrapolating with Eq. (52)], such an extrapolation (if not carried out too far) gives correct θ values since the errors in the E_M and β_0 values compensate each other.

It is seen from Fig. 5 that in the adsorption from 16% alcohol the linear sections become shorter at increasing lutidine concentrations. This means that the γ values increase. From the figure, at concentrations $c = 17.6$, 24.4, 40, and 52.7 mM the γ values are 1.22, 1.42, 1.53, and 1.57, respectively (for $E_M = -0.70$ V, and $\beta_0 = 115$ M^{-1}). The increase of the γ value may be the result of displacement of the adsorption region toward more negative potentials, where higher values were observed [265], or may be caused by the smaller effect of the alcohol on the attraction factor at increasing lutidine: alcohol ratios. It must be mentioned that in the presence of alcohol the attraction factor is much smaller. Therefore, at the same lutidine concentration the surface coverage by lutidine and the rate of approach to the adsorp-

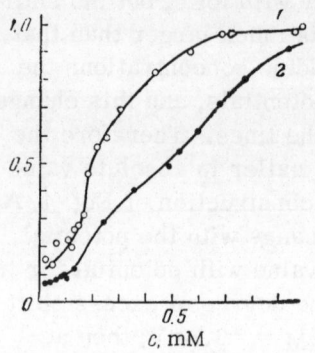

tion equilibrium are much smaller in alcoholic solutions than in aqueous solutions. A partial explanation of this phenomenon is that the adsorption equilibrium is reached in aqueous solutions at much lower θ values than in 16% ethyl alcohol solution (compare Figs. 4 and 5).

Fig. 6. Polarographic ("instantaneous") adsorption isotherms of 5-bromo-2-acetyl-thiophene from alkaline solutions in 4% (1) and 40% (2) methyl alcohol. m = 2.33 mg/sec, t = 0.17 sec.

The evaluation of the adsorption isotherm using the charging currents can be made even when the adsorbed compound enters an electrochemical reaction relatively easily and the range of the residual current occupies only a narrow potential region on the polarogram ahead of the wave of the depolarizer. To determine the adsorption isotherm in this case, the ratio of the value of the residual current to that of the discharge wave of the depolarizer is measured at a single potential as a function of the depolarizer concentration in the solution. Furthermore, using Eq. (51), the instantaneous θ value is calculated at different c values. Then a graph of θ vs. c is constructed. In Fig. 6 the "instantaneous" adsorptions isotherms of 5-bromo-2-acetylthiophene measured in 4 and 40% methyl alcohol in a supporting electrolyte consisting of 0.1 N CH_3COOK and 0.1 N KOH are shown as examples. The curves were obtained at -0.90 V for the 4% solution and at -0.95 V (S.C.E.) for the 40% methyl alcohol solutions at 25°C, respectively [348].

In the method just described for the evaluation of the adsorption characteristics, the effect of the adsorbed compound on the faradaic component of the residual current was not considered. Most likely, due to the errors thus committed, the adsorption isotherms of Fig. 6 do not pass through the zero point of the coordinates. As has already been mentioned, at sufficiently low drop periods, and when the oxygen is carefully removed from the test solution, the faradaic current component is not large. Thus the error in the determination of adsorption characteristics caused by change of the faradaic component can be neglected in many polarographic calculations.

In several cases polarography gives the exact value of the capacity component of the residual current and thus the integral capacity of the dropping electrode. This can be done when the specific integral capacity is independent of the drop period of the dropping electrode, i.e., in the absence of surface-active compounds in the solution (C can be determined at $\theta = 0$) and also when an excess of surface-active compounds is present at surface coverages close to complete coverage. For the determination of the integral capacity, the residual current value is measured as a function of the life of the drop [349] or as a function of the drop period at constant flow rate of mercury (the drop period is regulated either by a little hammer, falling periodically on the capillary, or by a propeller-driven scraper [350]). A plot of $i_{res}/t^{1/6}$ vs. $t^{-1/2}$ is constructed; it is linear since [compare with Eqs. (47) and (48)]

$$i_{res} = i_c + i_D = 0.0085 m^{2/3} t^{-1/3} C \Delta E + 607 m^{2/3} t^{1/6} (\Sigma D_i^{1/2} c_i n_i), \qquad (54)$$

where ΔE is the potential, measured from the electrocapillary zero in the given solution.

From the slope of this plot the integral capacity C can be calculated.

The adsorption characteristics of compounds on mercury electrodes can also be determined by so-called adsorption prewaves (or adsorption postwaves), which will be discussed in the following section.

D. THE EFFECT OF ADSORPTION OF THE DEPOLARIZER IN REVERSIBLE REDOX SYSTEMS; ADSORPTION PREWAVES AND POSTWAVES

In the polarography of organic compounds, reversible redox systems are formed in several instances, and on the polarograms small additional steps can be observed that, according to potentiometric data, should not appear. Such a wave was first observed by Brdička and Knobloch [351] on the polarograms of riboflavin. Müller [352], independently of the previous investigators, found a similar wave on the polarograms of α-oxyphenazine. He showed

by special experiments that this wave cannot be caused by the reduction of some impurities in the solution, and he explained its appearance by the existence of an unknown modification (or tautomer
form) of the compound investigated. Brdička also found a similar
wave on the polarogram of methylene blue [353] and assumed that
the appearance of such waves is caused by adsorption phenomena.
On this assumption he developed the theory of adsorption waves
[278].

If the adsorption of one of the components of the redox system is much larger than the adsorption of the other, then according to Brdička the adsorption energy changes the potential at
which the electrode process takes place. If the product of the
electrode reaction is adsorbed, the liberated adsorption energy
promotes the electron transfer in the forward electrochemical reaction. This means that the electrochemical process takes place
at a more positive potential and on the polarographic curve an adsorption prewave appears. If the starting material is strongly adsorbed, then an adsorption postwave can be observed on the polarographic curves [278].

Senda et al. [354] investigated the adsorption prewave
in solutions of flavinomononucleotide by the method of polarography with superimposition of alternating voltage (Breyer's method).
They observed that the capacity current of the supporting electrolyte is reduced at substantially more negative potentials than the
reduction wave, as compared to the decrease of the residual current at potentials more positive than the region of potentials corresponding to the adsorption prewave. This confirms the conclusion of Brdička that for an adsorption prewave the reduced form
of the redox system is more strongly adsorbed than the oxidized
form.

To develop equations that relate the height of the adsorption
wave to the determining factors [278], let us consider polarographic curves of reduction processes showing an adsorption prewave.
In this case the reduced form of the redox system is more strongly adsorbed (equations for the other case can be obtained by an
analogous technique).

The experimental results show that at low concentrations of
the depolarizer in the solutions only a single wave appears on the

polarographic curves at potentials where at higher concentrations the prewave is observed. The height of this prewave increases in proportion to the concentration of the compound that is being reduced. With increasing concentration of the depolarizer the prewave height increases and reaches a limit. Usually, the main reduction wave begins to increase only after this has occurred. The sum of limiting currents of both waves will be proportional to the concentration of the compounds being reduced in the solution. According to Brdička, the maximum height of the adsorption prewave (i_a) is determined by the quantity of particles in the reduced form of the redox system that can be adsorbed at the given electrode surface. The instantaneous adsorption current is determined by the quantity of particles adsorbed during unit time. Under conditions favoring the attainment of maximum adsorption current, i.e., when the adsorbable products of the electrode reaction are present in excess at the electrode, the instantaneous adsorption current is evidently proportional to the surface of the dropping electrode increased per unit time (ds/dt). It is easy to show that

$$ds/dt = {}^2/_3 \cdot 0.85 m^{2/3} t^{-1/3}. \tag{55}$$

If A is the electrode surface covered by one adsorbed particle of the electrode reaction product, the instantaneous limiting adsorption current can be given as

$$i_{(a)\,\text{inst}} = \frac{{}^2/_3\,0.85 m^{2/3} n F}{A\,t^{1/3}}\,, \tag{56}$$

where n is the number of electrons taking part in the electrochemical reaction and F is the Faraday number.

From Eq. (56) it follows that as the drop grows, the instantaneous limiting adsorption current decreases owing to the decreased rate of formation of new surface. This was first shown by Brdička on current intensity vs. time curves measured during the life of single drops [278].

The mean value (during the drop life) of the maximum possible current of the adsorption prewave is determined by the average surface of the drop electrode [compare with Eq. (23)] formed in unit time. Expressing the average limiting current of the prewave (i_a) in μA, A in Å , and m in mg/sec, i_a can be calculated using

$$i_a = \frac{13.5nm^{2/3}}{At^{1/3}}.$$ (57)

From Eq. (57) several characteristic properties of the limiting adsorption current (i_a) can be determined. First of all, i_a is
independent of the depolarizer concentration. With increasing
height of the mercury column (h_{Hg}) above the orifice of the dropping electrode, i_a increases proportionally (because m changes
proportionally and t is inversely proportional to the h_{Hg} value).
Finally, i_a is larger for those compounds whose particles occupy
a smaller surface area in the adsorbed state in the monolayer
(because more such particles can be absorbed on the same electrode surface). Two further properties of the adsorption current
must be mentioned. With increasing temperature the adsorptivity
of materials decreases. Therefore, at sufficiently high temperatures, the adsorption prewaves disappear. Furthermore, if strongly adsorbed compounds that displace the depolarizer from the
electrode surface are added to the polarographed solution, a decrease of the adsorption wave can be observed.

From the values in Eq. (57) only the A value cannot be determined directly. Equation (57) can therefore be used for the determination of A values from polarographic waves. Such deter-
minations of A values were made by a series of investigators.
Thus, for methylene blue, Brdička [278] determined an A $\approx$ 100 $\overset{\circ}{A}^2$
value, which corresponds closely to the molecular dimension.
Trifonov [355] investigated the anodic adsorption wave of sulfide
ions, and from its height calculated a value of A $\approx$ 14 $\overset{\circ}{A}^2$ for HgS
which is practically identical with the value calculated from the
molar volume of HgS.

The appearance of adsorption prewaves and postwaves is
quite common in polarography. Besides examples already mentioned, the adsorption prewaves in the polarography of pyocyanine
(α-methoxyphenazine) [356], chloranil (3,6-dichloro-2,5-dihy-
droxybenzoquinone) [357], perinaphthenon [358], triphenyltetrazoyl
chloride [359], quinidine [360], steroid hormones [361], and 5-
hydrooxynaphthoquinone [362] can be mentioned. It is interesting
to note that on the polarograms of cholestenone the adsorption prewave is approximately twice as large in acidic solution as in alkaline solution. The authors explained this difference [361] by the

different orientation of the adsorbed molecules at the electrode surface.

Adsorption prewaves appear in the polarography of the solution of certain heart glucosides containing aldehyde groups [363], of astofloxine [364], and also in polarographic measurements of certain anthraquinone derivatives in anhydrous acetic acid containing 10% H_2SO_4 [365].

Quite often adsorption prewaves can be observed in anodic processes, which lead to formation of surface-active products by the interaction of mercury ions (formed by the oxidation of the mercury of the electrode) with compounds present in the solution. This occurs, for example, in the case already mentioned of anodic polarization of the dropping electrode in the presence of sulfide ions [355] and also in the presence of certain mercaptans [366], barbituric acid [367], veronal (barbital) [368], derivatives of thiobarbituric acid [343, 369], and derivatives of barbituric acid [344].

With increasing temperature, the adsorption prewave of diethylaminoethyl mercaptan disappears [366]. This is explained [366] by the desorption of the product of interaction of this mercaptan with mercury from the surface of the electrode. From the maximum value of the height of this prewave, the area occupied by one particle of this product on the electrode surface (33.5 $\AA^2$) was calculated [366]. It was shown in the example of the waves in barbituric acid solutions [365] and veronal solutions [366], that the maximum height of the prewave increases when the drop period of the electrode is reduced, in accordance with Eq. (57). The prewave shows exceptionally high values (at sufficiently high concentrations of the derivatives of barbituric acid) at the streaming mercury electrode, which has a very high rate of new surface formation.

It must be mentioned that prewaves very similar to those found by Brdička can be observed when the electrode process is inhibited by a layer of adsorbed surface-active products of the electrode reaction (compare with section F of this chapter). This effect can apparently be observed to a certain degree in solutions of certain derivatives of barbituric acid [344].

It was already mentioned that at low depolarizer concentrations in the solutions, usually only one wave can be observed at the

potential corresponding to i_a on the polarographic curves. The
limiting current of this wave has the character of a diffusion current, and its value is proportional to the depolarizer concentration. At increasing depolarizer concentrations the limiting current
of this wave gradually loses its diffusion current characteristics
as it approaches its maximum value. Usually the main (second)
wave (i) appears on the polarograms only after the prewave (i_a)
has reached (or almost reached) its maximum value. In certain
cases, in the polarography of quinine, for example [370, 371], the
normal wave appears and increases with increasing quinine concentration at considerably lower concentrations than those at
which the prewave reaches its limiting value.

The simultaneous appearance and increase of the prewave
and main wave at higher concentrations of the compound reduced
in the solution was observed by Asahi [372] on the reduction polarograms of the protonated form of narcotine–N–oxide. The
height of the prewave increases in acid medium with increasing
N–oxide concentration (c) according to $i = abc/(1 + bc)$, and for
the electrode used by the author [372], $a = 1.25$ μA and $b = 2.67$
mM^{-1} (for m = 0.817 mg/sec and t = 3.94 sec). The A value corresponding to narcotine is 12 $\mathring{A}^2$ in acidic solution, while in alkaline solution, where the maximum height of the prewave is much
lower, $A = 50$ $\mathring{A}^2$ and $\Gamma_\infty = 3.3 \cdot 10^{-10}$ moles/cm^2. Thus the
narcotine molecules and similarly those of cholestenone [361] are
oriented differently on the electrode surface during adsorption
from acidic and alkaline solutions.

The simultaneous appearance of the adsorption prewave and
the main wave at very low depolarizer concentrations takes place
only if the products of the electrochemical reaction are adsorbed
to a small extent. In this case not all product particles formed
are adsorbed on the electrode surface. Instead, a certain portion
diffuses back into the solution, thus decreasing the height of the
adsorption prewave. At the same time as they are transported by
diffusion to the electrode surface they cause the appearance of the
main wave (which corresponds to the normal reversible electrode
process in the absence of adsorption of electrode products). The
simultaneous appearance of the prewave and the main wave on the
polarographic curves is apparently a very frequent case. If the
reaction products are very strongly adsorbed, only an insignificant

portion of the product (at incomplete coverage of the surface) passes back into solution. Therefore, the current of the main wave is much smaller than the current of the prewave, and at sufficiently low concentrations only one wave (i_a) can be seen on the polarographic curves. In these cases the limiting current of the prewave (i_a) increases linearly with the depolarizer concentration (c) until the maximum value of the adsorption current is attained. The plot of the height of wave i_a as a function of concentration thus consists of two linear sections. The first line passes through the zero point of the coordinates and its slope is equal to the proportionality factor of the Ilkovič equation for the given compound. The second line is parallel to the abscissa at a distance corresponding to the maximum adsorption current (after the maximum value is attained the height of the prewave does not change when the depolarizer concentration is further increased). A sharp break in the i vs. c plots is rarely observed. Usually, instead of a sharp break, a rounding up is observed with more or less smooth transition from one line to the other. With increasing surface coverage by the reaction product the conditions become less favorable for its adsorption, and an increasing proportion of the product diffuses into the solution. This is the cause of the absence of a sharp break. It is interesting to note that with veronal solutions, whose curves show no rounding up on the i vs. c plots for the adsorption prewave on the dropping mercury electrode, the plot changes into a continuous curve, reminiscent of the form of a Langmuir isotherm if the polarographic curve is recorded using a streaming mercury electrode (compare Figs. 2 and 11 in [368]).

For certain thiobarbituric acid derivatives the adsorptivity of products decreases in alkaline solutions compared to acidic solutions [369].

If the adsorptivity of the products of a reversible electrochemical reaction is not very high, i.e., if both the prewave and the main wave appear on the polarographic curves at incomplete coverages of the electrode surface by adsorbed material, then the adsorption isotherm of the electrode product can be determined from the dependence of the wave height on depolarizer concentration (by the Brdička theory [371]). If one assumes that the diffusion coefficients of the depolarizer and product are equal, the concentration of the reaction product (c_S) at the electrode (in the elec-

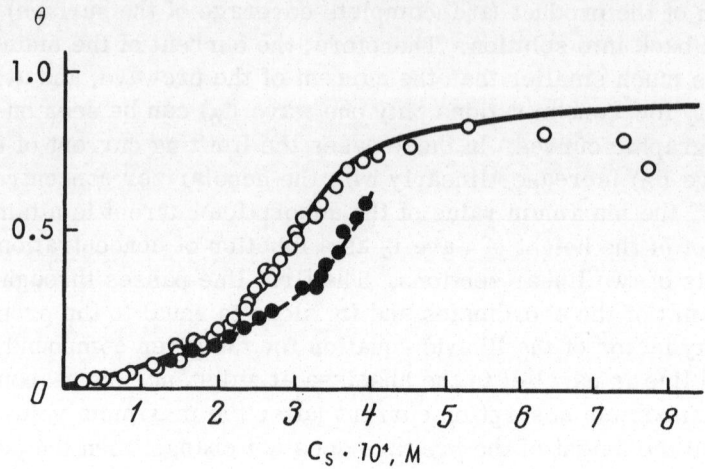

Fig. 7. The fractional coverage of the dropping electrode surface as a function of dihydroquinine concentration at the electrode.

trode layer volume) can be calculated from the following:

$$c_s = \frac{i_D - i}{\varkappa} = \frac{\varkappa c - i}{\varkappa},$$ (58)

where i_D is the diffusion current of the depolarizer (corresponding
to the total height of the prewave and main wave), i is the current
corresponding to the height of the adsorption prewave, and $\varkappa$ is
the constant of the Ilkovič equation (the proportionality factor between the diffusion current and depolarizer concentration).

The fractional coverage of the electrode surface by the product θ at a given concentration (c), and consequently, at a given
value c_S [compare with Eq. (58)], is equal to the ratio of the height
of the prewave and its maximum value: $\theta = i_a / (i_a)_{max}$. The relationship of θ and c_S gives the adsorption isotherm. Figure 7
shows the adsorption isotherm of the reduction product of quinine
(dihydroquinine) at −1.40 V, determined by this method [371]. The
continuous curve in this figure is the theoretically constructed
Frumkin adsorption isotherm with $\beta = 6 \cdot 10^2$ M^{-1} and $\gamma = 1.7$; the
points represent the experimental data.

Figure 7 shows that the points closely follow the curve, corresponding to the Frumkin isotherm. On the same figure, full
circles show the dependence of θ on c_S, obtained in the same solution but at twice the mercury column height above the electrode,

which means a smaller drop time. Thus the desorption characteristics determined by this method depend on the drop time of the dropping electrode. A certain decrease of adsorptivity of dihydroquinine at reduced drop periods is apparently caused by the same effect as the adsorptivity of reaction product in the interaction of veronal with mercury, if the dropping electrode is replaced by a streaming electrode [368]. The nature of this effect has not been clarified.

If the starting material is more strongly adsorbed than the product of the reversible electrode reaction, the adsorption wave appears after the main wave [278]. Such an adsorption wave can be called a postwave. The properties of adsorption postwaves are exactly the same as those of the adsorption prewaves.

An example of an adsorption postwave is the third wave on the polarographic curves of copper (II) thiocyanide reduction [374]. Adsorption postwaves are also the second waves in the reduction of anthraquinone-1-sulfonic acid [375] and anthraquinone-1,5-disulfonic acid [376].

Adsorption postwaves also were observed in the reduction of rhodamine B in buffer solutions [377] and during the reduction of nitrobenzene-3-sulfonic acid in acid medium [378].

E. THE EFFECT OF THE ADSORPTION OF COMPOUNDS NOT TAKING PART DIRECTLY IN THE ELECTRODE PROCESS ON THE RATE OF ELECTRON TRANSFER

Usually, the adsorption of compounds not taking part directly in the electrochemical reactions decreases the rate of these reactions. Exceptions are observed when the adsorbed compounds change the ψ_1-potential (the potential at a distance of the ionic radius from the electrode surface) in such a way that the effective potential drop between the electrode and discharging particle increases in the presence of the added compound. If ions of the same sign as the charge of the electrode in the given potential region take part in the electrode process, the electrostatic repulsion between the electrode and discharging particle is also decreased.

Kolthoff and Barnum [379] apparently first observed the inhibition of the electrode process (reduction of cystine) when surface-active compounds – thymol, camphor, gelatine, and methylene blue – were added to the polarographed solution. Fiala [380] noticed the shift in the oxygen reduction wave to more negative potentials upon addition of dyes of the eosin group to the solution. Wiesner [381] studied the effect of eosin dyes on the reversible waves of certain quinones, hydroquinones, and ascorbic acid. He found when the surface-active dye is added to the solution, above a certain concentration the height of the reversible reduction wave begins to decrease, and at more negative potentials an additional wave appears corresponding to the irreversible reduction of the same compound. For anodic processes only a decrease of the reversible wave was observed on addition of surface-active compounds. The decrease in the reversible wave, as shown by Wiesner [381], depends on the concentrations of the surface-active compounds, their nature, the temperature at which the polarogram is measured, and on the characteristics of the dropping electrode. Wiesner proved that the cause of the height reduction of reversible waves is the adsorption of surface-active agents on the mercury electrode.

A major emphasis was placed by M. A. Loshkarev and A. A. Kryukova on clarifying the effect of surface-active compounds on electrode processes. Loshkarev and Kryukova [382] explained the inhibition of electrode processes in the presence of adsorbed compounds by the difficulty in passage of the depolarizer through the adsorbed layer to the electrode surface. They treated this phenomenon as a special case of chemical polarization. Loshkarev and Kryukova showed that the layer of adsorbed material inhibits the electrode process of reduction of metal ions more if the atomic weight of the ion is smaller and its charge higher, or if the field intensity of the discharging ion is larger.

When the desorption potential of the surface-active materials is reached their inhibiting effect ceases. The presence of adsorption layers on the electrode surface was proved by the measurement of the electrical double-layer capacity [383]. The investigation of the electrode double layer led to the determination of some properties of the adsorbed layers [383].

Heyrovsky also observed the inhibition of discharge of Pb^{2+} cations in the presence of larger quantities of pyridine, fatty acids, and alcohols [318] and gave another explanation. According to Heyrovský the discharge of polyvalent cations takes place by transfer of one electron with consecutive dismutation, for example,

$$Cd^{2+} + e^- \rightarrow Cd^+$$
$$2Cd^+ \rightarrow Cd^{2+} + Cd^0.$$

The simultaneous transfer of several electrons on a single cation is improbable, according to Heyrovský. The adsorbed compounds apparently inhibit only the second stage [384], the dismutation, and do not affect the transfer of the first electron. Nevertheless, Loshkarev and Kryukova proved that to a certain extent even the first electron-transfer process itself is inhibited in the discharge of simple ions. They found that many organic surface-active agents increase the overpotential of hydrogen and also substantially inhibit the dismutation process of titanium and vanadium ions [385]: $V(III) + e^- \rightleftharpoons V(II)$ and $Ti(IV) + e^- \rightleftharpoons Ti(III)$. In these cases only a single electron is transferred; thus, according to Heyrovský's viewpoint, inhibition should not be observed. Somewhat later Loshkarev and Kryukova [386] showed that tribenzylamine and tetrabutyl ammonium sulfate even inhibit the reduction of such ions as Ag^+ and Tl^+, whose discharge is not affected by gelatine, camphor, or similar compounds [383-385]. At lower temperatures the inhibiting effect of adsorbed layers increases substantially [386, 387]. At low temperatures even camphor inhibited the discharge of Ag^+ and Tl^+ [386].

The potential interval in which the inhibiting effect of additives can be observed is determined by their adsorption region. If ionic-type adsorbents are used, the inhibiting effect can be studied at potentials far from the zero charge potential. Thus, if salicylic acid is added to the solution (surface-active anion) an inhibition can be observed at potentials more positive than -0.2 V (S.C.E.) [387].

Loshkarev and Kryukova repeatedly emphasized that the inhibiting action of the layers is not a result of reduced diffusion of discharging particles but is caused by an added activation barrier [388]. The inhibiting effect of adsorbed layers is much larger than

the effect of the change of the ψ_1-potential. Loshkarev and Kryukova showed [389] that upon addition to acidic solutions, tetrabutyl ammonium sulfate or tribenzylamine cause a sharp shift of
the cystine reduction wave to negative potentials, while the simultaneously occurring change of the absolute value of the negative
ψ_1-potential should shift the cystine reduction wave toward positive potentials. The inhibition of the electrode process of cystine
reduction in acidic solution is much larger when cationic surface-
active compounds (tetrabutyl ammonium sulfate, tribenzylamine)
are added than on addition of neutral compounds such as naphthol,
eosin, or camphor [389].

In the inhibition of the electrode process the exchange current decreases. Miller and Pleskov [390] first observed the reduction of exchange currents in presence of adsorbed compounds.
Stromberg and co-workers [391-394], determined that an addition
of surface-active compounds (camphor and gelatine) to the solution
causes a splitting of reversible anodic—cathodic waves in the
measurement of polarograms with dropping amalgam electrodes.
In this process the cathodic and anodic waves shift in opposite directions from the reversible potential of the system. At the same
time the exchange currents decrease substantially.

The study of Zagainova and Stromberg [393] dealing with the
simultaneous effect of two surface-active materials (camphor and
gelatine) on the reversible electrochemical reaction taking place
on a cadmium amalgam electrode in a cadmium salt solution is
very interesting. The gelatin, up to a concentration of 0.1%, does
not affect the rate of the electrode process, while camphor shows
a strong inhibiting action. Addition of gelatin to a solution saturated with camphor suppressed the inhibiting effect of camphor,
and at sufficiently high gelatin concentrations, completely eliminates its inhibiting action. The authors [393] explain this effect by
a change in the structure of the double layer in the presence of
gelatine which results in an increase in the rate of the electrochemical reaction. It can also be assumed that the gelatin, which
by itself does not inhibit this process, displaces the camphor from
the electrode surface and thus removes the inhibiting effect caused
by camphor.

Stromberg and his co-workers, using electrocapillary curves
of mercury in camphor solutions, determined the dependence of

adsorption on the camphor concentration and the electrode potential. These data allow prediction of the effect of the adsorbed surface-active material layer on the electrode process at different potentials and also clarify why the inhibiting action of the electrochemical reaction begins to decrease gradually, even before the potentials corresponding to desorption are attained. This later phenomenon was pointed out by Loshkarev [395] in 1950.

Bonting and Aussen [396] observed a shift of the polarographic reduction wave of oxygen and zinc to negative potentials in the presence of surface-active compounds, and showed that this effect is caused by formation of an adsorbed layer on the electrode surface.

V. V. Losev investigated the exchange currents between zinc in amalgam and a zinc sulfate solution, using radioactive indicators. He found that small additions of tetrabutyl ammonium reduced the exchange current by two orders of magnitude [397]. Losev also showed that if the concentration of the indifferent salt ($MgSO_4$) is increased, the inhibiting effect of surface-active compounds (tetrabutyl ammonium and tribenzylamine) on the anodic and cathodic processes increases. In this work he used amalgam electrodes (0.3 to 0.5 atom %) of zinc and cadmium [398]. The increased inhibiting effect upon addition of magnesium sulfate to the solution was observed only when the solution was not saturated with respect to the surface-active compound. These observations were explained by the salting-out effect of the salts which increased the adsorption of surface-active compounds. With solutions saturated with the adsorbed compound, addition of magnesium sulfate does not change the adsorption and does not affect the electrode process.

Laitinen and Subcasky [399] investigated the effect of n-octyl alcohol, β-naphthol, and camphor on the transfer coefficient and on the heterogeneous rate constant (k) of the electrode process. This investigation was made on the reduction of Ni^{2+}, Cu^{2+}, and Sn^{4+} on the dropping mercury electrode under conditions where concentration polarization did not affect the results. In the presence of surface-active compounds, the rate constant (k) of Ni^{2+} reduction decreases. In the potential-determining step two electrons are transferred, while in the absence of surface-active compounds only the first electron takes part in the potential-determining step. In the presence of camphor at potentials corresponding to its desorption, a sharp current increase can be observed. On the tin(IV) wave a minimum appears in acid medium on addition of camphor, and the current in the minimum decreases nearly to zero with increasing camphor concentrations.

Řiha [400] found that a polymeric compound of pyridine [401] can almost entirely inhibit the discharge process of cobalt(II) ions. The inhibition of this electrochemical reaction over a certain potential range was shown by the absence of radioactivity in mercury collected under the dropping electrode, when a Co^{60} salt was added to the solution. The inhibition of Co^{2+} ion discharge by a layer of polymer product, formed in the reduction of pyridine, was also indicated by the absence of the cobalt ion discharge wave on i vs. E curves repeatedly measured on a stationary electrode surface. These measurements were made on a hanging mercury drop in a solution containing pyridine and cobalt ions [400]. It is interesting to note that this polymeric product inhibits the discharge of other ions, such as Cd^{2+}, Pb^{2+}, and Cu^{2+} to a much smaller degree than Co^{2+} [400].

Josepovits showed [402] that surface-active compounds, in the region of potentials corresponding to their adsorption, change the slope of the reduction wave for several organic compounds. On the basis of this effect it was possible to distinguish between the type (anionic or cationic) of active groups of the adsorbed compound.

In early work dealing with the effect of surface-active compounds, the kinetics of their accumulation on the surface of the electrode was not taken into account.

In one of.the previous sections it was mentioned that the accumulation of adsorbed materials on the electrode is limited by the rate of their diffusion to the electrode surface. Koryta [287] was the first to quantitatively relate the inhibiting action of surface-active compounds to the rate of formation of the adsorbed layer on the electrode surface. From experimental data on the shift of the reversible polarographic wave of quinone, hydroquinone, and phthiocol by surface-active dyes (eosin, erythrosin, bengal red), Koryta concluded that the rate of formation of the adsorbed layer is limited by the rate of diffusion of the adsorbing material from the solution to the electrode. Furthermore, he found that the inhibiting effect on the electrode process becomes apparent only at a certain coverage of the electrode surface by the adsorbing compound.

Schmid and Reilley [403] reported on investigations showing that the adsorption of surface-active compounds lowers the current intensity vs. time (i vs. t) curve taken during the life of a single drop. Using as an example the reduction wave of the copper complex with ethylenediamine tetra-acetic acid, they showed that the decrease of the i vs. t curves upon addition of surface-active material can be observed under conditions corresponding to the splitting of this wave. These authors [403] also showed that the adsorption of eosin is limited by diffusion and corresponds to the equations derived by Koryta [287]. With inhibition by a camphor layer, the formation of the layer is slower than the diffusion process. The authors [403] assumed that in the adsorption of camphor the limiting stage is not the diffusion but the adsorption itself. Apparently the adsorption equilibrium begins to take effect in this case. This was not taken into consideration by Koryta [287].

Delahay and co-workers [288-290] discussed the inhibition of electrode processes by surface-active materials on the basis of kinetic analysis of the adsorption with diffusion control and also included the effect of adsorption equilibrium.

The effect of adsorption equilibrium can be neglected only when the compound is strongly adsorbed and when the reverse process of desorption can be disregarded nearly up to complete coverage of the surface by adsorbed particles. Weber, Koutecký, and Koryta [404] calculated the effect of adsorbed surface-active compounds on the rate of the electrode process for similar cases. They assumed that the covered and uncovered portions of the surface can be characterized by their respective rate constants and that thus the observed rate of the electrode process is a linear function of the electrode coverage. The authors [404] determined the dependence of the instantaneous current on the lifetime of the drop, on the concentration of the surface-active compound, and on the rate of the electrode reaction both at the covered and uncovered electrode surface. Kůta and Smoler [405] compared experimentally determined i vs. t data that were measured during the life of the "first" drop (which means conditions under which concentration changes of the depolarizer at the electrode, caused by electrolysis on previous drops, are absent [19]), with the calculation results of Weber, Koutecký, and Koryta [404]. They found that on the i vs. t curves taken in presence of adsorbing compounds, a decrease of the current can be observed with increasing time, in some cases nearly to i = 0. For a strongly adsorbed uncharged compound the i vs. t curve is close to zero almost to the last section and follows the calculated curve closely. In the last section, the drop in the current is substantially lower. This is caused, in our opinion, by the effect of desorption, which becomes significant at high (nearly complete) coverages of the surface. If surface-active cations are added to the solution, the experimental i vs. t curves corresponding to the discharge of positively charged depolarizers are substantially below the calculated curves. The current decrease in this case is caused by a decrease in the rate constant of the electrode process at uncovered portions of the surface, due to a smaller negative value of the ψ_1-potential [405]. The instantaneous and average current in the presence of charged and uncharged adsorbing surface-active compounds was calculated [406] for irreversible discharge. The effect of change of ψ_1-potential and its influence on the rate of the electrode reaction was included in this calculation. These calculations were also made for currents that are affected by adsorbed electrochemically inactive compounds and for reversible electrode processes [407]. Finally, they were carried out for the case when the adsorption itself is the slow stage [407].

Kůta and Smoler [408] gave a review of the work dealing with the effect of surface-active compounds on the form of i vs. t curves. As has already been mentioned, during the adsorption of surface-active ions, the i vs. t curves also are affected by the change of the ψ_1-potential of the electrode. The work of Kůta and Smoler, dealing with the form of i vs. t curves for this case, will be discussed later.

Recently, the effect of interaction of the adsorbed compound and depolarizer on the electrode process was discussed. This interaction results in the formation of a nonadsorbed, electrochemically inactive product [409]. The i vs. t curves corresponding to this case are similar in form to the curves obtained for inhibition by an adsorbed compound [409]. In our opinion, it is very hard to realize such a case when a nonadsorbed product is formed as a result of chemical interaction of the adsorbed compound with the depolarizer.

The very interesting work of L. Holleck and co-workers must be mentioned. They investigated the mechanism of reduction of nitro compounds and compounds containing carbonyl groups. The effect of added adsorbing compounds on the polarographic waves and on i vs. t curves was studied. In 1951 Holleck and Exner [410] showed that the reduction waves of a series of nitro compounds in the presence of surface-active materials (tylose, gelatin, agar-agar) split into two parts in alkaline solution. The wave remaining at the potential of the original wave decreased until it reached the height corresponding to a one-electron diffusion current. On the basis of detailed investigation of this one-electron wave in alkaline solution in presence of adsorbed compounds, Kastening and Holleck [411] partly proved that these compounds do not inhibit the transfer of the first electron in the reduction of nitro compounds. This transfer is reversible. But these compounds substantially inhibit the reduction of the primary products. The investigation of the i vs. t curves in the reduction of nitro compounds [412] by Kastening and Holleck led to the conclusion that in strongly alkaline solutions the kinetics of inhibition by surface-active compounds is determined by the rate of their adsorption.

In a further study [413], based on the analysis of the form of i vs. t curves in the reduction of p-nitrochlorobenzene in presence of tylose, it was proved that the anion radical of p-nitrochlorobenzene formed by transfer of one electron reacts with the tylose. In the same work [413] it was stated that the inhibiting effect of surface-active compounds on the second stage of nitrocompound reduction is apparent only when anion radicals are formed that are not protonated rapidly. An inhibition of the electrode process and splitting of the wave was also observed on the polarographic curves of nitronaphthols on addition of tylose, camphor, or triphenylphosphine oxide to the polarographed solution [414]. A doubling of the reduction wave of nitrobenzene disulfonic acid was observed in acid solution upon addition of surface-active leuko-bases of anthraquinone and of the sulfonic acid ester of anthraquinone [415].

The inhibiting effect of surface-active compounds depends to
a certain extent on their nature. Holleck, Kastening, and Williams
[416] attempted to give a quantitative comparison of the effect of
surface-active agents. This comparison was based on that concentration of the surface-active material causing the same shift
of the half-wave potential of the upper portion of the first reduction step of p-chloronitrobenzene (from −0.79 to −1.1 V) in alkaline solution. The same inhibiting effect was found for the following concentrations of the adsorbed compounds in the solution (in %):
2.0 cyclopentanoneoxime; 2.0 benzaldoxime; 1.0 n-valeronitrile;
0.4 benzonitrile; 0.1 p-tolunitrile; 0.01 dibenzylsulfonic acid; 0.01
diphenylsulfoxide; 0.0045 dibenzylsulfide. But these figures, as
has been mentioned by the authors [416], are only qualitative: the
dependence of the half-wave potential on the logarithm of the concentration of the surface-active compound ($c_{S.A.}$) is not linear.
The $E_{1/2}$ vs. $\log c_{S.A.}$ curves for different compounds intersect one
another. This is the result of different adsorption isotherms of
the compounds investigated.

The inhibiting effect of different compounds can depend on
the nature of the inhibited electrochemical reaction. The relative
inhibition effect often changes due to the specific interaction between the adsorbed compound and the depolarizer (compare, for
example, with [413]). Furthermore, the relative inhibition changes
because of the different adsorptivity of surface-active compounds
at different potentials. This latter effect becomes pronounced if
the effect of a given series of compounds is studied on reactions
taking place at different potentials. For adsorption of compounds
of similar structure, their inhibiting effect on the same electrochemical process increases for compounds having a higher surface activity. Thus, the inhibiting action of tetrasubstituted n-
alkyl ammonium salts sharply increases with increased chain
length; the latter determines the adsorptivity of their cations. As
an example, the shift of the first reduction wave of the methyl
ester of the p-chlorophenylpropionic acid may be mentioned. This
shift can be observed on replacement of a potassium chloride solution by tetrasubstituted ammonium salts (the wave corresponds to
the reduction of a triple bond into a double bond) [417]. At a sufficiently high concentration of the tetrasubstituted salt, the first and
second waves on the polarographic curves merge even though the
difference of the half-wave potentials of these two waves in the

absence of surface-active agents may be as much as 0.5 V. The authors [417] give interesting data showing how the concentration of the tetrasubstituted ammonium cation with unbranched chains required for the merging of the two waves decreases with increasing length of the radical. For C_1 [which is $(CH_3)_4NCl$] the salt concentration must be more than 5 M; for C_2 [which corresponds to $(C_2H_5)_4NCl$], ~1; for C_3, ~0.2; for C_4, ~0.01; for C_5, ~4 · 10^{-4} M; and for C_6 and C_7 the salt concentration must be only 2 · 10^{-4} M.

Besides the work of Stromberg already mentioned, Müller and Lorenz [418] also investigated the electrochemical discharge — dissolution processes of Zn^{2+} and Cd^{2+} on amalgam (Zn and Cd) electrodes. They showed that in presence of tetrasubstituted ammonium salts of butyric acid, tertamyl, and n-hexyl alcohol, the relative decrease in exchange current is directly proportional to the degree of coverage (θ) of the electrode surface by the adsorbed compound. In their work the θ value was determined from differential capacity curves. However, Rek [419] found that the exchange current in a solution of iron oxalate is a linear function of the surface coverage only in the presence of n-hexyl alcohol. For compounds such as thymol, polyoxyethylene esters of lauric acid, and sodium dodecyl sulfate, the dependence of the exchange current on θ deviates from the linear relationship.

In the work already mentioned [418], an impulse galvanostatic method was used for the investigation of the kinetics of electrochemical processes in the presence of surface-active agents. Many authors have used chronopotentiometry for the study of the inhibition of electrode processes in the presence of surface-active agents (for example, [420-423]). Dračka [421] developed equations for the calculation of transition time for semi-infinite linear diffusion. These equations can be used when the adsorbed layer on the electrode hinders the transport of material to the electrode surface and also inhibits the antecedent chemical reactions. Dračka found that the inhibiting effect depends on the ratio of the thickness of the inhibiting layer to that of the diffusion or reaction layer. The inhibition of the electrode process by a layer of adsorbed material results in a decrease in the wave on regular polarograms and in a decrease of the $i_0 \tau^{1/2}$ value at increasing i_0 [422]. To study the inhibiting effect of the layer when the thickness and the properties of the layer remain constant with

time and also remain constant if the electrode potential and the
composition of the solution changes, Fischer [423] used an electrode covered with a cellophane membrane in chronopotentiometric investigations of different electrode processes. This cellophane membrane formed a "homogeneous" medium of known
thickness [424]. The inhibiting effect under these conditions apparently has a different nature from that encountered in investigations by Loshkarev and Kryukova, in which the inhibiting layer had
molecular dimensions. Fischer's [423] experimental data show
that the inhibiting effect for the diffusion and for electrode processes with antecedent chemical reaction caused by the cellophane
membrane, is correctly described by the equations given by
Dračka [421].

   A decrease of the rate constant of reversible or pseudore-
versible electrochemical reactions, caused by addition of adsorbable compounds to the solution, sharply reduces the peaks on the
polarograms measured by Breyer's method with superimposition
of alternating voltage [425-427]. The inhibition of electrode processes in the presence of adsorbed compounds also changes the
form of curves taken by oscillographic polarography. The oscillo-
grams for the reduction of nitrobenzene in the presence of camphor [428] confirmed Holleck's views that the adsorbed compound
inhibits the further reduction of the intermediate product formed
by transfer of one electron on the nitro-compound molecule.

   The case when the inhibiting effect of a compound present in
the solution is decreased or completely eliminated by addition of
a second surface-active compound is very interesting. This was
mentioned in the previous section where the simultaneous effect
of gelatin and camphor on the anodic−cathodic waves on the cadmium amalgam electrode in Cd^{2+} salt solutions [393] was discussed. In an acidic buffer solution of Prideaux−Ward containing
the surface-active phenylacetic acid, the half-wave potential of the
first wave in reduction of 4-nitropyrrole-2-carboxylic acid is
much more negative than in other buffer solutions at the same pH
that do not contain phenylacetic acid. This acid inhibits the reduction of the given compound [429]. An addition of small concentrations − up to 1 mM − of isomeric 5-nitropyrrole-2-carboxylic
acid to the polarographed solution shifts the 4-nitro-isomer wave

toward positive potentials [430]. It was shown that this effect is caused by the adsorption of the 5-nitropyrrole carboxylic acid, which by itself does not inhibit the reduction of the 4-nitro-isomer but instead displaces the phenylacetic acid from the electrode surface [429]. It is interesting to note that if a series of oscillographic single-sweep i vs. E curves is recorded during the life of a single drop, the reduction curves of 4-nitropyrrole-2-carboxylic acid in phenylacetic acid containing buffer solution and in the presence of small quantities of 5-nitro-isomer shift to positive potentials as the dropping electrode grows. This phenomenon is due to the gradual accumulation of 5-nitropyrrole-2-carboxylic acid on the electrode [429].

5-Nitropyrrole-2-carboxylic acid also eliminates the inhibiting action of phenylacetic acid in the reduction of several other 4-nitropyrrole derivatives containing different substituents (CHO, CH = NOH, CN) in the 2-position [429].

It must be mentioned that the splitting of the wave, the reduction of the wave height, and appearance of separate steps (false waves) due to introduction of adsorbable compounds into the polarographic solution is not caused only by inhibition of the electrochemical reaction. Kryukova [431] pointed out that the phenomena also can be caused by the damping of the tangential movement of the mercury surface of the electrode under conditions when a maximum of the second kind is formed [5]. An increase in current flow as a result of tangential movement of the mercury surface was observed by Frumkin and Bruns [432]. The dampening of this movement was studied in detail by the Soviet scientific school of A. N. Frumkin: by Kryukova and Kabanov [433-436], and the quantitative theory of this phenomenon was given by Frumkin and Levich [437]. In the adsorption region of surface-active compounds the tangential movement decreases. Therefore, the supplementary current caused by convection of the solution is reduced. Beyond the desorption potential of the adsorbed compound the tangential flow again appears, resulting in an increase of current or appearance of false waves. A detailed description of these phenomena with numerous examples can be found in the book by Kryukova, Sinyakova, and Aref'eva [5] and in the monograph of Frumkin, Bagotskii, Iofa, and Kabanov [7].

F. INHIBITION OF ELECTRODE PROCESSES BY
ADSORPTION OF DEPOLARIZER AND
ELECTROCHEMICAL REACTION PRODUCTS;
FORMATION OF ADSORPTION PSEUDO-
PREWAVES

The physical nature of inhibition of electrode processes by adsorption of the depolarizer or products of electrochemical reactions is the same as for the adsorption of added compounds. But the effects found in the adsorption of depolarizer or products show certain special properties that must be discussed separately.

The first case of strong inhibition of the electrode process by the adsorption of electrochemically active compound was described recently by Laviron [438]. In Britton–Robinson buffer solutions in the pH interval from 4 to 10, a sharp increase can be observed on the polarographic curves of the reduction of isomeric di-(2',2")- and di-(4',4")-dipyridyl-1,2-ethylene, suddenly after the start of the wave nearly up to the level of limiting diffusion current and corresponding to transfer of two electrons. For di-(2',4")-pyridyl-1,2-ethylene this wave form is only found in the pH interval 4-6. The sharp current increase on the polarographic curves of these compounds is a result of elimination of inhibition caused by the adsorption of the depolarizer itself, and the elimination corresponds to the beginning of desorption of the depolarizer. The electrocapillary curves of solutions of these compounds indicate a strong adsorption of the depolarizer and a certain adsorption of the reaction products [438]. The desorption of the products, occurring for each isomer (in which the pyridine ring is attached at position 3) at a potential of about −1.5 V causes a new, smaller increase on the polarographic curves. This increase is due to the removal of inhibition by reaction products. In alkaline solution a very high maximum can be observed [438]. It must be mentioned that the current increase at a potential of −1.5 V also can be caused (in part or entirely) by removal of the damping action exerted on the mercury surface tangential flow by adsorbed products. The desorption of the products removes this inhibition. The electrode used in this work [438] had an m value of 4 mg/sec. Such a high value, according to Kryukova [5], must cause tangential movements (appearance of a maximum of the second kind).

It is interesting to note that in the polarographic investigation of dipyridyl-ethylene compounds in a solution containing 50% ethyl alcohol, the described phenomena could not be observed [439]; this is most probably a result of the lower adsorptivity of dipyridylethylene compounds on the mercury electrode from aqueous alcohol solutions.

The inhibiting effect on the electrode process by adsorbed depolarizer and the electrochemical passivity of the depolarizer in the adsorbed state can be explained in our opinion as follows. Large organic molecules adsorbed on the mercury electrode surface can be oriented in different ways. At a given electrode potential the most favorable orientation of the molecules from the standpoint of adsorption energy can be unfavorable for the electrochemical process. For example, the electrochemically active centers of the adsorbed molecules can be comparatively far away from the electrode surface. The mutual orientation of adsorbed particles that is encountered quite often in the adsorption of large organic molecules, results in complete coverage of the electrode by a dense adsorbed layer of the compound. Within this layer a reorientation of the molecules becomes impossible, and the layer makes the passage of depolarizer molecules from the solution volume to the electrode surface difficult. With an increasingly negative potential of the electrode, the adsorptivity of the molecules decreases. The adsorbed layer becomes less dense, and a certain potential is reached where the adsorbed depolarizer molecules can be reoriented. At this potential a sharp increase of current can be observed, corresponding to the complete breakdown of the inhibiting adsorption layer.*

Several other cases of inhibition of the electrode process by adsorbed depolarizer (autoinhibition) are known. Thus, Zhdanov and Pozdeeva [440] noted a sudden current increase on the reduction polarograms of diphenylcyclopropenone in acidic medium at low alcohol concentrations caused by depolarizer desorption and

---

*It must be mentioned that this explanation is not entirely correct for dipyridylethylene, in which the π-electron system extends over the entire molecule. Electron transfer on this system can therefore take place at any orientation of the molecules on the electrode surface. Other explanations of the inhibiting effect at complete coverage of the electrode surface by the depolarizer have been given by Frumkin, Dzhaparidze, and Tedoradze (see page 266).

removal of inhibition of the electrode process. With increasing diphenylcyclopropenone concentrations the potential at which the wave increases discontinuously becomes more negative [440]. This is due to the shift of desorption potential to more negative potentials.

If increased pressures are imposed on the polarographed solution, the adsorptivity of organic compounds on the electrode increases. For example, in polarographic curves when unbuffered potassium chloride solution was used as a supporting electrolyte in the presence of pyridine, a decrease in current can be observed on the i vs. t curves togehter with a distortion in their shape with pressure increasing above 1000 atm. This is most probably caused by the inhibition of the electrode process by an adsorbed layer of depolarizer (or perhaps reaction products) [441].

The unprotonated forms of actinomycin-C_1 and chloro-actinomycin-C_1 give reversible waves in their reduction at the dropping mercury electrode only at concentrations $<5 \cdot 10^{-4}$ M. Inhibition effects appear at higher concentrations of these depolarizers, due to adsorption and the reduction becomes irreversible [442, 443].

The inhibition of electrode processes due to adsorption of products of electrochemical reactions is encountered much more frequently. Thus, in the polarography of the system uranyl [with uranium(VI)]−cupferron, in the electrode reaction the insoluble complex of uranium(IV) with cupferron is formed, which inhibits the electrode process [444]. An addition of ethyl alcohol to this solution dissolves the complex, and removes the inhibition of the electrochemical reaction. The formation of strongly adsorbed narcotine on the dropping mercury electrode during the reduction of its N-oxide results in the appearance of a minimum on the wave [372]. The minimum is located in the potential region close to the electrocapillary zero (from −0.42 to −1.02 V at a concentration of 1 mM of narcotine-N-oxide). At increasing concentrations of the N-oxide, the potential region corresponding to the current decreases, becomes wider, and the decrease of the current (minimum) becomes more pronounced [372].

Reaction product adsorption in the reduction of p-chloro-nitrobenzene and p-nitrotoluene in alkaline solution on rotating

gold, silver, and platinum disk electrodes, results in the waves of
the nitro compounds splitting into two if 15% methyl alcohol is
added to the solution [445]. A similar effect can be observed on
the mercury dropping electrode in the presence of surface-active
agents. The inhibition of a reversible electrode process by products of the electrochemical reaction in the reduction of quinoline
ethyl iodide results in distortion of the shape of the wave as shown
by logarithmic analysis [446]. The surface area occupied by one
product particle was calculated assuming that the distortion begins at a complete coverage of the electrode surface by products.
The product is formed by transfer of an electron on the N-ethyl
quinoline, and occupies an area of $37\,\mathring{A}^2$. Inhibition by electrode
products is often observed in the reduction of aromatic aldehydes
and ketones (which will be discussed in Chapters VIII and X of this
book) and in many other cases.

The degree of inhibition of the electrode process depends on
the coverage of the electrode surface by surface-active products.
Greatest inhibitions are usually observed at close to full coverages. Very often, if surface-active products are formed at the
dropping electrode that inhibit the electrode process, the formation of two waves can be observed. The first wave at small current values (and therefore at small coverages of the electrode surface by adsorbed products) corresponds to the electrode process
proceeding almost without inhibition. At the same time, the second
wave is caused by the process accompanied by inhibition at full
coverage of the electrode surface, or corresponds to the removal
of inhibition since the desorption potential of the surface-active
compound was already attained [438].

For products with sufficiently high adsorptivity, the electrode surface coverage is determined by the equations given by
Brdička. Therefore the first wave on polarograms with inhibition
of the process by electrode products is similar in character to the
adsorption prewaves of Brdička. Thus, at very low depolarizer
concentrations only one wave can be observed. This is the first
wave, and it increases with increasing depolarizer concentrations
to a certain limit. Then the second wave appears and begins to
increase. The maximum limiting current of the first wave changes
linearly with increasing height of the mercury column above the
orifice of the dropping electrode, and its value is usually close to

that which can be calculated by Brdička's equation (57). At higher
temperatures, when the adsorptivity of the products is lower, and
also when surface-active agents are added to the solution, this
wave may often disappear. The adsorption prewave of Brdička is
caused by a promotion of the reversible electrode process due to
the energy gain resulting from adsorption of electrode products.
Therefore, the prewave precedes the main electrode process
(whose half-wave potential is close to the redox potential of the
system). At the same time, the first wave on the polarograms,
corresponding to processes inhibited by products, is caused by
the uninhibited (or nearly uninhibited) electrolysis of depolarizer
particles. The following second wave corresponds to the discharge
of the same particles, but inhibited by a layer of reaction product
adsorbed on the electrode. Therefore, the adsorption waves for
inhibition of the process by reaction products are not prewaves
but rather are the main (normal) waves. Taking into consideration their size and character, these waves can be called adsorption pseudo-prewaves.

The difference between adsorption pseudo-prewaves and the
true prewaves of Brdička was pointed out by Schmid and Reilley
[403], Zhdanov and Frumkin [447], and also by Stromberg [343].
This difference is caused by the nature of the processes that result in prewaves and is expressed by the difference of certain
properties of the pseudo-prewaves from the characteristic properties of the waves of Brdička. Thus Zhdanov and Frumkin [447]
showed that the adsorption wave on the reduction polarograms of
the tropylium ion does not disappear on the addition of surface-
active agents, as is the case for prewaves of Brdička. The wave
even increased due to displacement of the reaction product from
the electrode surface. The reaction product, ditropylium, inhibited the electrode process, while the added adsorbable compounds did not affect the transfer of electrons. Similar phenomena
were observed by other investigators (see, for example, [393, 448,
449]). If the electrode reaction product is not adsorbed entirely,
then similarly as for waves of Brdička, the main wave appears
and increases even before the prewave reaches its maximum
value [371], determined by Brdička's equation. At the same time,
the height of the adsorption pseudo-prewave for incomplete adsorption of the product can thus exceed substantially the maximum
value calculated by Brdička's equation (57) on the basis of the area

occupied by the adsorbed product particle. Consequently, as pointed out by Zhdanov and Frumkin [447], the height of the pseudo-prewaves cannot always be used for the calculation of the electrode surface area covered by one adsorbed particle.

Pseudo-prewaves caused by inhibition of electrode processes are encountered quite frequently. Thus pseudo-prewaves are characteristic for the reduction polarograms of pyridinium ion derivatives, containing a carbonyl group in position 3 [438], as, for example, for 3-acetyl- and 3-formylpyridines in acidic solutions or for their n-alkyl-substituted salts (see also [450]). On the single-sweep i vs. E oscillographic curves of these compounds, two peaks can be observed which correspond to a pseudo-prewave and a consecutive wave. When the same i vs. E curve is recorded for a second time on the same drop, only one peak can be found. It corresponds to the second wave, since the electrode surface is covered by products and the pseudo-prewave is not formed [438]. A pseudo-prewave also can be observed on the polarograms of 2,6-diformylpyridine in acidic medium [438]. An inhibition of the electrode process by adsorbed reaction products was found in the polarographic examinations of 2- and 3-pyrrole derivatives containing electronegative substituents in position 5 [451]. In polarography using a hanging drop, two peaks are obtained only in the first measurement of the polarograms. When measurements are repeated on the same drop, the electrode surface is already covered by the product, and only one wave remains corresponding to the second wave on regular polarograms. With increasing temperature the height of the pseudo-prewave increases due to lower adsorptivity of products. For example, for 2-nitro-5-formylpyrrole oxime at temperatures above 37°C, only one wave can be observed [451]. A pseudo-prewave was observed in the reduction of trivalent arsenic compounds in acidic solutions [452, 453]. This wave is caused by inhibition of the electrode process by an arsenic(0) layer [453]. Calculation of the arsenic atom radius from the height of the pseudo-prewave gives a value of 0.98 Å [453], which is entirely reasonable.

It has been already mentioned that prewaves of Brdička can be observed during anodic polarization of dropping mercury electrodes in the presence of compounds that react with mercury ions. Under the same conditions (if a dense layer is formed that inhibits

the electrode process) adsorption pseudo-prewaves also can appear. Thus, during the anodic polarization of the mercury elec-
trode in cysteine solution, due to the RSH + Hg $\rightleftharpoons$ HgRS + H$^+$ + e$^-$
reaction a compact layer is formed, inhibiting the further rever-
sible electrode process, which appears on the polarograms as a
pseudo-prewave [454]. If the temperature is increased to 90°C,
the pseudo-prewave disappears, due to desorption of the HgRS
layer [455]. The pseudo-prewave also disappears when a strongly adsorptive compound like dithioglycolic acid, which prevents
the adsorption of HgRS particles [454], is added to the solution.

It has been shown on the basis of the area under the current
vs. time curve measured during the cathodic dissolution of the
layer, that the product layer formed by anodic polarization of the
mercury electrode in a cysteine solution is monomolecular [454].
The thickness of the layer is independent of the cysteine concentration as well as of the duration of the anodic polarization of the
electrode. In other instances the height of the anodic pseudo-prewave corresponds to less than complete monomolecular coverage
of the electrode surface by adsorbed particles. Thus, calculations
based on the height of the pseudo-prewave observed in anodic polarization of isomeric p- and m-nitrophenylhydrazone solutions,
yield 250 and 670 Å^2 for the areas occupied by one adsorbed product particle for the para and meta isomers, respectively [456].
The nature of this effect is not clear as yet.

Anodic prewaves are also observed in the formation of
strongly adsorptive products, due, most likely, to the inhibition of
the electrode process (which means pseudo-prewaves) in the polarography of inorganic depolarizers. Examples are the complex
cyanoferrate [Fe (II) complex] and hexathiocyanidochromate
[Cr (III) complex] ions [458].

The question of the nature of the prewave observed due to
formation of adsorptive products in a reversible electrode reaction, i.e., the question of whether the prewave is an adsorption
prewave of the Brdička type caused by promotion of the electrode
process, or whether it is a pseudo-prewave caused by inhibition
of the electrode process by a product layer cannot be answered by
the methods of classical polarography.

Breyer's method, with superimposition of an alternating voltage, can be used for the investigation of electrode processes: in particular when combined with a measurement of the capacity of the double layer by a bridge technique or with a vector polarograph, it enables us to differentiate between the reversible and irreversible process. Thus it can be distinguished whether the adsorption step corresponds to a Brdička prewave, or whether it is a pseudo-prewave caused by an inhibition of the irreversible process. For inhibition of a reversible electrode process the pattern is more complicated. On the curves, two pseudocapacity peaks can be observed that can correspond to the Brdička prewave and to the main wave, respectively. The same two peaks may correspond to the reversible pseudo-prewave and the subsequent peak of desorption of reaction products at a potential at which the second wave would normally appear on the polarographic curves (of the inhibited process). In such a case the Brdička prewave and the pseudo-prewave can be distinguished by the different dependence of the reaction pseudocapacity peak height and the peak caused by the adsorption–desorption process on the frequency of the applied alternating voltage.

The methods developed by Breyer permit a relatively simple gathering of qualitative data on the adsorptivity of starting materials or reaction products and on the changes caused by different factors. Thus, for example, in the polarography of flavinmononucleotide solutions [459], two peaks corresponding to the prewave and the main wave are observed on the Breyer polarograms. With increasing pH the peak centers shift toward negative potentials, and at the same time, the peaks move closer to each other. This can be explained by increased adsorptivity of the oxidized form as compared to the reduced form. This also is indicated on the Breyer polarogram by the greater decrease of the capacity current preceding the positive branch of the curve [459]. The increased adsorptivity with increasing pH is caused by the displacement of the wave from relatively positive potentials ($E_{1/2} = -0.12$ V measured against the S.C.E. at pH 2) toward potentials of maximum adsorptivity.

In the reduction of p-formylbenzoic acid (terephthal-aldehydic acid) on the dropping mercury electrode, at pH < 4.5, an adsorption prewave can be observed on the polarograms that has the

characteristics of a Brdička-type prewave. On the Breyer polarograms the height of this prewave in strongly acid solutions is much
decreased due to inhibition of the electrode process by adsorbed
depolarizer. If the solution pH is increased due to conversion of
the undissociated acid particles into the anionic form (pK_A = 4.47),
which is adsorbed to a much smaller extent, the height of the
Breyer peak corresponding to the prewave increases substantially
[460]. An inhibition of the electrode process by adsorbed starting
material or by reaction products results in an increase of the
Breyer peak corresponding to the first (main) wave at a rate
slower than that of the increase in the concentration of the starting material in the solution [460].

Biegler [461] recently discussed the inhibition of electrochemical reactions by adsorption of products.

Adsorption steps representing pseudo-prewaves can be observed on polarograms even when an added surface-active compound and not the reaction products are adsorbed on the electrode
if the concentration of this compound is of the same order as that
of the depolarizer. At sufficiently high adsorptivity of such a compound the coverage of the electrode surface by this compound can
be given by the equations derived by Brdička and Koryta. The appearance of an adsorption pseudo-prewave of this type was described by Bozyk [462]. If thymol ($\sim 2 \cdot 10^{-3}$ M) is added to a solution of glutathione the basic anodic wave is displaced to positive
potentials, and, where the wave was without thymol, only a smaller
step remains. The height of this step depends on the rate of the
electrode coverage by adsorbed thymol.

The examples given in this section about the effect of adsorption of surface-active agents on electrochemical processes do not
cover all investigated cases. The objective of this review has been
to show the basic characteristics of the effects of surface-active
materials — the great variety and complexity that exists in this
area. The physical nature of the effect of adsorbed compounds on
electrode processes is given in several excellent reviews by
Frumkin [463-465]. In his most recent publications [466, 467],
Frumkin reported new information on the adsorption of simple inorganic anions and cations and also on organic compounds (depolarizers among others) and on the effect of their adsorption on
electrode processes.

Chapter IV

Electrode Processes with Antecedent Protonation Reaction

A. PROTONATION IN THE ELECTROCHEMICAL REDUCTION OF ORGANIC COMPOUNDS

1. General Remarks

In electrochemical processes at the mercury electrode involving organic compounds, hydrogen ion addition or release usually takes place (an example of the very few electrode processes in which hydrogen ions do not take part is the reduction of aliphatic derivatives of vicinal dihalogen compounds: after the transfer of two electrons, two halogen ions are released and a multiple bond is formed between the carbon atoms [468-472]). On electrodes having a small hydrogen overvoltage the hydrogen atoms adsorbed on the electrode can take part in the electrode process (see, for example, [473]).

In the electrochemical reduction of organic compounds at the mercury cathode, hydrogen ions can be added to organic molecules both before they accept electrons or in successive steps. Addition of protons anteceding the electrochemical step always affects the kinetics of the electrode process. The addition of protons to the products of the electrochemical reaction (which means after the electron transfer) indirectly affects a reversible electrochemical process only if the rate of electron transfer (in the forward and reverse direction) is higher than the rate of the other stages of the electrode process.

We will discuss in detail the effect of protonation on processes with reversible and irreversible electrochemical step.

2. Electrode Processes with a
Reversible Electrochemical Step

The electrode potential in a reversible redox system is determined by the Nernst equation [compare with Eq. (5)] and the half-wave potential is close to the redox potential of the system E_0 [compare with Eq. (8)]. It is known from potentiometric data (compare, for example, [474]) that the electrode potential is a function of pH if hydrogen ions take part in the redox reaction.

The equilibrium electrode potential in a reversible electrochemical system in which n electrons and ν protons are transferred:

$$R \looparrowright ne^- + \nu H^+ \rightleftarrows RH_{\nu}^{(\nu-n)},\qquad\qquad \text{(XVI)}$$

is determined by equation

$$E = E_0 + \frac{RT}{nF}\ln\frac{[R][H^+]^{\nu}}{[RH_{\nu}^{(\nu-n)}]}.\qquad\qquad (59)$$

At the half-wave potential, where $[R] = [RH_{\nu}^{(\nu-n)}]$ (at the electrode surface), on the basis of Eq. (59) (disregarding the differences in diffusion and activity of the reduced and oxidized form)

$$E_{1/2} \simeq E_0 + \frac{RT}{nF}\ln[H^+]^{\nu} \simeq E_0 - \frac{\nu RT}{nF}\,pH.\qquad\qquad (60)$$

If hydrogen ions take part in the electrochemical reaction and if one or both components of the redox system are acids that can dissociate under experimental conditions, the dependence of the half-wave potential on the pH becomes more complicated [475]. Thus, for example, if in the reduction of R a product that can dissociate as a di-basic acid (with dissociation constants K_1 and K_2) is formed, then taking into consideration the decrease of the concentration of the reduced form of the redox system resulting from its dissociation, on the basis of Eq. (59) the following equation can be derived:

$$E_{1/2} \simeq E_0 - \frac{RT}{nF}\ln\frac{[H^+]^{2-\nu}}{([H^+]^2 + K_1[H^+] + K_1K_2)},\qquad\qquad (61)$$

since the fraction of undissociated di-basic acid in its entire analytical concentration (c) is equal to

$$\frac{[RH_2]}{c} = \frac{[H^+]^2}{[H^+]^2 + K_1[H^+] + K_1K_2} \cdot \tag{62}$$

Usually in reversible organic redox systems n = ν = 2 (for example, for derivatives of benzoquinone – hydroquinone, p-benzoquinone diamine – p-phenylenediamine, nitrosobenzene – phenylhydroxylamine), and therefore Eq. (61) can be given in a simpler form:

$$E_{1/2} \simeq E_0 - \frac{RT}{2F} \ln \frac{1}{[H^+]^2 + K_1[H^+] + K_1K_2} \cdot \tag{63}$$

The analysis of Eq. (63) shows that in strong acid medium, when $[H^+]^2 \gg K_1[H^+] > K_1K_2$, the dependence of the half-wave potential on pH is given by a straight line with a slope of -59 mV/pH (at 25°C). If the values of K_1 and K_2 differ from each other sufficiently, then in the pH range for which $pK_1 < pH < pK_2$, the dependence of the half-wave potential on pH is also linear, but the slope is -29 mV/pH, one half the previous value in strong acid medium. Formally, if $pK_1 < pH < pK_2$ in the discussed system, it can be considered that two electrons and one proton are transferred (which means n = 2, ν = 1). Therefore, in this case, Eq. (60) can be applied, from which it follows that $\Delta E_{1/2}/\Delta pH = -29$ mV/pH. In sufficiently alkaline medium, where pH > pK_2, the reduction product is completely dissociated, which means that formally ν = 0, and hence it follows from Eq. (60) that the half-wave potential is independent of pH. The same result can be obtained from Eq. (63) if we disregard the first two terms as compared with the third term in the denominator of the fraction in the logarithm. Thus the general relation between the half-wave potential and pH is expressed by three linear sections, which are interconnected by smaller curved sections. It must be mentioned that the pH values of the intersection points of the linear sections correspond to the pK values for dissociation of the reduced form (compare, for example, with [476]).

If the oxidized form of a reversible redox system is a polybasic acid, then the dependence of the half-wave potential on pH also is expressed by several straight-line sections. But in this case, with increasing pH the slope of each subsequent section will be larger than that of the previous section. For example, let us

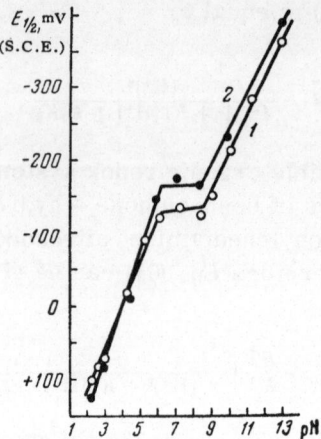

Fig. 8. Dependence of the half-wave potential of reduction for (1) tetramethylpiperidone N-oxide, and (2) tetramethylpiperidine N-oxide on pH (in citrate—phosphate and borate buffer solutions).

discuss the case of the reversible reduction of a protonated particle without participation of a proton, when the electrochemical step proper is

$$RH^+ + e^- \rightleftarrows \dot{R}H. \tag{XVII}$$

The fraction of protonated RH^+ particles depends on the hydrogen ion concentration in the solution [compare with Eq. (28)]. Therefore, the half-wave potential can be given by [475]:

$$E_{1/2} \simeq E_0 + \frac{RT}{nF} \ln \frac{[H^+]}{K_1 + [H^+]}. \tag{64}$$

In strongly acidic solution, if $[H^+] \gg K_1$, the half-wave potential is independent of pH; this means that all particles are present in the protonated form. In the pH region for which $K_1 \gg [H^+]$, the half-wave potential changes linearly with pH, and if n = 1, then $\Delta E_{1/2} / \Delta pH = -60$ mV.

If both the oxidized and the reduced form of the systems are acids, then breaks are found on the $E_{1/2}$ vs. pH graph close to each value of pK. If after the break the absolute value of $\Delta E_{1/2} / \Delta pH$ increases, then the given break is caused by the dissociation of the oxidized form; if this value decreases, then the break is caused by

the dissociation of the reduced form. Many examples of this relation between the half-wave potential and pH can be found in the
literature (see, for example, [477]). Figure 8 shows the dependence of the half-wave potential on pH for the reduction wave of
tetramethylpiperidone and tetramethylpiperidine N-oxide free
radicals [478]. In neutral and alkaline medium the waves are
close to reversible. Therefore, the breaks on Fig. 8 can be explained by acidic dissociation of the components of the redox system (in the original work [478] a somewhat different explanation
is given for the break in the alkaline region). It must be mentioned that the reduction of N-oxide radicals is complicated by kinetic factors [478] which appear due to incomplete reversibility
of the electrochemical stage.

3. Protonation Preceding a Slow
Electron Transfer: Quasidiffusion Waves

If the electrochemical stage itself of the process is irreversible, the electrode potential is no longer determined by the concentration of the oxidized and reduced form at the electrode, but
rather by the kinetics of the electron transfer. Due to electrostatic factors (see Chapter V), the rate of electron transfer from
the cathode to the particles increases sharply, if the charge of the
particles is made more positive (less negative) under identical
conditions. The protonation of particles results in an increase of
their positive charge

$$R + BH^+ (AH) \underset{\rho a}{\overset{\rho}{\rightleftarrows}} RH^+ + B (A^-) \tag{XVIII}$$

(BH$^+$ and AH are the acidic components of the solution and B and
A$^-$ are the conjugated bases); for this reason the protonation of
the particles promotes their electrochemical reduction.

For kinetic currents limited by the rate of protonation of
anions or uncharged bases (compared with section D-2 of Chapter
II), the height of the kinetic wave is determined by the kinetics of
addition of hydrogen ions. When the proton-donor ability of the
solution is increased (for example by increasing its buffer capacity or decreasing its pH) the rate of protonation (ρ) increases,
the height of the kinetic wave also increases [compare with Eq.
(25)]. At a sufficiently high rate of protonation (ρ) the height of

the kinetic wave tends toward the level of the diffusion current [$i_{lim} \rightarrow i_D$, compare with Eq. (25)]. The limiting current of such a quasidiffusion wave practically ceases to depend on pH and shows the properties of a diffusion current. If the acidic dissociation constant of the depolarizer remains much larger than the concentration of hydrogen ions in the solution even when the pH value is decreased (which means that $pK_A \ll pH$), the antecedent protonation remains in effect without influencing the limiting current but instead influences the rate of the electrochemical reaction, which means the potential value at which the reduction process takes place. We will discuss this case in detail, using for simplicity the concept of reaction layer.

If the electrochemically active form of compound RH^+ is formed in a protonation reaction (XVIII) taking place in the solution, then the current at any point of the kinetic wave can be given by

$$i = nsF\mu\rho([R]_s - \sigma[RH^+]_s).$$ (65)

For the limiting kinetic current the equation is [see Eq. (12)]

$$i_{lim} = nsF\mu\rho[R]_{s,\,lim}.$$ (66)

where $[R]_{s,lim}$ is the concentration of the inactive (unprotonated) form of the depolarizer in the reaction layer under conditions of the limiting current.

Expressing the $[R]_s$ and $[R]_{s,lim}$ concentrations by the diffusion current and the constant of the Ilkovič equation [see Eq. (14)] and subtracting Eq. (65) from (66), we obtain the equation

$$i_{lim} - i = nsF\mu\rho\left(\frac{i - i_{lim}}{\varkappa} + \sigma[RH^+]_s\right).$$ (67)

Combining with Eq. (15) we obtain

$$[RH^+]_s = \frac{(i_{lim} - i)\,c}{\sigma i_{lim}}.$$ (68)

Based on the theory of slow discharge [7, 8], the current intensity [see Eqs. (1) and (3)] can be given by equation

$$i = nsF\,[RH^+]_s\,k^0_{el}\exp\left(-\frac{\alpha n_a EF}{RT}\right).$$ (69)

Substituting into Eq. (69) the expression for $[RH^+]_S$ from Eq. (68), we obtain the equation for the kinetic wave with antecedent protonation proceeding in the solution volume close to the electrode surface

$$E = E_{1/2} - \frac{RT}{\alpha n_a F} \ln \frac{i}{i_{\lim} - i} , \qquad (70)$$

where

$$E_{1/2} = \frac{RT}{\alpha n_a F} \ln \frac{ns F k_{el}^0 c}{\sigma i_{\lim}} .$$

An equation similar to Eq. (70) was given first by Tanaka and Tamamushi [479] and also by Tur'yan [152].

Under conditions of the quasidiffusion current ($i_{\lim} \rightarrow i_D$), Eq. (70) becomes identical to Eq. (9). The only difference is that the expression for the half-wave potential of quasi-diffusion waves with antecedent protonation contains the additional factor σ which takes into consideration the effect of solution pH. This factor does not appear in Eq. (10) for the half-wave potential of ordinary irreversible waves with antecedent protonation. If the Ilkovič equation is considered [Eq. (6a)], the half-wave potential of quasidiffusion waves can be given by

$$E_{1/2} = \frac{RT}{\alpha n_a F} \ln \frac{0.81 k_{el}^0 \, t^{1/2}}{\sigma D^{1/2}} = b \, \log \frac{0.81 k_{el}^0 \, t^{1/2}}{\sigma D^{1/2}} . \qquad (71)$$

From Eqs. (26) and (71) the dependence of the half-wave potential for quasidiffusion waves on pH can be given by

$$\frac{\Delta E_{1/2}}{\Delta \, pH} = -\frac{2.3 RT}{\alpha n_a F} = -b. \qquad (72)$$

Thus the half-wave potentials for quasidiffusion waves at pH values larger than the pK of the depolarizer changes linearly with the solution pH, and the slope of the line representing this function is equal to the slope of the wave in $\log[i/(i_D - i)]$ vs. E coordinates.

The literature describes very many cases of similar change of half-wave potentials as a function of solution pH for waves, the wave height of which can be determined by the Ilkovič equation. Apparently, in all these cases the protonation of depolarizer par-

ticles precedes the electron transfer [480]. Very many examples
of this type of electrode processes can be given. For instance,
an antecedent protonation can be found in the reduction of different azomethine derivatives in acid solution [160, 161, 481], di-
phenylphosphoryl hydrazones [482], amidoximes of aromatic and
aliphatic acids [483], nitrosamines [484], aromatic nitrosohydroxylamines [159, 485, 486], thiobenzamide derivatives [487], for
several substituted 6-azauracils [488], amides of aromatic carboxylic acids [489], amine N-oxides [372, 490-493], azulene [494],
tropolone [495, 496], tropone [496], cyclopentadienylide of pyridinium and guajazulene [497], different pyridine derivatives [498-500],
quinoline derivatives [501, 502], isoquinoline derivatives [502],
pteridine [503], sydnones [504], sydnonimides [505], pyridoxine
and its several derivatives [506], tetrazolium salts [507, 508], different derivatives of imidazole and thiazole [509], and salts of iso-
benzpyrilium derivatives [510]; furthermore, in the reduction of
ethylenimide rings, attached to the phenyl ring of benzhydroquin-
one [511], N-substituted imides of maleic acid [512], phenylhydroxylamine [513], azobenzene [174, 175], aminoketones [514],
pyrrolidinetrione [515], and ketosteroides [361, 516]. The electrochemical reduction of carbonyl groups of aromatic aldehydes and
ketones also is preceded by their protonation in acid media [517];
only the consideration of antecedent protonation made possible the
explanation of the effect of various substituents on the half-wave
potentials of reduction waves for derivatives of benzaldehyde in
acid solutions [517]. A preliminary protonation takes place in the
reduction of trifluoroacetylacetonyl thiophene [518].

An antecedent protonation takes place in nonaqueous media
by addition of proton donors in the reduction of aldehydes [519]
and ketones [520]. The same behavior is shown by the intermediate reduction products of quinones and several other compounds
[521] and also for hydrocarbons with conjugated double bonds [522].

In the majority of the enumerated cases the waves with antecedent protonation in aqueous solutions seem to be quasidiffusion
waves. These waves decrease in height at sufficiently high pH
values and acquire the properties of kinetic waves limited by the
rate of proton addition.

At sufficiently low pH, when the pH reaches (or becomes
smaller than) the pK value of the acid, the half-wave potential of

the wave ceases to depend on pH. In this case the solution already contains a large quantity of protonated particles, and the $[RH^+]_S$ value [see Eqs. (28) and (14)] can be given by

$$[RH^+]_s = \frac{[H^+]}{[H^+] + K} \frac{(i_D - i)}{\varkappa}. \tag{73}$$

Substituting the $[RH^+]_S$ value from Eq. (73) into Eq. (69) and putting i equal to $i_{D/2}$, we obtain the general equation of half-wave potentials of quasidiffusion waves which is correct for any pH region:

$$E_{1/2} = \frac{RT}{\alpha n_a F} \ln \frac{0.81 k_{el}^0 t^{1/2}}{D^{1/2}} + \frac{RT}{\alpha n_a F} \ln \frac{[H^+]}{[H^+] + K}. \tag{74}$$

It can be seen from Eq. (74) that for $[H^+] > K$ the half-wave potential of the wave is independent of pH. An equation similar in form to Eq. (74) was derived by Laviron [438]. He used this equation to study the dependence of the half-wave potential on pH in strong acid medium in the polarographic investigations of derivatives of pyridine [438], imidazole, and thiazole [509]. It must be mentioned that the value of the acidic dissociation constant of the protonated form of the depolarizer RH^+ can be determined from the dependence of the half-wave potential of the wave on pH over the pH interval in which the half-wave potential becomes independent of pH. This means the region of transition of the wave from quasidiffusional to diffusional; the latter is limited by the rate of diffusion of RH^+. At $[H^+] \gg K$, Eq. (74) becomes practically identical to Eq. (10).

At sufficiently high pH (usually if alkaline solutions are used) the half-wave potentials of the waves also become independent of pH. In alkaline media waves of compounds that are reduced in acid media with antecedent protonation correspond either to the reduction of species formed by protonation by water, or to a reduction in an unprotonated state. Only the wave height of the second type is independent of pH in alkaline medium (for a wave accompanied by protonation the pH affects the σ value, and therefore the half-wave potential too). Thus, in the polarography of α-bromobutyric acid in alkaline medium, the $C-Br$ bond is split in the anion without protonation [523]. Therefore, the half-wave potentials of α-bromobutyric acid are in strongly acidic and alkaline media independent of pH (in strongly acid media the splitting of the C − Br bond takes place in the undissociated acid) and changes with pH in the intermediate region. Consequently, two waves should appear on the polarographic curves of α-bromobutyric acid in weakly acid and neutral solutions. One wave, of kinetic nature, would correspond to the reduction of the bromine of the undissociated acid. The second would correspond to the reduction of the bromine on the anion of the α-bromobutyric acid. Actually, only one wave can be observed on the polarographic curves, and the height of this wave is independent of pH [523]. It is possible that this wave consists of two parts, which cannot be distinguished on the polarograms. The merging of waves, probably also takes place in many other cases, when on the polarograms only a single wave can be observed. This wave has a height equal to the diffusion current, and both protonated and unprotonated particles are reduced at the same time.

Recently, the first wave on the polarograms of acetophenone was successfully resolved into two waves by selection of proper conditions. The first wave shows a kinetic character and corresponds to the reduction of protonated acetophenone molecules. The second wave corresponds to the reduction of unprotonated molecules. The derivative polarograms of acetophenone are given in Fig. 9. Three maxima can be clearly differentiated: the first and second correspond to the reduction of protonated and unprotonated acetophenone; the third maximum corresponds to the further reduction of formed radicals [524]. By comparing curves a and b in Fig. 9 we can see that at increasing pH, the first wave decreases substantially and the second increases.

For many compounds an S-shaped form of the dependence of half-wave potentials on pH can be found [475], particularly in the polarography of nitro compounds. It is known that nitro compounds can accept the first electron even in a proton-free medium [119, 525-527]. Therefore, they can be reduced in media of low proton-donor activity (for example, in alkaline solutions) without previous protonation. In weakly acidic medium the reduction of nitro compounds is preceded by their protonation, but the character of the protonation is different from that encountered for quasidiffusion waves. The dependence of the half-wave potentials on pH for quasidiffusion waves is not determined by the kinetics of the antecedent protonation, but only by the protolytic equilibrium by which

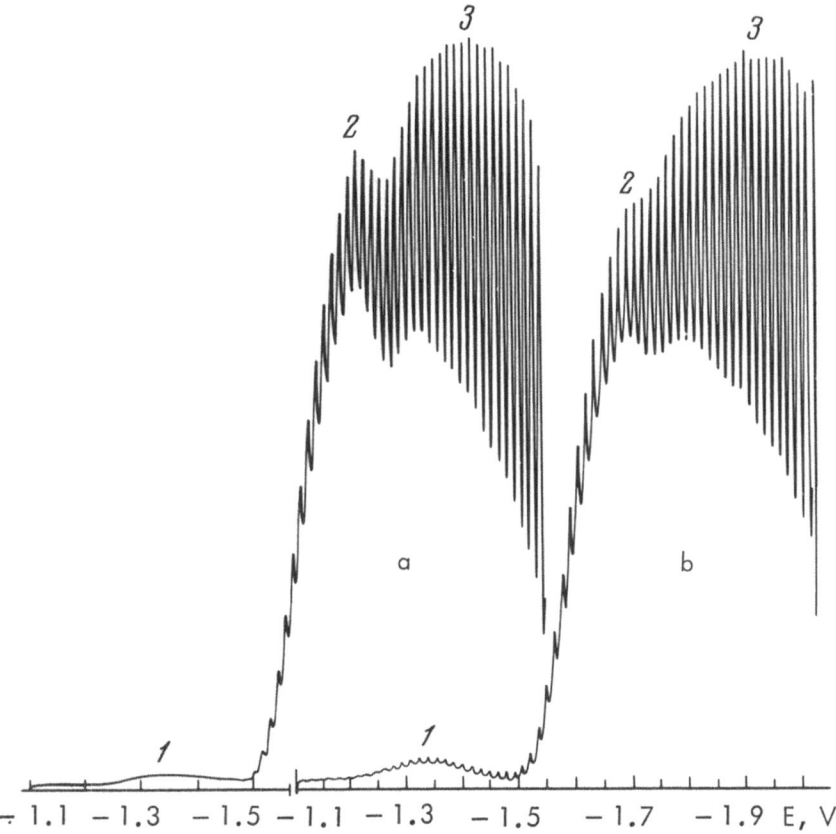

Fig. 9. Derivative polarograms ∂i/∂E vs. E for the reduction of acetophenone in buffer solutions at pH 7.8 (a) and 7.2 (b). (1) Reduction of the protonated molecule; (2) reduction of the unprotonated molecule; (3) reduction of the alcohol radical.

the concentration of electrochemically active particles is decreased due to dissociation. Therefore, in equations expressing changes in half-wave potentials [Eqs. (71) and (74)] not the kinetic constant (ρ) but the equilibrium constant $(\sigma$ or $K)$ is included. For nitro compounds the half-wave potentials of their reduction wave in acid solution depends not only on the pH of the solution but also on the nature of the buffer solution and its buffer capacity [528]. This means that the half-wave potential is affected by the kinetics of antecedent protonation.

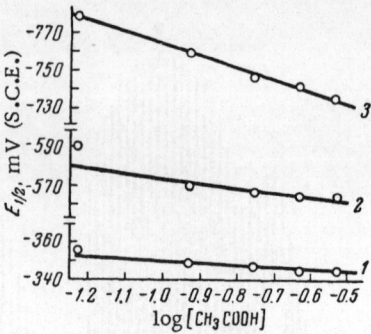

Fig. 10. Dependence of half-wave potentials of nitro compounds on the logarithm of the concentration of undissociated acetic acid in a solution of pH 4.60. (1) For the nitrobenzene wave (c = 0.263 mM); (2) for phenylnitromethane (c = 0.172 mM); (3) for 1-nitropropane (c = 0.368 mM).

Figure 10 shows the dependence of the half-wave potentials of the reduction waves of three nitro compounds on the concentration of undissociated acetic acid and acetate buffer solutions (the composition of the solutions was acetic acid + potassium acetate + potassium chloride). The pH and ionic strength of these solutions was kept strictly constant (by addition of potassium chloride instead of potassium acetate). As can be seen from Fig. 10, with increasing buffer capacity of the solutions the half-wave potentials of the waves for all three nitro compounds investigated shift to positive potentials. The half-wave potentials are displaced nearly linearly with the logarithm of the concentration of undissociated acetic acid in the solution. It must be pointed out that the buffer capacity was not exhausted during the polarographic measurements, since the intensity of the diffusion current always was less than 1.5 μA, and the concentration of undissociated acetic acid in the solution of smallest buffer capacity was 59 mM. The greatest shift of half-wave potentials was observed for the wave of 1-nitropropane ($\Delta E_{1/2}/\Delta \log[CH_3COOH] \approx 60$ mV). The shift of half-wave potentials for phenylnitromethane (~21 mV) was much smaller. The smallest shift was observed for nitrobenzene (~14 mV). These data indicate that in acidic solutions the rate of the reduction process can be determined to a certain degree by the kinetics of protonation of the nitro group. The protonation takes place not only by interaction with H_3O^+, but also under the influence of the acidic components such as undissociated acetic acid molecules in acetate solutions.

The shift of half-wave potentials with changing buffer capacity, which shows the effect of the kinetics of antecedent protonation, as might be expected, decreases with increasing reversibility of the electrochemical stage. The latter is equivalent to an increased rate of electron transfer in the series from nitropropane to nitrobenzene. The degree of reversibility can be determined to a first approximation by the reciprocal slope of logarithmic graphs of the waves in E vs. $\log[i/(i_D - i)]$ coordinates. The slope of these lines is independent of the buffer capacity and for nitrobenzene is ~50 mV; for phenylnitromethane, ~66 mV; and for 1-nitropropane, ~87 mV [528].

The nature of buffer components also affects the half-wave potentials to a great extent. Thus the half-wave potentials measured in a phthalate solution at pH 4.6 and an ionic strength of 0.5 at a monophthalate concentration of 23 mM are, for nitrobenzene, −0.398 V; for phenylnitromethane, −0.649 V; and for nitropropane, −0.811 V (vs. S.C.E.). These values are substantially different from those determined in acetate buffer solutions (see Fig. 10).

It must be mentioned that the antecedent protonation in the reduction of nitro compounds in acid solutions explains the negative ρ constant in the Hammett equation [529]. This equation was developed for the nitro compounds.

To explain the effect of the kinetics of antecedent protonation on the half-wave potentials of the nitro compounds in weakly acidic medium, it can be assumed that, due to the partial reversibility of the first electron transfer and also due to other as yet not understood features of the reduction of nitro compounds, the observed wave corresponds to the reduction of both protonated and unprotonated particles. This was the case in the polarographic investigation of α-bromobutyric acid already discussed. The ratio of protonated and unprotonated particles in the layer at the electrode affected the half-wave potentials. This ratio depends on the protonation kinetics of the nitro group.

The rate of antecedent protonation can also affect the half-wave potentials of quasidiffusion waves. This happens when the

limiting kinetic current is still somewhat different from the diffusion current (i_{lim} is, for example, 95 to 98% of i_D). Under these
conditions a shift of the half-wave potential of the 2-acetylthio-
phene semicarbazone reduction toward positive potentials was observed when the buffer capacity of the acetate solution increased
at pH 4.65 [480].

B. ANTECEDENT CHEMICAL REACTIONS AT THE SURFACE AND IN THE VOLUME; THE EFFECT OF ADSORPTION OF REACTION COMPONENTS IN THE ELECTRODE LAYER ON KINETIC CURRENTS

Up to now, in the discussion of electrode processes limited
by the rate of antecedent chemical reaction, it has been assumed
that these reactions proceed in a certain reaction volume close to
the electrode surface. In developing equations for kinetic currents,
Koutecký, Brdička, Hanuš, Delahay, and others also started with
the assumption that the antecedent chemical reactions proceed in
a certain volume of the solution in the neighborhood of the electrode surface. Nevertheless it has been shown [479, 480, 530,
531] in many cases — mainly when organic molecules take part in
the antecedent reactions — that these reactions proceed not only in
the solution volume but also on the surface of the electrode with
participation of adsorbed particles. The observed kinetic current
in such cases is equal to the sum of two components, "surface"
and a "volume" component. These current components character-
ize the processes taking place at the electrode surface and in the
volume reaction layer. The majority of organic compounds show
a substantial surface activity at the mercury—solution interface
at mercury potentials that are not too far from the zero charge
point (section A, Chapter III). For this reason a great majority of
kinetic waves with participation of organic depolarizers contain a
surface current component.

Let us discuss now the role of the adsorbed inactive depolarizer form in limiting the kinetic current. For simplicity, let
us deal with the case where the depolarizer is completely regenerated from the reaction products and its diffusion does not affect
the electrode process, i.e., the case of a pure kinetic current. We
will compare the quantities of electrochemically active species

that can be formed in the limits of the reaction layer volume and on the electrode surface from the adsorbed inactive form.

Let us assume that the thickness of the protonation reaction layer (see page 9) has an order of magnitude of 50 Å (in most real cases μ < 50 Å). If the concentration of the electrochemically inactive form of depolarizer R in the solution is 0.5 mM (which is the concentration normally used for the compound investigated in polarography), then the quantity of R in the reaction layer on a 1-cm^2 electrode surface area is: $5 \cdot 10^{-7}$ cm $\cdot$ 1 cm^2 $\cdot$ $5 \cdot 10^{-7}$ g-mole/cm^3 = $2.5 \cdot 10^{-13}$ g-mole.

Let us also assume that R is weakly adsorbed on the electrode and, thus, the surface coverage by R is under experimental conditions only 0.5%. It should be mentioned that such a small coverage cannot be observed by methods usually applied, e.g., by the measurement of surface tension or capacity of the electrical double layer. The potential region in which such an adsorption can occur is far outside that of the desorption peaks. But even at such small electrode coverages the quantity of adsorbed compound R on 1-cm^2 surface area (Γ) is $2.5 \cdot 10^{-12}$ g-mole (assuming that $\Gamma_\infty = 5 \cdot 10^{-10}$ g-mole/cm^2). Thus even for those extremely unfavorable conditions the quantity of adsorbed compound R on the electrode is larger than the quantity that is within the limits of the reaction layer volume by one order of magnitude. If we further consider that the surface coverage is often much larger than 0.5% and that the thickness of the reaction layer is smaller than 50 Å, the ratio of quantities adsorbed and present in the reaction layer is much larger.

Let us assume that the rate constant of the antecedent chemical reaction proceeding in the solution volume and that on the electrode surface (with participation of adsorbed R) have the same order of magnitude (actually the reaction with adsorbed material usually proceeds at a higher rate). On the basis of the previous considerations it can be concluded that even for a relatively low adsorption of R, the antecedent reaction takes place mainly on the electrode surface, thus causing a surface wave [480], and not in the reaction layer volume. This is particularly applicable to reaction with antecedent protonation, in which case, as will be shown later (see page 165), the concentration of the second component of the electrode reaction (the proton donor) is also increased at the

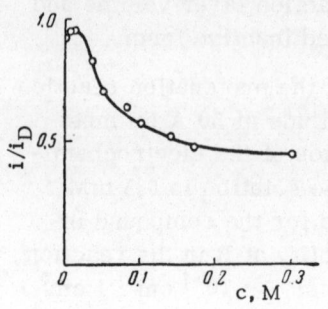

Fig. 11. Effect of tetraethylammonium benzenesulfonate on the ratio of the kinetic current to diffusion current for the reduction of maleic acid in citrate—phosphate buffer solution at pH 5.0.

electrode compared to the bulk solution. The majority of polarographic waves limited by similar processes are apparently either entirely of surface character or have a mixed surface—volume character.

The increased reaction rate at the electrode caused by adsorption of the inactive depolarizer form is also caused by increased reactivity of the adsorbed particles in the field of the electrode, as compared with the reactivity of some particles in the limits of the reaction layer volume (compare with section H of Chapter V).

Thus, if we disregard the effect of the adsorption and calculate the rate constants of the reactions at the electrodes by the equations obtained for purely volume kinetic waves the rate constants obtained for antecedent reactions will usually be much larger. Several rate constant values given in Brdička's review [532] for the recombination of acids with hydrogen ions are higher than the maximum possible value (close to $1 \cdot 10^{11}$ liters/mole · sec) [533, 534]. This is the rate constant of reaction for hydrogen and hydroxyl ions [535]. The rate constant values determined without consideration of adsorption from polarographic kinetic currents are usually higher than values determined by nonpolarographic methods (compare the data in [532] and [536, 537]).

Figure 11 shows the change of the ratio of the kinetic current to the diffusion current on addition of a surface-active compound (tetraethylammonium benzenesulfonate) to the solution [531]. The kinetic current value in this case is limited by the rate of addition of protons to the monoanions of maleic acid; a substantial adsorptivity of monoanions on the electrode results in the appearance of a surface current component. The displacement of the maleic acid monoanions from the electrode surface by the strongly adsorbed tetraethylammonium cations, if their concentration is sufficiently high, leads to a decrease of the surface component of the kinetic current. If we consider, as indicated in the Koutecký equation (25) the value of the apparent protonation rate constant in the absence and in the presence of tetraethylammonium benzene-sulfonate, we find that on its addition, the rate constant decreases by more than 100-fold.

The apparent increase of the rate constant of antecedent re-action due to the adsorption of reagents and the appearance of a surface component of the kinetic current results in a substantial decrease of the "polarographic" constants of acid dissociation [46, 137] (see page 26) compared with their true values. Therefore, surface kinetic currents can be observed even at a pH that is eight to nine units larger than the pK value of the discharging acid. This was observed, for example, in the reduction of pyridine-N-oxide on a dropping mercury electrode [492]. The dissociation constant of the protonated form of pyridine N-oxide is pK = 0.8 [538]. The polarographic wave characteristic of the reduction of the proton-ated form of N-oxide can be observed up to a pH of 9 [490-493]. The polarographic, apparent dissociation constant, i.e., the pH value at which the kinetic current is equal to one half of the diffu-sion current, is 7.5 [490]. Thus the difference between pK' and pK is 6.7, while according to Eq. (29) for a purely volume current, even if we take $k_{H^+} = 10^{11}$ liters/mole $\cdot$ sec, this difference can-not exceed 4-4.5 units.

At higher temperatures the adsorptivity of the components of antecedent reactions decreases. This results in a decrease of the surface current fraction in the total limiting current of the wave. If the surface current fraction is sufficiently high, then its decrease at higher temperatures can be compensated to a certain degree by the increased reaction rate of the antecedent reaction. This apparently explains the low value for the change of kinetic currents with temperature observed occasionally [162]. This cor-responds to an apparent activation energy for the antecedent reac-tion that is close to zero or even negative [162].

The quantity of adsorbed material on the electrode surface increases with adsorption time if the adsorption equilibrium is not yet attained. The experimental results of Delahay [183] can be ex-plained by the accumulation of electrochemically inactive (unpro-tonated) azobenzene on the electrode and by the increase of the sur-face component of the kinetic current. He found an apparent in-

crease of the rate constant of the protolytic reaction between
chloroacetic acid and azobenzene at longer durations of the experiments. The rate constants were computed from the reduction
current of protonated azobenzene.

C. PROTONATION REACTIONS IN THE VOLUME

1. Volume Character of Kinetic Currents Limited by the Recombination of Maleic Acid Dianions

Owing to significant adsorptivity of organic compounds on
the mercury electrode, examples of purely volume kinetic current
with participation of organic compounds are very rare. The current of the third wave on the polarographic curves of maleic acid
has been studied in greatest detail. This wave can be observed in
buffer solutions in a pH range of 8-11. It is limited by the protonation rate of the maleic acid dianion, which is converted into the
electrochemically active monoanion. The latter is reduced into
the succinic acid anion.

The relatively small size of this dianion, the two charges,
and the relatively high negative potentials (−1.5 V vs. S.C.E.) at
which this wave can be observed almost exclude the possibility of
adsorption of this dianion under these conditions. For the same
reason, surface components of the kinetic current cannot appear.
The absence of surface components in the third wave of the maleic
acid polarographic curves was proved by special experiments [531].
Introduction of a surface-active compound into the solution, which
reduces the adsorptivity of the compound investigated, should remove the surface component of the kinetic current. Upon introduction of different surface-active materials, the decrease of the third
wave could never be observed on the polarograms of maleic acid
reduction [531]. Figure 12 shows the effect of increasing concentrations of two surface-active compounds − tetraethylammonium
ions and isobutyl alcohol − on the height of the third wave for maleic acid. It can be seen from the figure that the kinetic current
does not decrease, as was the case, for example, with the second

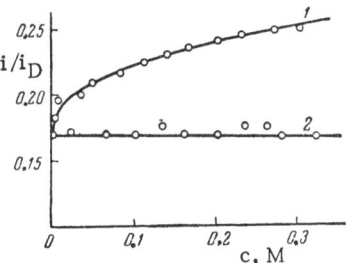

Fig. 12. Effect of surface-active compounds on the
limiting kinetic current controlled by the recombination of dianions of maleic acid in the reaction volume,
in veronal (barbital) buffer solutions at pH 9.0. (1)
Tetraethylammonium benzenesulfonate; (2) isobutyl
alcohol.

wave of the same acid (see Fig. 11). In fact, an increase in the kinetic current was even observed on addition of tetraethylammonium benzenesulfonate. The reasons for this increase will be discussed in section D of Chapter V. Therefore, the third wave on
the polarograms of maleic acid is purely volume in character, and
will be used for the detailed investigation of protonation reactions
taking place within the reaction volume of the solution.

2. Calculation of Rate Constants

from Polarographic Data Measured

in Buffer Solutions

The rigorous method of Koutecký was used for the calculation of rate constants of antecedent volume reactions from limiting kinetic current values. Corrections were made for the sphericity of diffusion. By this method, measuring the ratio of the limit-
ing kinetic current to the diffusion current $[i_{lim}/i_D = \overline{F}(\chi_1)]$, and
using the tables given by Weber and Koutecký [35], the value of the
χ_1 function can be determined:

$$\chi_1 = \sqrt{\frac{12\rho t}{7\sigma}} .$$ (75)

From this the rate constant (ρ) of the first-order reaction can be determined. The correction for the sphericity of diffusion [78] is made by adding to the $F(X_1)$ value a small component $\xi \bar{H}_C(X_1)$, which depends on X_1 [in the form of a certain function $\bar{H}_C(X_1)$] and on the characteristics of the dropping electrode, expressed by the ξ coefficient.

The calculation of ρ is made in the following steps. From the experimental data, using a plot of the $i_{lim}/i_D = \bar{F}(X_1)$ function constructed from the tables given by Weber and Koutecký [35] an approximate value of X_1 is determined. This preliminary value is then used to determine the correction for the sphericity of diffusion; the $H_C(X_1)$ values are determined from the table given by Koutecký and Cižek [78], and the ξ coefficient is determined by Eq. (45) in [78].

$$\xi = 50.4 \, D^{1/2} m^{-1/3} t^{1/6}, \tag{76}$$

where D is the diffusion coefficient of the inactive depolarizer form.

The ξ value usually does not exceed 0.10; the $\bar{H}_C(X_1)$ value is within the limits from 0.05 to 0.20. Therefore, the $\xi \bar{H}_C(X_1)$ correction is usually smaller than 0.015 or, expressed differently, rarely exceeds 7% of the i_{lim}/i_D value. The correction is used for the determination of the corrected $F(X_1) = i_{lim}/i_D + \xi H_C(X_1)$ value. From this value the corrected X_1 is calculated. Then the ρ value is calculated using Eq. (75).

The diffusion currents in this investigation of maleic acid were determined experimentally from wave heights in sufficiently acid solutions (pH ~ 3). Under these conditions the current is limited only by the diffusion of the depolarizer. For maleic acid, the wave in acid medium corresponds to the diffusion current of undissociated molecules. Brdička and Hanuš showed [137] that the diffusion coefficient of the dianion of maleic acid is somewhat lower than that of the undissociated acid, and therefore the i_D value of maleic acid multiplied by a correction coefficient of 0.837 (according to [137]) was used in the calculation. The diffusion coefficients, which are needed for calculation of the sphericity correction by Eq. (76), were determined by the Ilkovič equation from the limiting diffusion currents.

3. Determination of the Protonation

Rate Constants of Maleic Acid Dianions

under the Influence of the Components

of the Buffer System

The i_{lim}/i_D ratio for the third wave on the polarograms of maleic acid at constant pH and ionic strength of the solution depends strongly on the nature of the acid buffer solution and its concentration (as was determined from the effect of the buffer capacity). Under equal conditions, the i_{lim}/i_D value increases with increasing buffer capacity. This means that the overall rate of the antecedent reaction (ρ) increases.

The ρ value represents in this case the overall rate constant of the dianion protonation as affected by all acids (proton donors) present in the solution [see Eq. (30)]. Therefore, by changing the concentration of just one acidic component of the buffer solution, the specific protonation rate constant affected by this acidic component can be determined on the basis of the ρ-value change.

In the investigation of maleic acid dianion protonation almost all buffer solutions contained only one acidic component (not including hydrogen ions and water). This means that the buffer solutions consisted only of the salt of one acid. Only potassium (and ammonium) salts were used to prepare the solutions. In all solutions the ionic strength was $\mu = 1.0$ (by addition of calculated quantities of potassium chloride).

The undissociated acid concentration in the solution was calculated from the total concentration of the buffer system, the solution pH, and the dissociation constant of the given acidic component.

Figure 13 shows, for example, the dependence of ρ on the NH_4^+ ion concentration in ammoniacal buffer solutions. It can be seen from the figure that for solutions of the same pH, the experimental points fall on a straight line. The slope of this line ($\partial \rho / \partial [NH_4^+]$) is equal to the specific protonation rate constant of maleic acid dianions by ammonium ions ($k_{NH_4^+}$).

The protonation rate constants for boric acid, veronal, and glycine (Figs. 14, 15) were similarly determined. All these data were determined at 25°C with an electrode of the following characteristics: m = 1.46 mg/sec, t = 0.30 sec.

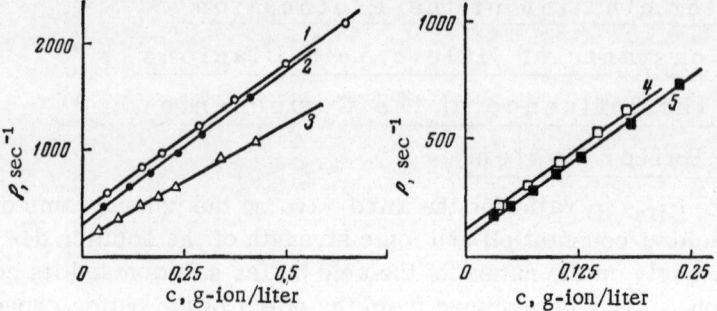

Fig. 13. The protonation rate constant of maleic acid dianions (ρ) as a function of NH_4^+ ion concentration in ammoniacal buffer solutions at several pH values. (1) 8.8; (2) 9.1; (3) 9.3; (4) 9.5; (5) 9.7.

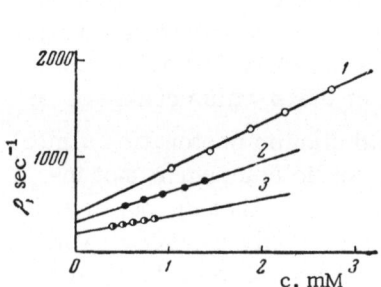

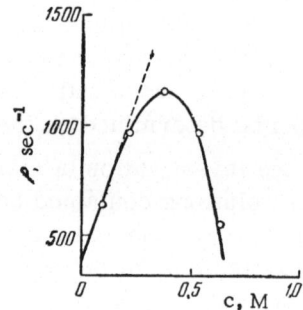

Fig. 14. The protonation rate constant of maleic acid dianions (ρ) as a function of un-ionized veronal (barbital) concentration at several pH values: (1) 8.8; (2) 9.1; (3) 9.3.

Fig. 15. The effect on undissociated glycine concentration on ρ, the protonation rate constant of maleic acid dianions at pH 9.0.

It can be seen from Fig. 15 that, if the undissociated glycine concentration is increased above ~0.15 M, the increase in ρ becomes slower and at glycine concentration above 0.4 M, the ρ value decreases. This is most probably caused by blocking of the dropping electrode surface by an adsorbed glycine layer, which inhibits the electrode process (compare with section E of Chapter III).

To calculate the specific rate constant of protonation of maleic acid dianion by glycine, only the first (linear) section of the graph given in Fig. 15 was used.

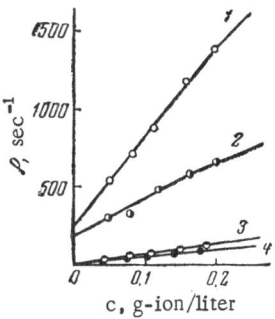

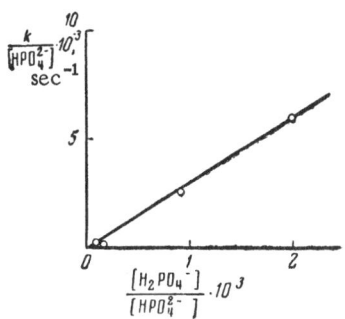

Fig. 16. The dependence of ρ on the total concentration of mono- and divalent phosphoric acid anions in phosphate buffer solutions at several pH values: (1) 9.1; (2) 9.4; (3) 10.2; (4) 10.5.

Fig. 17. The ratio of the total protonation rate constant by phosphoric acid anions in phosphate buffer solutions and the HPO_4^{2-} ion concentration as a function of the ratio $[H_2PO_4^-]/[HPO_4^{2-}]$ [according to Eq. (78)].

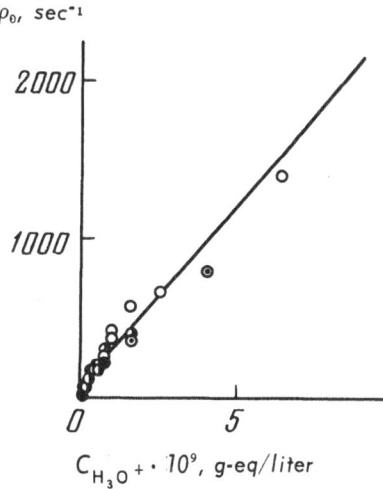

Fig. 18. The dependence of ρ_0 values on the hydrogen ion concentration at 25 °C in buffer solutions of different composition.

TABLE 1. Protonation Rate Constants (in liter/mole · sec) for Maleic Acid Dianions by Certain Characteristic Proton Donors

Electron donor	p	q	15°C		25°C		35°C		45°C		E, kcal per mole	A	$\Delta s^{\neq}$, entropy units
			K_A	k_A	K_A	k_A	K_A	k_A	K_A	k_A			
H_3O^+	1	1	55.55	$2.1 \cdot 10^{11}$	55.67	$2.3 \cdot 10^{11}$	55.84	$2.5 \cdot 10^{11}$	56.05	$2.6 \cdot 10^{11}$	2.1	$6.6 \cdot 10^{12}$	-1
$H_2PO_4^-$	2	3	$3.3 \cdot 10^{-7}$	$1.8 \cdot 10^6$	$4.0 \cdot 10^{-7}$	$2.8 \cdot 10^6$	$4.7 \cdot 10^{-7}$	$3.0 \cdot 10^6$	$4.9 \cdot 10^{-7}$	$3.9 \cdot 10^6$	4.3	$3.2 \cdot 10^9$	-17
Veronal	2	1	$1.3 \cdot 10^{-8}$	$3.2 \cdot 10^4$	$1.6 \cdot 10^{-8}$	$8 \cdot 10^4$	$2.5 \cdot 10^{-8}$	$1.5 \cdot 10^5$	$3.3 \cdot 10^{-8}$	$2.8 \cdot 10^5$	12.0	$3.2 \cdot 10^{13}$	3
H_3BO_3	3	1	$1.3 \cdot 10^{-9}$	$3.5 \cdot 10^3$	$1.6 \cdot 10^{-9}$	$4.5 \cdot 10^3$	$1.8 \cdot 10^{-9}$	$5.5 \cdot 10^3$	$2.1 \cdot 10^{-9}$	$9 \cdot 10^3$	7.1	$6.3 \cdot 10^8$	-20
NH_4^+	1	1	$2.7 \cdot 10^{-10}$	$1.4 \cdot 10^3$	$5.6 \cdot 10^{-10}$	$2.8 \cdot 10^3$	$1.1 \cdot 10^{-9}$	$4.4 \cdot 10^3$	$2.1 \cdot 10^{-9}$	$1.1 \cdot 10^4$	12.4	$3.2 \cdot 10^{12}$	-3
HPO_4^{2-}	1	4	$4.1 \cdot 10^{-12}$	$2.5 \cdot 10^2$	$4.1 \cdot 10^{-12}$	$3.5 \cdot 10^2$	$4.4 \cdot 10^{-12}$	$3.0 \cdot 10^2$	$4.1 \cdot 10^{-12}$	$3.5 \cdot 10^2$	—	—	—
H_2O	1	1	$8.1 \cdot 10^{-17}$	0.77	$1.8 \cdot 10^{-16}$	1.2	$3.7 \cdot 10^{-16}$	1.7	$7.2 \cdot 10^{-16}$	2.2	7.0	$5.6 \cdot 10^4$	-39

In the investigation of the dependence of ρ on the buffer capacity of phosphate solutions, only the total interaction constant for both donors, HPO_4^{2-} and $H_2PO_4^-$, can be directly determined from the slope of the dependence of ρ on the phosphate buffer concentration (Fig. 16). This can be expressed by

$$k = k_{HPO_4^{2-}} [HPO_4^{2-}] + k_{H_2PO_4^-} [H_2PO_4^-]. \tag{77}$$

The known change in the concentration of each phosphate ion with the pH of the solution was used to determine the individual specific constants. Equation (77) can be given in the form

$$k/[HPO_4^{2-}] = k_{HPO_4^{2-}} + k_{H_2PO_4^-} \frac{[H_2PO_4^-]}{[HPO_4^{2-}]} . \tag{78}$$

If we plot the ratio of the total constant and the concentration of divalent anion as a function of the $[H_2PO_4^-]/[HPO_4^{2-}]$ ratio, the pK_A'' value of the second dissociation step of phosphoric acid can be determined from the solution pH. Then, from the slope of the line plotted in this manner (Fig. 17), the value of $k_{HPO_4^{2-}}$ can be obtained. Finally, from Eq. (77) the value of $k_{H_2PO_4^-}$ can be calculated. The data, determined by this method for $k_{H_2PO_4^-}$ and $k_{HPO_4^{2-}}$, are given in Table 1.

By extrapolation of the linear relationships giving the dependence of ρ on the concentration of the acidic buffer component (the lines represented in Figs. 13-16) to zero concentration the ρ_0 values were determined. These values give the protonation rate by hydrogen ions and water, i.e.,

$$\rho_0 = k_{H_2O} [H_2O] + k_{H^+} [H^+]. \tag{79}$$

A plot of the ρ_0 values, determined in different buffer solutions, at a series of pH values as a function of hydrogen ion concentration can be fitted by a straight line (Fig. 18). From the slope of the line the rate constant of protonation by hydrogen ions k_{H^+} can be calculated (see Table 1).

It must be mentioned that the data given in Table 1, as well as some other values given in this book, are somewhat different from the data given by us in our previous publications. The explanation of this is that in calculating the values given in the book, more exact constant values were used (mainly dissociation constants of acids) determined in much later work. Thus, for example, in [531], in calculating the interaction constant for dianions with boric acid, through an oversight the concentration of tetraboric acid was used instead of the concentration of the boric acid. Therefore, the value given in [531] for $k_{H_3BO_3}$ was much higher.

4. Development of Equations for Kinetic Volume Waves Observed in the Polarography of Salts of Weak Acids in Unbuffered Solutions

According to Eq. (79), an extrapolation of the ρ_0 values (on a plot as shown in Fig. 18) to zero [H^+] concentration should give the rate constant for protonation by water (k_{H_2O}). But the errors committed in the extrapolation (in the determination of the ρ_0 values) and the scatter of the points in a plot similar to that given in Fig. 18 does not permit a sufficiently exact determination of the k_{H_2O} value by this method. Another method, based on the determination of polarographic waves in unbuffered solutions, was developed [540] for the determination of the k_{H_2O} value.

Besides being useful in the direct determination of the rate constant for protonation by water kinetic currents in unbuffered media can also be used for the study of protolytic reactions under such conditions when buffer solutions cannot be used. This is the case, for example, for acids reducible at negative potentials in supporting electrolytes containing tetrasubstituted ammonium salts or in nonaqueous solvents.

Let us consider the kinetic current of an exactly neutralized salt of a weak acid and strong base in an unbuffered solution. If the electrochemically active form is the undissociated RH acid, the reduction process follows the protonation of the R^- acid anions (salt hydrolysis) at the electrode surface

$$R^- + H_2O \underset{k_2}{\overset{k_1}{\rightleftarrows}} RH + OH^-. \qquad \text{(XIX)}$$

We will consider only the case where the amount of salt hydrolysis in the solution does not exceed 1%, i.e., we will deal with salts of only moderately weak acids (pK < 8). In this case the equilibrium of reaction (XIX) is shifted toward the left and the RH concentration in the solution is negligible compared to the $[R^-]$ concentration.

The only proton donor in the case under discussion seems to be water. The concentration of hydrogen ions is very low (the pH of solutions of weak acids is of the order of 8-9) and the buffer capacity is negligible. Therefore, even if all hydrogen ions that are transported to the electrode from the solution by diffusion react with R^-, the kinetic current component, which is equal to the diffusion current of hydrogen in the given solution, is extremely small. It is so small that its value can be neglected in comparison with the current determined by the rate of reaction (XIX) in the forward direction. If in the reduction of RH to AH n electrons and x hydrogen ions are used

$$RH + ne^- + xH_2O \xrightarrow{\text{el}} AH + xOH^- + P^{x-n}, \qquad (XX)$$

where P represents the side products of the electrode process (for example, halogen ions in the reduction of halogen-substituted organic acids).

If pK_A of the acid AH formed in the electroreduction is smaller or equal to the pK value of the starting acid, then AH dissociates at the electrode surface

$$AH + OH^- \rightleftarrows A^- + H_2O. \qquad (XXI)$$

If ν is the number of hydrogen ions used, the total process can be given by

$$R^- + ne^- + \nu H_2O \rightarrow A^- \text{ (or AH)} + \nu OH^- + P^\nu{}^n. \qquad (XXII)$$

From this equation it can be seen that as a result of the electrode process, hydroxyl ions are formed. These shift the equilibrium (XIX) to the left and reduce the rate of formation of electrochemically active particles, which means that the electrode process is auto-inhibited. Formally, the process is similar to the autocatalytic process discussed by Brdička. This later process is characterized by a kinetic wave limited by the rate of dehydration of formaldehyde in unbuffered solutions (see page 36).

Let us assume that R^- and RH are not adsorbed on the mercury at potentials corresponding to the reduction of RH, which means that reaction (XIX) takes place only in the reaction volume. Then for the average (over the life of the drop) limiting kinetic current, the equation derived by Koutecký and Weber [Eq. (25)] or Eq. (24) can be used. This equation was derived using the reaction layer concept of Brdička, Wiesner, and Hanuš, and differs from Eq. (25) only in the coefficient preceding the square root (0.81 instead of 0.886). For neutral and alkaline unbuffered solutions the first-order rate constant (ρ) is equal to

$$\rho = k_1 [H_2O], \tag{80}$$

and the σ value is

$$\sigma = \frac{[R^-]_s}{[RH]_s} = \frac{K_A}{[H^+]_s} = \frac{K_A[OH^-]_s}{K_W}, \tag{81}$$

where K_A is the dissociation constant of the investigated RH acid; K_W is the ion product of water; and the s subscript indicates that the concentration of compounds is given at the electrode surface.

The $[OH^-]_s$ concentration at the electrode surface, as can be seen from (XXII), increases at higher currents. If the removal of OH^- can be described [541] by the Ilkovič equation, with a proportionality factor of $\varkappa_{OH^-}$ ($\varkappa_{OH^-} = i_{D,OH^-}/[OH^-]_s$), then the $[OH^-]_s$ concentration can be given by

$$[OH^-]_s = iv/n\varkappa_{OH^-}. \tag{82}$$

Substituting the appropriate values from Eqs. (80)-(82) in Eq. (25) after simple rearrangement, we obtain

$$i_{\lim}^{3/2} = Q (c - i_{\lim} \varkappa), \tag{83}$$

where

$$Q = 0.886 \left(\frac{k_1 [H_2O] \, t K_W n \, \varkappa_{OH^-}}{K_A v} \right)^{1/2} \varkappa. \tag{84}$$

The calculation using the Brdička−Wiesner−Hanuš method gives a similar expression for Q [541], except that the numerical coefficient is 0.81.

The Koutecký–Weber equation (25) or Eq. (24) is correct if the concentration of proton donors and that of the conjugated bases with the proton donors is constant within the thickness of the reaction layer μ. For the given case this latter is [541]

$$\mu = \sqrt{D/k_2 \, [OH^-]_s} = \sqrt{\frac{D\varkappa_{OH^-} n}{k_2 i v}},\tag{85}$$

where D is the diffusion coefficient of RH and k_2 is the rate constant of the reverse reaction (XIX).

Therefore, Eq. (83) is strictly correct only if the thickness of the reaction layer (μ) is much smaller than the OH^- ion diffusion layer, and the OH^- concentration can be taken as constant within the limits of the μ layer and equal to $[OH^-]_S$. It can be seen that this condition is fulfilled at higher current intensities according to Eq. (85). The experimental results [541] showed that this condition is met exactly for current intensities that are higher than a certain "transition" value, which is for electrodes with regulated drop formation equal to a few tenths of a microampere.

To develop the equation of the wave, the method based on the reaction-layer concept will be used. From the balance of the process rates in the electrode layer it follows that

$$i = nsF\mu \, (k_1 \, [R^-]_s \, [H_2O] - k_2 \, [RH]_s \, [OH^-]_s);\tag{86}$$

$$i_{lim} = nsF\mu_{lim} k_1 \, [R^-]_s \, [H_2O],\tag{87}$$

where μ_{lim} is the reaction-layer thickness according to Eq. (85) under limiting current conditions; s is the average electrode surface, and F is the Faraday number.

The i and i_{lim} values are much smaller than i_D, especially under such conditions when Eq. (83) can be correctly applied. Therefore, $[R^-]_S$ remains almost constant along the wave. Taking this into consideration, and comparing Eqs. (86) and (87), we obtain

$$i = \frac{i_{lim} \mu}{\mu_{lim}} - nsF\mu k_2 \, [RH]_s \, [OH^-]_s.\tag{88}$$

From Eq. (85) it follows that $\mu/\mu_{lim} = (i_{lim}/i)^{1/2}$ and

$$i_{lim}^{1/2} - i^{1/2} = nsFk_2 \, [RH]_s \, [OH^-]_s \, i^{1/2},$$

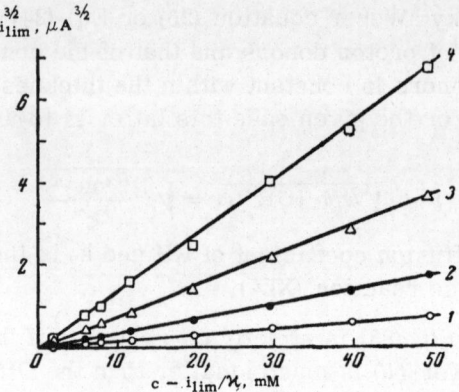

Fig. 19. The $i_{lim}^{3/2}$ value of kinetic waves for maleic acid dipotassium salt solutions as a function of $(c - i_{lim}/\varkappa)$ in a supporting electrolyte of 1.0 M potassium chloride at the following temperatures: (1) 15°C; (2) 25°C; (3) 35°C; (4) 45°C.

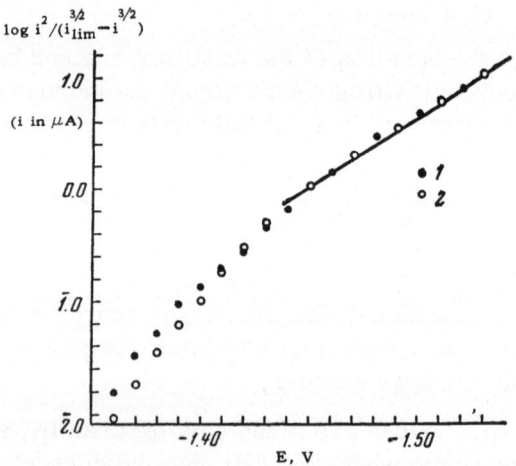

Fig. 20. The dependence of the current function according to Eq. (90) on potential for two concentrations of maleic acid dipotassium salt at 25°C: (1) 4.84 mM; (2) 32.7 mM.

and with Eqs. (82) and (85) we find

$$[RH]_s = \frac{i_{lim}^{3/2} - i^{3/2}}{insF} \cdot \left(\frac{n\varkappa_{OH^-}}{\nu k_2 D} \right)^{1/2}. \qquad (89)$$

Substituting the $[RH]_S$ value from Eq. (89) into Eq. (69) of the theory of slow discharge [in Eq. (69) the concentration of the electrochemically active depolarizer form at the electrode is given by $[RH^+]_S$], we obtain the equation giving the form of the kinetic reduction wave for weak acids in unbuffered solutions of their salts

$$E = \varepsilon_0 - \frac{RT}{\alpha n_a F} \ln \frac{i^2}{i_{lim}^{7/2} - i^{3/2}}, \tag{90}$$

where ε_0 is a characteristic potential independent of the acid salt concentration. Its value is given by

$$\varepsilon_0 = \frac{RT}{\alpha n_a F} \ln k^0_{el} \left(\frac{n \varkappa_{OH^-}}{\nu k_2 D} \right)^{1/2} 10^3, \tag{91}$$

if the current in Eq. (90) is expressed in μA.

5. Kinetic Waves of Maleic Acid Dianions in Unbuffered 1 M Potassium Chloride Solution; Determination of the Protonation Rate Constant of Maleic Acid Dianions by Water

The relationships given in the previous section were confirmed for volume kinetic waves in solutions of maleic acid dipotassium salt on a background of 1 M potassium chloride [540].

In Fig. 19 is given the plot of $i_{lim}^{3/2}$ against $(c - i_{lim}/\varkappa)$ relationship, measured with a dropping electrode (having a small glass plate for the regulated removal of drops) with the following characteristics: m = 0.828 mg/sec, t = 0.28 sec. From Fig. 19 it follows that Eq. (83) expressed the experimental data well. To determine the $\varkappa$ value, diffusion waves of maleic acid were measured with the same electrode in a buffer solution at pH ≈ 3 (with an ionic strength of 1.0). The value determined by this method was corrected for the difference of diffusion coefficients of the acid and its anions. The $i_{lim}/\varkappa$ value is much smaller than c (mainly at high c values) and therefore expression (83) is practically identical in its form with the analogous equation for volume catalytic waves of hydrogen in unbuffered media [541] (see page 259). From the slope of line 2 in Fig. 19 the rate constant of protonation for maleic acid dianions by water was determined according to Eq. (84) — the hydrolysis rate constant of potassium dimaleate. The value at 25°C is $k_1 = 1.2$ liters/mole $\cdot$ sec. In the calculation it was assumed that n and ν in (XXII) are 2; the $\varkappa_{OH^-}$ value of 2.05 $\mu A/mM$ was calculated from the Ilkovič equation, with a diffusion coefficient of OH$^-$ ions in 1 N potassium nitrate equal to $2.26 \cdot 10^{-5}$ cm^2/sec [542].

Figure 20 shows a plot of the current function given by Eq. (90) versus potential for the waves corresponding to two concentrations of the dipotassium salt [540]. It can be seen from the graph that at sufficiently high current values the experimental points fall on a straight line, given by Eq. (90) with $\varepsilon_0 = -1.45$ V and a reciprocal slope value of $\sim$75 mV.

Nearly the same slope was obtained on plots of the reduction wave for maleic acid monoanions. These measurements were made in borate buffer solutions with pH 9-9.5. It must be mentioned that at small current values the $\log [i^2 / (i_{lim}^{3/2} - i^{3/2})]$ vs. E curve, constructed from the experimental data, lies below the calculated data (see Fig. 20).

6. The Relationship Between the Acid Strength and Its Proton Donor Effect

The rate constant of the proton transfer process from the acid (donor) to the base (acceptor) is determined by the properties of both the donor and the base-acceptor. If we consider a series of acids, which give their protons to the same base, then the rate constants of these protolytic reactions can be used as a measure of the proton-donor ability of these acids.

Such a series is given, for example, in Table 1 by the specific protonation rate constants of maleic acid dianions, which play the role of the base (proton acceptor) in the reaction with various acids. Table 1 shows that the rate of proton transfer to the maleic acid dianion increases with the K_A value of the acid (from the bottom of Table 1 upward) showing that the rate constants of protolytic reactions and the dissociation constants of the acid reactants, follow the Brönsted equation (see, for example, [167]).

The Brönsted relation, with correction for the so-called statistical effect, is given by

$$k_A = pG^1 \left(\frac{q}{p} K_A \right)^\alpha, \tag{92}$$

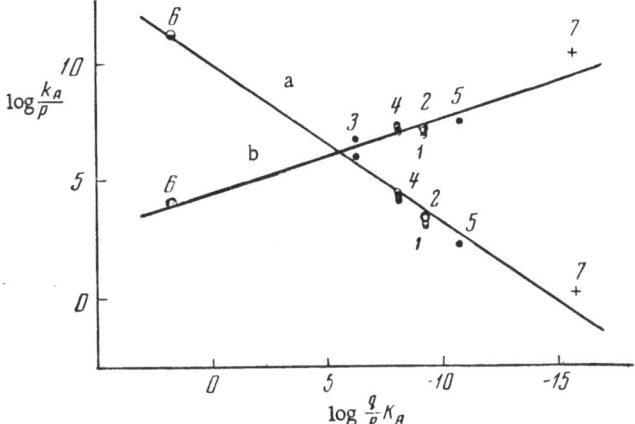

Fig. 21. The dependence of specific protonation rate constants of maleic acid dianions on the dissociation constants of the corresponding acids (proton donors) (a) and specific dissociation rate constants of maleic acid monoanion by different bases (b). (1) Boric acid; (2) NH_4^+; (3) $H_2PO_4^-$; (4) veronal (barbital); (5) HPO_4^{2-}; (6) hydroxyl ion; (7) water.

where $\alpha < 1$ and G_1 is a constant; p and q are statistical factors, which take into consideration the number of protons that can be removed from the different atoms of the acid, and the number of places at the different atoms of the base where the protons can be accepted [167]. If several protons can be removed from the same atom of the acid, as for example for H_2O or NH_4^+, then p is taken as unity. The p and q values for acids and conjugated bases used in calculating the protonation of maleic acid dianions are given in Table 1.

By plotting $\log k_A/p$ as a function of $\log (q/p)K_A$, the line a in Fig. 21 was obtained for protonation constants of maleic acid dianion. This line can be described by the Brönsted equation (92), with $\alpha = 0.68$ and $\log G_1 = 10.0$.

From the k_A value and from the acid dissociation constant (K_A) of the proton donor and the reacting base (maleic acid dianion), the rate constant of the reverse reaction (k_B) can be calculated. In the given case this process is proton splitting from the maleic acid monoanion by the base, conjugated with the acid of the buffer solution

$$k_B = k_A \frac{K''_{MA}}{K_A},$$ (93)

where K''_{MA} is the acid dissociation constant of maleic acid mono-anion.

The k_B value is larger if the strength of the reacting base increases (or if the conjugated acid becomes weaker). In Fig. 21 the line b characterizes the dependence of the k_B values [determined by Eq. (93)] on the K_A value of the acids conjugated with the reacting bases.

The correctness of the Brönsted relationship was proved for many protolytic reactions (compare, for example, with [543-545], and also see Chapter 14 in Shatenshtein's monograph [167]).

Wiesner and co-workers [164] first discussed the application of the Brönsted relationship for protonation rate constants of weak acids determined from polarographic data. Kemula, Grabowski, and Bartel [162] applied this relationship to the calculation of the protonation rate constant of p-dimethylaminobenzaldehyde. But these investigators did not take into account the surface nature of protonation, and therefore the given rate constants are apparently higher than the true values.

The Brönsted relationship was used [492] to prove the surface nature of protonation for pyridine-N-oxide.

7. The Effect of Temperature on the Protonation Rate of Maleic Acid Dianions

For purely volume antecedent reactions, investigation of kinetic waves at several temperatures in buffer solutions of different composition and buffer capacity permits the determination of the temperature dependence of specific rate constants. The activation energy of the separate protonation reactions can also be determined.

The kinetic waves for recombination of maleic acid dianions were investigated [546] at 15, 25, 35, and 45°C in different buffer solutions of 1.0 ionic strength (addition of potassium chloride) by the method described.

In all buffer solutions investigated, an increase of the absolute kinetic current value of the third wave can be observed on the polarographic curves of maleic acid with increasing temperature. The ratio of kinetic current and diffusion current (i_{lim}/i_D) also increases. The overall rate constant values (ρ) calculated from the i_{lim}/i_D data for buffer solutions of given composition at constant pH were a linear function of the buffer capacity. Figures 22 and 23 show examples of the dependence of ρ on the concentration of undissociated acid components in the same buffer solutions at different temperatures. The pH values of the solutions at the corresponding temperatures are given in the figures. These figures show that the protonation rate increases substantially with temperature. The partial rate constants for protonation, determined from the slopes of the lines of the type given in Figs. 22 and 23, are given in Table 1.

The ordinate intercepts of the ρ vs. acid concentration lines determine the ρ_0 value. This is the total rate constant of interaction of maleic acid dianion with water and hydrogen ions [see Eq. (63)]. By plotting the ρ_0 values as a function of $[H^+]$, the k_{H^+} value can be determined (see Fig. 18).

The straight lines on the plot given in Fig. 18 pass through the $k_{H_2O}[H_2O]$ value, which can be found on the ordinate. These values were determined from experiments in unbuffered 0.1 M potassium chloride solutions at different temperatures (compare with Fig. 19). The $k_1 = k_{H_2O}$ values are given in Table 1.

The specific protonation rate constant data determined at different temperatures for maleic acid dianions were used to construct the lines according to the Brönsted equation (92). It was found that over the entire temperature interval investigated (from 15 to 45°C for every 10°C) the experimental values lie on straight lines of the Brönsted type. The slope of these lines (the α value) was found to be independent of the temperature within the experimental error ($\alpha = 0.63$).

The constant slope is the result of compensation of the increased value of the specific constant for weaker acids (right-hand portion of Fig. 21) by an increased dissociation constant at higher temperatures (see the k_A and K_A values at different temperatures in Table 1). Therefore, at higher temperatures, the lines corre-

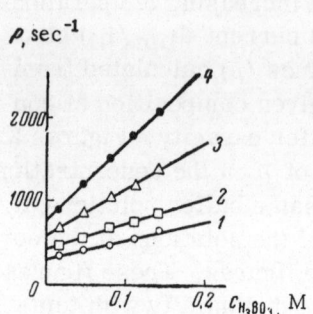

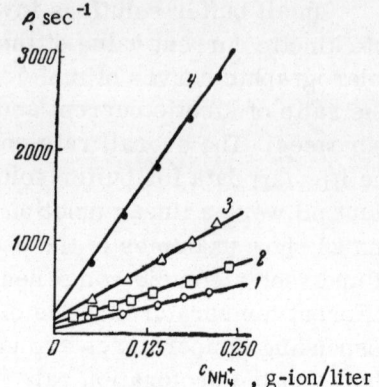

Fig. 22. The dependence of ρ on the concentration of undissociated boric acid at several temperatures. (1) 15°C (pH 9.08); (2) 25°C (pH 9.0); (3) 35°C (pH 8.92); (4) 45°C (pH 8.84).

Fig. 23. The dependence of ρ on the ammonium ion concentration at several temperatures. (1) 15°C (pH 10.08); (2) 25°C (pH 9.7); (3) 35°C (pH 9.45); (4) 45°C (pH 9.17).

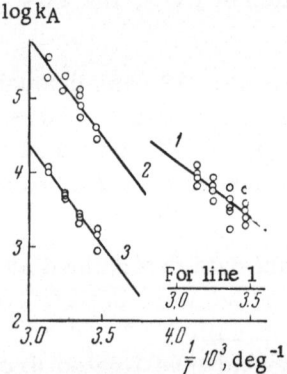

Fig. 24. The quantity $\log k_A$ as a function of $1/T$ for the protonation of maleic acid dianions with different agents. (1) Boric acid; (2) undissociated form of veronal; (3) ammonium ions.

sponding to the Brönsted equation are only somewhat displaced parallel with respect to each other.

The constancy of the α coefficient in the Brönsted equation with a temperature change is apparently of general character. The same was observed for Brönsted-type lines at different temperatures for the dissociation and recombination rate constants of nitro compounds and also for the protonation of pyridine (see page 259).

It must be mentioned that the highest rate constant value — for the reaction of maleic acid dianions with hydrogen ions — is close to the highest value theoretically possible. According to Debye [534] and Onsager [533], this is equal to the rate constant of diffusion-limited interaction of the hydrogen ion and anion in the solution.

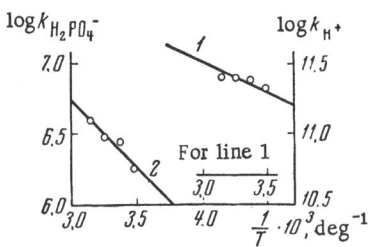

Fig. 25. The quantity $\log k_A$ as a function of $1/T$. (1) Veronal (barbital) ions; (2) $H_2PO_4^-$ ions.

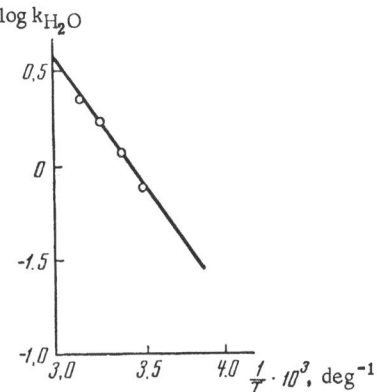

Fig. 26. The quantity $\log k_{H_2O}$ as a function of $1/T$.

Figures 24–26 give the logarithms of the specific rate constants for maleic acid dianions with different proton donors as functions of $1/T$. It can be seen from the figures that the scatter of the points is very large. The straight lines, corresponding to the Arrhenius equation, were constructed by the method of least squares. The scatter is particularly large for the interaction with divalent HPO_4^{2-} anion. Therefore, the activation energy for this process could not be determined from the experimental data. The determined experimental activation energies (E), the pre-exponential factor of the Arrhenius equation (A), and the calculated activation entropy [547] data are given in Table 1.

Table 1 shows that the activation energies of the processes investigated are quite low.

It is known that small experimental activation energies are characteristic for processes limited by diffusion [547], where E is usually of the order of 4–5 kcal/mole; for hydrogen ions this value is even lower [547]. The very small experimental activation energy value for the interaction of maleic acid dianions with hydrogen ions (about 2 kcal/mole) again confirms that the rate of this process is diffusion limited. This is also indicated by the activation entropy, which is close to zero [547].

It is interesting to compare the value of the pre-exponential factor with the change in the product of charges of the reacting particles ($z_{dianion} \cdot z_{proton\ donor}$). If this product increases, the value of the pre-exponential factor, in accord with theory [547], decreases (see the first three lines in Table 1). It must be men-

·tioned that due to the large scatter in the plot of $\log K$ vs. $1/T$, the accuracy of the determined values of E and A is very low, and therefore only qualitative conclusions can be drawn on the basis of their comparison.

8. Certain Other Processes with Fast Antecedent Volume Protonation Reaction

It has already been mentioned that because of the significant adsorptivity of organic compounds at the mercury electrode, the kinetic waves limited by the rate of recombination of organic acids usually have a large surface component. One of the most thoroughly investigated examples of pure volume currents is the limiting current of the third wave of maleic acid. The results of these investigations were discussed in the previous sections. Besides this process, which was limited by the rate of recombination of maleic acid dianions, other electrode processes are known for which it was determined that the antecedent protolytic reaction takes place mainly in the reaction-layer volume. Thus, for example, a protonation reaction taking place in the reaction volume limits the reduction current for N-methylpiperidine-N-oxide in neutral and weakly alkaline media [492]. From the dependence of the limiting current on the buffer capacity of borate solutions at several pH values, the protonation rate constants of N-methylpiperidine-N-oxide by undissociated boric acid $(k_{H_3BO_3} = 1.6 \cdot 10^3$ liters/mole · sec) and hydroxonium ions $(k_{H^+} = 7 \cdot 10^9$ liters/mole · sec) (at 25°C) were calculated [492]. The kinetic wave in pyridine-N-oxide solutions contained a significant surface component, which indicated a much higher adsorptivity of pyridine-N-oxide than for N-methylpiperidine at corresponding conditions. This difference in adsorptivity is caused partly by the potential of reduction: pyridine-N-oxide is reduced at less negative potentials than N-methylpiperidine (by 150-200 mV, according to [490, 492]). Furthermore, the planar structure of the pyridine ring, under identical conditions, is seemingly more favorable for the adsorption of pyridine-N-oxide.

Volume kinetic waves were also observed in the polarography of alkaline dinitroethane [548, 549] and nitroacetic ester [549, 550] solutions.

Based on the dependence of the wave height on the concentration of buffer components, the rate constants of protonation of dinitroethane by NH_4^+ and H_3BO_3 [$1.1 \cdot 10^7$ and $1.5 \cdot 10^7$ liters per mole $\cdot$ sec, respectively (at 25 °C)] were determined. The rate constants were also determined for the protonation of the nitroacetic acid ester by veronal (50 liters/mole $\cdot$ sec), by the acidic form of tris-(oxymethyl)aminomethane, and by H_3BO_3 [12 and 2 liters per mole $\cdot$ sec, respectively (at 25°C) [550].

Polarography of these nitro compounds in unbuffered solutions of potassium salts in a supporting electrolyte consisting of 0.5 M potassium chloride gave protonation rate constants for dinitroethane anions by water ($8 \cdot 10^3$ liters/mole $\cdot$ sec) and for nitroacetic ester ($4 \cdot 10^{-3}$ liter/mole $\cdot$ sec) [549].

Chapter V

The Effect of the Double-Layer Structure on Electrode Processes

A. INTRODUCTION

A. N. Frumkin was the first to consider the effect of the double-layer structure on the kinetics of electrochemical processes in developing the theory of slow discharge and hydrogen overpotential [7, 8, 551]. Frumkin's views were later fruitfully employed by himself and by his co-workers to explain the phenomena observed in the electrochemical reduction of oxygen [552], anions [463-466, 551-559], and several neutral compounds. Recently, the ideas developed by Frumkin have become widespread among electrochemists (see, for example, [420, 560-562]); several publications have appeared in which the effect of the double-layer structure on electrode processes was discussed. These later cases dealt with electrode processes limited by antecedent chemical reactions [563-569]. The structure of the electrical layer and its effect on the kinetics of electrode processes were summarized in several reviews, among which the outstanding reviews of Damaskin [570], Parsons [571], and of Nürnberg and Stackelberg [572] must be mentioned.

The effect of the double-layer structure on irreversible polarographic waves is generally caused by three basic factors: (1) changes of the effective potential drop between the electrode and discharging particle, (2) different ion concentration at the electrode surface as compared with the bulk solution, and (3) adsorptivity of compounds. Levich [573, 574] also considers the effect of potential gradient within the limits of the diffusion part of the double layer on the movement of charged particles.

We will discuss the effect of double-layer structure on different electrode process types.

B. IRREVERSIBLE ELECTRODE PROCESSES WITH PARTICIPATION OF UNCHARGED COMPOUNDS WITHOUT ANTECEDENT PROTONATION

The effect of double-layer structure on discharge of uncharged particles is caused by the change of the effective potential drop between the electrode and discharging particle, which determines the rate of the electron transfer process.

Due to the action of electrostatic forces at the electrode surface, which is immersed in a solution of supporting electrolyte, the concentration of those ions increases that have a charge opposite to that of the electrode. Some of these ions are directly on the electrode surface and form a so-called dense (Helmholtz) part of the double layer; other ions, which are at a certain distance from the surface, form its diffusion part. The potential drop between the electrode and a point in the bulk solution takes place in these parts of the double layer. If we assume that the active center of a particle entering an electrochemical reaction is at the interface of the Helmholtz and diffusion part of the double layer (which means that the center is at the so-called Helmholtz surface, which is about one hydrated ion radius away from the electrode surface), then the electrode potential drop taking place in the diffusion part of the double layer will not affect the rate of electron transfer. Thus, the potential drop in the diffusion part of the double layer, which is usually designated by ψ_1, reduces the potential drop between the electrode and the discharging particle. The effective potential drop is equal to $E - \psi_1$. Therefore, if the ψ_1 potential changes, the half-wave potential of the wave also must change, and for the uncharged depolarizer

$$\Delta E_{1/2} \simeq \Delta \psi_1. \tag{94}$$

If the ions of the indifferent electrolyte do not show a specific adsorptivity at the electrode, the absolute value of the ψ_1 potential decreases with increasing supporting electrolyte concentration, and with increasing valence of the ions, forming the double layer. A specific adsorptivity of ions causes a sharp decrease of the ψ_1 potential, which will be discussed later.

For solutions of 1 -1 valent electrolytes in the absence of specific adsorptivity, the ψ_1 value can be calculated [7] with sufficient precision using

$$\varphi_a = \psi_1 + \frac{1}{C}\sqrt{\frac{2DRTc}{\pi}}\, sh\left(\frac{\psi_1 F}{2RT}\right), \tag{95}$$

where φ_a is the potential of the electrode measured from the electrocapillary zero; C is the integral capacity of double layer; D is the dielectric constant of the solution; c is the total (molar) concentration of $1-1$ valent electrolytes.

At potentials sufficiently far from the zero charge, the ψ_1 value can be expressed approximately by

$$\psi_1 \simeq \text{const} + \frac{RT}{F}\ln c - \frac{2RT}{F}\ln \varphi_a, \tag{96}$$

For aqueous solution at 25°C, if c is expressed in M, the constant is about -0.06 V.

Levin and Fodiman [575] first related the change of the half-wave potentials of reduction waves for organic compounds (halogen derivatives of aromatic hydrocarbons) caused by changes of the composition of the supporting electrolyte with the change of the ψ_1 potential. Later Rogers and co-workers [561] discussed the effect of ψ_1 changes on the half-wave potentials (and also on the slope of logarithmic analysis of the waves). A quantitative investigation of the effect of potassium chloride concentration on the half-wave potentials of the reduction wave of several iodo derivatives showed [40] that the half-wave potential changes nearly the same way as the ψ_1-potential value, as is indicated in Eqs. (94) and (95) (see Fig. 27).

An even better agreement between the half-wave potentials and ψ_1 can be found for the second waves of aldehydes and ketones [480]. In Fig. 28, the points show the half-wave potentials of the first (a) and the second waves (b) for benzaldehyde (1), acetophenone (2), 2-acetothiophenaldehyde-4 (3), and 2,2-dithienyl ketone (4) in acetate buffer solutions at a pH 4.65 and $[CH_3COOH] \approx 0.1$ M, at different ionic strength (0.1 M CH_3COOK + KCl). The second waves correspond to the irreversible reduction of free radicals without participation of hydrogen ions in the potential-determining step [576, 577]. According to the theory, with increasing ionic

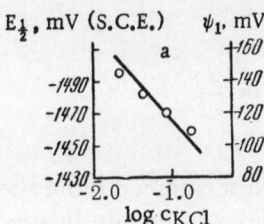 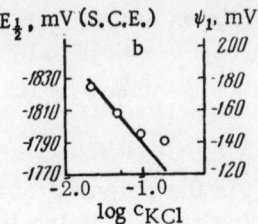

Fig. 27. The half-wave potentials (points) and also the ψ_1-potentials (lines) as functions of potassium chloride concentration for the waves of (a) iodomethyldimethyl phenylsilane and (b) primary isobutyl iodide.

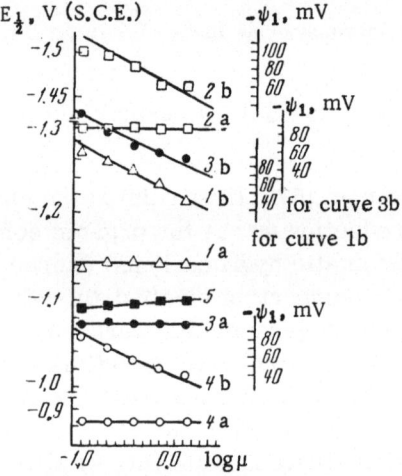

Fig. 28. The half-wave potentials (points) for the first (a) and second (b) reduction waves of several compounds at different ionic strength in acetate buffer solution $(0.1 \text{ M } CH_3COOH + 0.1 \text{ M } CH_3COOK + x \text{ KCl})$ and pH 4.65. The b lines show the change of ψ_1-potential as a function of K^+ concentration. (1) Benzaldehyde; (2) acetophenone; (3) 2-acetothiophenaldehyde-4; (4) 2,2-dithienyl ketone; (5) 2-acetylthiophene.

strength of the solution, the half-wave potential becomes more positive. The lines 1b–4b in Fig. 28 were constructed using Eq. (95), assuming that $C = 18 \ \mu F/cm^2$, $D = 78$, and the potential of the electrocapillary maximum is -0.45 V (S.C.E.). The lines express the change of ψ_1 potential with increasing salt concentration. It can be seen from Fig. 28 that the change of the half-wave potential follows quite well the change of ψ_1 potential (the small change of the diffusion potential for liquid compounds was disregarded here and in the following).

It must be mentioned that for $c_{KCl} < 0.1$ M the second wave on the polarograms of acetophenone is almost masked by the background discharge current. With increasing c_{KCl} the second wave shifts toward positive potentials and is easily distinguished. For benzaldehyde with increasing c_{KCl}, the second wave begins to merge with the first wave. Therefore, at $c_{KCl} \geq 0.3$ M for the $E^I_{1/2}$ and $E^{II}_{1/2}$ (i.e., for the half-wave potential of the first and second wave, respectively) the $E_{1/4}$ and $E_{3/4}$ values of the resulting wave were accepted, since the height of the first and second wave for benzaldehyde are equal under these conditions.

Due to the high degree of reversibility of the electrochemical step of the process, the double layer does not affect the half-wave potentials of the first waves of aromatic aldehydes and ketones.

Stradyn' and Teraud [578] found a change in half-wave potentials with increasing ionic strength of the solution, corresponding to the change of the ψ_1 potential. This was determined for the second wave of furfural in acidic solutions and for the entire wave in neutral and alkaline solutions.

Bezuglyi, Mel'nik, and Dmitrieva [579] showed that the half-wave potential of the reduction wave for β-acetyltetraline in a buffer solution at pH 8.45 becomes more positive with increasing ionic strength of the solution. The $\Delta E_{1/2}/\Delta \log c_{K^+}$ value is close to 36 mV. This value is much smaller than that corresponding to the ψ_1-potential change (the $E_{1/2}$ of the wave is close to -1.6 V vs. S.C.E.); the discrepancy between $\Delta E_{1/2}$ and $\Delta \psi_1$ was explained [579] by the effect of the double-layer structure on the rate of antecedent protonation, which can be always observed under these experimental conditions (the half-wave potential of β-acetyltetralin close to pH 8.5 is somewhat dependent on the solution pH).

A sharp increase of the ψ_1 potential can be observed for supporting electrolytes containing ions which undergo a specific adsorption of the electrode surface. This does not correspond to Eqs. (95) or (96). The direction of the ψ_1-potential change depends on the sign of the adsorbed ion: cations shift the ψ_1 potential to positive values, anions toward negative values. It is well known that many inorganic anions show a very high specific adsorptivity at the mercury electrode [291, 293, 580, 581]. The adsorptivity of the halide ions increases sharply in the series from F^- to I^-, and the shift of the ψ_1 potential toward negative values increases under identical conditions in the same order also. In the potential region corresponding to the adsorption of halide ions (at positive and not very high negative charges of the surface), a shift of the half-wave potentials of the irreversible waves for uncharged compounds occurs toward negative potentials due to the negative shift of the ψ_1 potential on introduction of halide salts into the polarographic solution. Thus, Lothe and Rogers [582] showed that the half-wave potentials of the first wave of carbon tetrachloride, which is in the potential region corresponding to the adsorption of these ions, becomes more negative if the chlorides taken as supporting electrolyte are exchanged for bromides, and further for thiocyanates and iodides. A shift of the half-wave potentials of the first wave of carbon tetrachloride toward negative potentials can also be observed at increasing potassium iodide concentrations. At the same time, the half-wave potential of the second wave, which lays outside of the adsorption region of iodides, becomes more positive [582] due to the decrease in the absolute value of the negative ψ_1 potential, which is determined in this case by the potassium ions according to Eq. (96).

C. THE EFFECT OF THE DOUBLE LAYER
ON ION DISCHARGE

If the discharging particles are ions, which means that they bear positive or negative charges, then because of the electrostatic action of the electrode field, their concentration in the electrode layer must be different from their concentration in the bulk solution. The concentration change of the ions close to the electrode surface, in the absence of specific adsorption of ions, follows the Boltzmann distribution. Therefore, the concentration of the cations and anions immediately at the electrode surface [7] can be given by

$$[R^z]_s = [R^z]_\delta \, e^{-z\psi_1 F/RT}, \qquad (97)$$

where s and δ subscripts correspond to the concentrations at the electrode surface and on the outer interface of the diffusion part of the double layer; z is the charge on discharging particles.

The change of the depolarizer concentration at the electrode affects the rate of the electrochemical reaction. We can substitute the depolarizer concentration at the electrode surface from Eq. (97) into Eq. (69) and express $[R^Z]_\delta$ from the Ilkovič equation as $[R^Z]_\delta = (i_D - i)/\varkappa$ (the thickness of the diffusion part δ of the double layer is much smaller than the diffusion space). Furthermore, the effective potential drop can be considered to be smaller by the ψ_1 value than the total potential drop [which means that in Eq. (69), E is replaced by $E - \psi_1$]. After a simple transformation we obtain the following for the change of the half-wave potentials of the ion discharge wave due to the change of the double-layer structure:

$$\Delta E_{1/2} \simeq \Delta \psi_1 \frac{\alpha n_a - z}{\alpha n_a}. \tag{98}$$

For uncharged particles (z = 0) it follows from Eq. (98) that $\Delta E_{1/2} \simeq \Delta \psi_1$, which means that the shift of half-wave potentials corresponds just to the change of the effective potential. This case was discussed in the previous section. The absolute value of z is usually larger than αn_a. Therefore, for the discharge of cations (z > 0) the signs of $\Delta E_{1/2}$ and $\Delta \psi_1$ are opposite. This means that the half-wave potential becomes more negative with decrease of the absolute value of ψ_1 (for example, at increased ionic strength of the solution). For anions, z < 0 and the $(\alpha n_a - z)/\alpha n_a$ value is positive and much larger than unity. Therefore, in accord with Eq. (98), a decrease of the absolute value of ψ_1 causes a large shift of half-wave potentials toward positive potentials. Experiments have completely confirmed these conclusions.

As was mentioned earlier, A. N. Frumkin and his co-workers first investigated quantitatively the effect of the double-layer change on the kinetics of hydrogen ion and other depolarizer electrolysis. Tur'yan successfully applied the conclusions developed by Frumkin to the investigation of the effect of supporting electrolyte concentration and nature on the reduction of Ni and Co(II) ions [583]. With increasing supporting electrolyte concentration, the half-wave potentials of the discharge wave for Co(II) and Ni(II) ions, in accord with Eq. (98), becomes more negative. A substantial shift of half-wave potentials to negative values was observed on addition of Ca^{2+} salts to the solution.

It must be mentioned that although Frumkin's views have been generally accepted and fruitfully utilized to explain the effect of the nature of the supporting electrolyte and its concentration on the kinetics of irreversible electrode processes, even very recently publications have appeared in which the authors intend in vain to explain the observed effects without considering the effect of the double-layer structure [for example, in work [584], in which the effect of different salts was studied on the reduction of Co(II) and Ni ions].

Several examples can be cited from the field of polarography of organic compounds that show the effect of the double-layer structure on the kinetics of electrode processes. The second of the well-known empirical rules, the Shikata—Tachi rule, states that under equal conditions a cation reduces more easily on the mercury electrode than uncharged molecules [585] and that it is even more difficult to reduce the anion. This is an expression of the effect of the double-layer structure. It was mentioned previously that an antecedent protonation, which increases the charge of the particles, makes the reduction easier. In the same way the formation of tetra-substituted ammonium cations in the interaction of amine derivatives with alkyl halides results in easier reduction. For example, the half-wave potentials of the waves for N-methyl derivatives of pyridine aldoximes, mainly in neutral and alkaline solutions, are more positive than the half-wave potentials of unsubstituted (on the pyridine nitrogen) derivatives of the same aldoximes [586]. Similar phenomena can be observed with imidazole derivatives [509], for the alkaloide protopine [587], and in many other cases.

It can be seen from Eqs. (96) and (98) that the magnitude and charge of the depolarizer can be determined from the character of the change of the electrolysis of the species studied caused by concentration changes of the supporting electrolyte. Frumkin and co-workers devoted a detailed investigation to this problem [588, 589].

The double-layer structure can be changed not only by changing the nature and concentration of the supporting electrolyte, but also by changing the electrode material. Thus thallium amalgam drop electrodes were used for the investigation of the electrode process kinetics [562, 590]; they permitted a displacement of the zero charge by 0.4 V compared with that of the mercury electrode.

The change of the ψ_1-potential value caused by changing the supporting electrolyte nature and concentration, is also expressed on the Breyer polarograms, which are measured with superimposition of an alternating current [591]. Thus, with increasing perchlorate concentration in the solution the peak height on polarograms for Zn^{2+} and the heights of the second peak for uranium reduction decreases, owing to a decrease in the concentration of these cations at the electrodes.

D. SOME PECULIARITIES OF ANION REDUCTION

The decrease of the negative ψ_1-potential value at the mercury cathode, caused by increased concentration of supporting electrolyte, makes the reduction of anions much easier [523, 556-559, 592-594]. The decrease of the ψ_1 potential is so large that the reduction waves of many organic anions, which are usually hidden by the electrolytic current of the supporting electrolyte, can be observed if the concentration of the supporting electrolyte is sufficiently high, and especially when polyvalent or surface-active cations are added to the solution [145, 151, 595, 596].

Electrostatic repulsion of the anions from the cathode surface not only causes a shift of their reduction wave toward negative potentials, but often results in a drop in the wave on the portion of the horizontal sections corresponding to the limiting current. This interesting phenomenon was observed by Kryukova [555] on the reduction wave of persulfate anion: with increasing negative polarization at potentials more negative than −0.5 V, the limiting current begins to decrease, reaches a minimum at about −1.0 V, and on further increase of the negative potential increases again nearly to the level of the diffusion current. If the supporting electrolyte concentration is increased, the minimum on the limiting current becomes less deep, and at sufficiently high concentrations of the supporting electrolyte, the minimum can no longer be observed. The minimum does not appear on the wave if there are even very small quantities of polyvalent cations present in the solution [555]. The troughs on the reduction waves for anions were quantitatively

explained by Frumkin and Florianovich [554]. They showed that
the trough is caused by the decrease of the electrode process rate
due to decreased anion concentration at the electrode when the
negative charge of the mercury surface increases close to the
point of zero charge. Above a certain negative potential the increased electron transfer rate becomes predominant and suppresses the repulsion effect. As a result the current increases
on the polarogram. It also was shown [554], that by calculating
the change of concentration at the electrode from Eq. (97) and by
using the ψ_1-potential value calculated by an equation given by
Bagotskii [597], theoretical curves that have a shape similar to
that observed on recorded polarographic curves can be obtained
if the constants are properly selected. The disappearance of the
trough at increasing supporting electrolyte concentrations or on
introduction of polyvalent cations is caused by the decrease of the
absolute value of the negative ψ_1 potential. A trough on the reduction curves for anions at very low supporting electrolyte concentrations can be observed also for rotating solid electrodes of lead
or cadmium [598] and also for the thallium amalgam dropping
electrode [590].

Sharp decreases in current caused by electrostatic repulsion
can usually be observed on reduction waves of di- and trivalent
anions [464-466]. Similar current decreases also were observed
on reduction polarograms of monovalent MnO_4^- anions [599], and
for the $[Co(NH_3)_4Cl_2]^+$ cation [600]. In the latter case the repulsion apparently cannot be caused by the charge of the entire ion,
but rather by the presence of a negatively charged group in the
particle [600]. It is not entirely impossible that electrode processes involving ions such as $[Co(NH_3)_4Cl_2]^+$ have a chemical step
preceding the electron transfer. Therefore, the trough on the wave
might be caused by the surface character of this step (see section
C, Chapter VI).

E. CHANGE OF pH IN THE ELECTRODE LAYER
AS COMPARED TO ITS VALUE
IN THE BULK SOLUTION

In any buffer system the magnitude of the charge of the acid
component is always larger than that of the conjugated base. There-

fore, the electrode field will affect the distribution of the acid and the basic components at the electrode differently [480].

Let us discuss that most common type of buffer system in which the acid is either uncharged (HA as, for example, in acetate or borate buffer solutions) or has a positive charge (BH^+ as in ammoniacal, amine, or pyridine solutions)

$$A^- + H^+ \rightleftarrows HA \text{ and } B + H^+ \rightleftarrows BH^+, \qquad \text{(XXIII)}$$

where A^- and B are the bases conjugated with the acids. If the components of the buffer system do not show a specific adsorptivity, then their concentration at the electrode surface can be determined by Eq. (97). If the capacity of the buffer solution is high enough, its consumption in the electrode layer can be neglected (due to consumption of hydrogen ions in the electrode process), and it can be assumed that the concentration of the buffer components on the outer interface of the double layer is equal to their concentration in the bulk solution. This means that $[BH^+]_\delta = [BH^+]_0$ and $[A^-]_\delta = [A^-]_0$. Furthermore, assuming that the dissociation constants of the acids in the diffusion part of the double layer are equal to their values in the bulk solution, the following equation can be given based on Eq. (97) [480]:

$$\left(\frac{[HA]}{[A^-]}\right)_s : \left(\frac{[HA]}{[A^-]}\right)_0 = \left(\frac{[BH^+]}{[B]}\right)_s : \left(\frac{[BH^+]}{[B]}\right)_0 = \frac{[H^+]_s}{[H^+]_0} = e^{-\psi_1 F/RT}, \qquad (99)$$

since for uncharged particles in absence of specific adsorption $[HA]_0 = [HA]_s$ and $[B]_s = [B]_0$.

Furthermore, since the charge of the acid components of the buffer system is always larger than the charge of the bases conjugated with the acids, the acid–base equilibrium at the cathode surface is displaced toward the acid form compared with the bulk solution. This shift can reach very significant levels. Thus, for $\psi_1 = -120$ mV, which can be encountered under the usual polarographic conditions [in an 0.1-M solution of a 1–1 electrolyte and at an electrode potential $E \approx -1.5$ V (S.C.E.)], the pH at the electrode is smaller than the pH in the bulk solution by as much as two units.

In electrode processes taking place after a preceding chemical protonation reaction [see scheme (XVIII)] the pH decrease at the electrode surface causes a higher protonation rate of compound R, which is converted into the electrochemically active cationic form RH^+. The largest pH shift, and consequently also the largest increase in the protonation rate, takes place in the thin layer immediately at the electrode surface.

F. THE EFFECT OF DOUBLE-LAYER STRUCTURE ON PROCESSES WITH ANTECEDENT PROTONATION

It has already been mentioned that the largest concentration change expressed by Eq. (97) takes place immediately at the electrode surface. As the distance away from the electrode and into the solution increases, the ionic concentration gradually approaches the value in the bulk solution. For this reason the effect of the double-layer structure on the chemical reaction preceding the electron transfer taking place with participation of ions is larger if a greater proportion of the reaction volume is within the limits of the diffusion part of the double layer [560, 565, 566]. In other words, the effect of the double layer increases with the ratio of the thickness of the reaction layer μ [see Eq. (16)] to the thickness of the diffusion layer of the double layer (δ). This latter can be expressed for aqueous electrolyte solutions at 25°C by

$$\delta = 3 \cdot 10^{-8} \left(\frac{1}{2} \Sigma c z^2 \right)^{-1/2}, \tag{100}$$

where δ is given in cm and c in M.

The influence of the double-layer structure on antecedent reactions with participation of ions is a complicated function of the μ/δ ratio [566, 567]. The mathematical expression becomes very simple for the two limiting cases: $\mu/\delta \gg 1$ and $\mu/\delta \ll 1$.

If the thickness of the reaction layer μ is much larger than δ, the concentration change of the ions in the diffusion part of the double layer barely affects their average concentration in the limits of the reaction layer. Therefore, for $\mu/\delta \gg 1$, the double layer does not show any effect on antecedent reactions taking place with participation of ions. In the opposite case, for $\mu/\delta \ll 1$, the entire reaction layer lies within the diffusion part of the double layer. It occupies a thin space immediately at the electrode surface, where the ionic concentration is determined by Eq. (97). By using this equation it is easy to consider the effect of the double-layer structure. The relationships are the same for surface processes when one of the reaction components is in an adsorbed state on the electrode surface and the other component is present in the solution as an ion.

It was shown earlier (see page 121) that the half-wave potentials of waves with antecedent protonation are pH-dependent. For this reason, in considering the effect of double-layer structure on irreversible waves, the pH change at the electrode compared to its value in the bulk solution also must be taken into consideration. Usually, the displacement of the half-wave potentials for waves in acid and neutral solutions is about 40-80 mV for each pH unit [475, 601] ($\Delta E_{1/2}/\Delta pH \approx -40$ to -80 mV). Therefore, according to Eq. (99), the wave shift caused by a change of the ψ_1 potential resulting from pH change only is

$$\Delta E_{1/2} = \frac{\Delta E_{1/2}}{\Delta pH} \frac{\Delta \psi_1 F}{2.3 R \Gamma},$$

This value is close to $-\Delta \psi_1$. For the general case of irreversible ion electrolysis with antecedent protonation in a very thin reaction layer or on the surface of the electrode, the change of the half-wave potential with ψ_1 can be given [480] by

$$\Delta E_{1/2} \simeq \Delta \psi_1 \left(\frac{\alpha n_a - z}{\alpha n_a} + \frac{\Delta E_{1/2}}{\Delta pH} \frac{F}{2.3 R T} \right). \tag{101}$$

With using Eq. (72), Eq. (101) reduces to:

$$\Delta E_{1/2} \simeq \Delta \psi_1 \left(\frac{\alpha n_a - z - 1}{\alpha n_a} \right). \tag{102}$$

Equation (101) is preferred to Eq. (102) because in Eq. (101) the physical significance of the last term is clearly expressed. Furthermore, Eq. (101) also can be used when Eq. (72) is not valid as, for example, for strongly acidic medium, for basic solutions, and for certain quasidiffusional surface waves (compare with sections C and E of Chapter VI).

Equations (101) and (102) show that in certain cases a change in ionic strength of the solution can cause a significant shift of the half-wave potentials. Indeed, for example, for the reduction wave of N-methylpiperidine N-oxide in an acetate buffer solution at

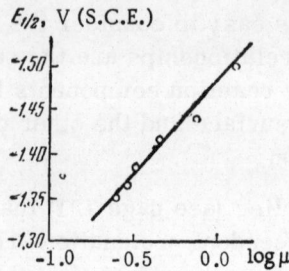

Fig. 29. Dependence of the half-wave potentials
for the reduction of N-methylpiperidine N-oxide
from the ionic strength of the solution in acetate
buffer at pH 4.65.

pH 4.6 a shift of the half-wave potential by 200 mV toward negative potentials (Fig. 29) was observed on tenfold increase of the
ionic strength of the solution (sodium chloride). This value is in
good agreement with that calculated by Eq. (101), since, in this
case, z = 1 (pK$_A$ = 5.12 for methylpiperidine N-oxide), $\alpha n_a \approx 0.3$,
$\Delta E_{1/2}/\Delta pH$ = −60 to −70 mV/pH unit [492].

For compounds that are present in un-ionized form in the solution, which is the case for the majority of organic compounds in
solutions not too acidic, z = 0. From Eq. (101) it follows that for
waves corresponding to processes with antecedent protonation with
participation of uncharged compounds (z = 0), the half-wave potential should depend little on the change of the ψ_1 value or on the
ionic strength of the solution, especially if $\Delta E_{1/2}/\Delta pH$ = −60 mV.
Indeed, for example, the half-wave potential of the reduction wave
for 2-acetylthiophene semicarbazone (line 5 in Fig. 28) at strictly
constant pH and buffer capacity of the solution is almost independent of the ionic strength. In this case the effect of the change in
potential drop on the change of ψ_1 nearly compensates the shift of
the half-wave potential caused by the (pH)$_S$ change.

In certain cases these two effects may not compensate each
other, but may instead be superimposed. This occurs, most probably, in the reduction of nitromethane in acid solution in presence
of iodide ions [560]. But for the clarification of this effect it is not
possible to limit the discussion to simple schemes of the double-
layer structure; the presence of two Helmholtz surfaces must also

be considered. Grahame [602] showed that owing to the greater deformation of anions, their centers are closer to the electrode surface in the adsorbed state than the centers of cations. Therefore, the ψ_1 potentials corresponding to the inner (ψ_1^i) and outer (ψ_1^o) Helmholtz surface must be differentiated. The displacement of the half-wave potential of the nitromethane wave with increasing potassium iodide concentration was compared [560] with the change of the ψ_1^i and ψ_1^o values. The latter values were calculated by Grahame [603] for several KI concentrations and different potentials on the basis of theoretical considerations developed by Ershler [604], and Esin and Shikhov [605]. Under the conditions of nitromethane polarography with increasing c_{KI}, ψ_1^i becomes more negative and ψ_1^o more positive, and the absolute values of $\Delta\psi_1^i$ and $\Delta\psi_1^o$ are nearly equal (see Fig. 6 in [560]). The observed shift of the half-wave potential of nitromethane takes place toward negative potentials, and the change of half-wave potentials is nearly twice as large as the $\Delta\psi_1^o$ value [560]. It is reasonable to assume, as was done by the authors of this work [560], that the shift of the half-wave potential is caused by the change of the effective drop $\Delta\psi_1^i$. The unexplained [560] larger change of the absolute value for $\Delta E_{1/2}$ than that for $\Delta\psi_1^i$ can be most probably ascribed to the increase of pH in the outer Helmholtz surface caused by the change of the ψ_1^o values toward positive potentials [480].

It must be mentioned that apparently the change of effective potential drop has a determining role in the change of the half-wave potential for nitromethane with the concentration of supporting electrolyte. The pH change in the electrode layer has a smaller value, most probably owing to the partial volume character of the antecedent protonation.

Figures 30 and 31 show the results of the investigation [606] of the effect of potassium chloride, cesium chloride, and potassium iodide concentration on the half-wave potential of the reduction wave of certain organic nitro compounds. The measurements were made in an acetate buffer solution containing 0.1 M potassium acetate, at constant solution pH (4.65). The reduction of nitrobenzene occurs at potentials corresponding to the positive branch of the electrocapillary curve. Therefore, the chloride anions show a predominant effect on the ψ_1 potential (and also on the half-wave potential). With increasing chloride ion concentration, the ψ_1 value becomes more negative and the half-wave potential must

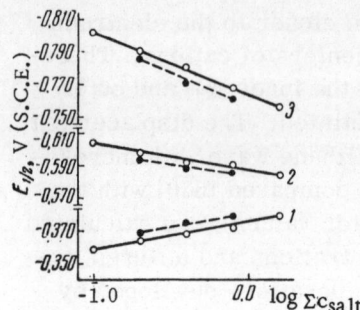

Fig. 30. Half-wave potentials of (1) nitrobenzene, (2) phenyl nitromethane, (3) 1-nitropropane as functions of total salt concentration (0.01 M CH$_3$COOK + x MCl) in the solution. Continuous lines, KCl; dotted lines, CsCl.

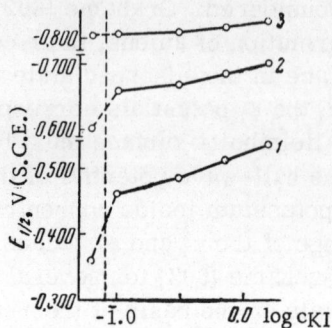

Fig. 31. Half-wave potentials of (1) nitrobenzene, (2) phenyl nitromethane, (3) 1-nitropropane waves as functions of potassium iodide concentration in the solution (the points given at the left of the dotted lines are half-wave potentials measured in absence of KI.

shift toward negative potentials due to the change of the effective potential drop. The reduction of nitro compounds in acid medium takes place with antecedent protonation, and if this protonation even partly takes place at the electrode surface or in a very thin reaction layer, then its rate must be affected by the pH change at the electrode surface. At negative ψ_1 values or at increasing absolute ψ_1 values, the pH decreases at the electrode and the half-wave potential (due to the pH change) must become more positive. The observed shift for the nitrobenzene wave, $\Delta E_{1/2}/\Delta \log \Sigma c_{salt} \simeq$ −16 mV (see Fig. 30), shows that the first of the mentioned effects is predominant. Therefore, the protonation in this case takes place both in the solution and within the surface layer.

Phenyl nitromethane is reduced at potentials corresponding to the left branch of the electrocapillary curve, where the ψ_1 values are influenced by the cations. Therefore, an increase of K$^+$ ions in the solution decreases the negative ψ_1 potential. (If we assume that the zero charge point is at −0.45 V, and the potassium chloride concentration is changed from 0.1 to 1.0 M, the ψ_1 value changes at the phenyl nitromethane reduction potential by about 25 mV.) As a result of this change, the half-wave potential must become more positive by the same value. The observed shift ($\Delta E_{1/2}/\Delta \log c_{K^+} \simeq$ +16 mV) is much smaller than the change of the ψ_1 potential. This can be explained by an increase in pH in the electrode layer at higher potassium chloride concentrations, which causes a shift of half-wave potentials toward negative potentials. The magnitude of this effect is relatively small and that indicates that the protonation reaction takes place not only at the surface, but also in the solution volume.

For the 1-nitropropane wave the observed $\Delta E_{1/2} / \Delta \log c_{K^+} \simeq$ +37 mV is almost equal to the change of the ψ_1 potential. Therefore, the pH change in the electrode layer does not affect the process. This can apparently be explained by the relatively small adsorptivity of 1-nitropropane, which results in a lower proportion of the protonation reaction taking place as a surface reaction.

As can be seen from Fig. 30, replacement of potassium chloride with cesium chloride barely affects the half-wave potential (compare with next section). This can be expected for half-wave potentials of such processes that take place at potentials close to the electrocapillary zero. An exchange of Cl^- for I^- ions at the same time results in a considerable shift of the half-wave potential to negative potentials (see Fig. 31). This effect is the result of the adsorption of I^- ions, which lead to a decrease of the ψ_1 potential. The largest shift of half-wave potentials takes place for nitrobenzene, most probably owing to larger adsorptivity of I^- ions at the positive potentials at which this wave can be observed. With increasingly negative potential the iodide adsorptivity decreases, and at the same time the ψ_1 value begins to be affected by K^+ ions. Therefore, for the phenyl nitromethane, and even more so for the 1-nitropropane wave, the effect of I^- ions is smaller (see Fig. 31).

For an electrochemical step taking place after a first-order chemical reaction [see, for example, scheme (XVIII)] a change in the concentration of starting material R (where among all reagents only R is an ion) in the reaction layer caused by the effect of the electrode field results in a proportional change of the concentration of the electrochemically active product of the chemical reaction. Consequently, the rate of the electrode reaction also changes. Thus for electrode processes with antecedent chemical reaction for which one of the reagents is an ion, the effect of the double-layer structure on the half-wave potential can be given by Eq. (98). In this case the z value in the equation is the charge of the electro-

chemically inactive R compound. If the compound in the solution
is present mainly in the un-ionized state, and its transition into the
ionized form takes place only before the discharge at the electrode
surface (for example, surface protonation occurs), then the electrostatic field of the electrode does not affect the distribution of
this material at the electrode, and therefore the value of z in Eq.
(101) is zero.

Grabowski and Bartel [564] first studied the effect of the
ionic strength of buffer solutions on the kinetic current with antecedent protonation reaction from the viewpoint of ψ_1-potential
changes. Their experimental data agreed well with the theory because of the surface character of the process investigated.

The effect of the double-layer structure on kinetic currents
was observed by many investigators. Grabowski [563] showed that
the disappearance of the minimum on the kinetic wave for $Hg(CN)_4^{2-}$
discharge, observed by Siekierski [607] on addition of potassium
chloride to the solution, is caused by the effect of the double-layer
structure on the antecedent $Hg(CN)_4^{2-} \rightleftharpoons Hg(CN)_3^- + CN^-$ reaction.
Siekierski's earlier observation [608] of the increase of the limit-
ing current for cyanide complexes of cadmium, and the decrease
of the minimum on the limiting current found on addition of potassium nitrate to the solution, also can be explained by the effect of
the double-layer structure.

G. THE EFFECT OF THE NATURE
OF SUPPORTING ELECTROLYTES

It was shown in the previous section that adsorbed anions
can significantly change the ψ_1 potential at not too high cathodic potentials. But certain simple inorganic cations also show substantial adsorptivity: in their presence the absolute value of the negative ψ_1 potential decreases much more than would be expected from
Eq. (96). The existence of such excessive adsorption on the mercury cathode for simple cations with large radius was shown by
Frumkin, Damaskin, and Nikolaeva-Fedorovich [609]. These
authors compared the effect of the cation of the supporting electrolyte on the differential capacity of the mercury electrode (observed by Grahame [610]) with the character of the effect of the
same cations on the rate of electrochemical reduction of anions

[464, 611]. Grahame, in his investigation of a series of alkali salts, found that the double-layer capacity increases from lithium to cesium. He explained this phenomenon by an increase of the capacity of the Helmholtz part of the double layer. But as was shown by Frumkin and co-workers [609, 558], in this case an increase of the absolute value of the negative ψ_1 potential can be expected, and indeed the increased rate of reduction of anions in a transition from lithium to cesium salts indicates that the negative ψ_1-potential value actually decreases.

The decrease of the absolute value of the negative ψ_1 potential in the series of salts from lithium to cesium is caused by the increased adsorptivity of their cations. This also explains the effect of the nature of these cations present in the supporting electrolyte solution, on certain electrochemical processes.

Herasymenko and Slendyk [612] first observed the effect of the cation nature on electrochemical processes. They found that if solutions containing lithium salts are replaced with cesium salt solutions, the rate of hydrogen ion discharge decreases significantly. From the viewpoint of the theory of slow discharges, this observation can be easily explained by the decrease of the absolute value of the negative ψ_1 potential. This decrease, according to Eq. (98), results in a shift of the potential in the cathodic direction at constant current density (since $\alpha n_a = 0.5$, $z = +1$). Analogous effects of the supporting electrolyte cation nature were observed by Izgaryshev and Ravikovich [613] in the electrodeposition of nickel and by Vasenin and Gorbachev [614] during the reduction of copper.

Ashworth [593] first studied the effect of supporting electrolyte nature and concentration on the reduction of organic compounds. He investigated in detail their effect on the second waves on polarograms of benzophenone and fluorenone in alkaline solutions. These waves are characteristic of the reduction of the anions formed in the first step of the process. These anions are $(C_6H_5)_2CO^-$ and $(C_6H_4 \cdot C_6H_4)CO^-$, respectively. On increase of the supporting electrolyte concentration (by alkali hydroxides) the half-wave potentials of these waves become more positive by approximately 180 mV on a tenfold concentration increase. A substantial displacement of the half-wave to positive potentials also can be observed if cations of larger radii are used. Thus, replacement of lithium hydroxide by sodium hydroxide at equal concentrations shifts the half-wave potential of the second wave for benzophenone by about 80 mV. Approximately the same shift can be observed if sodium hydroxide is replaced by potassium hydroxide. A somewhat larger shift was observed (about 100 mV) toward positive potentials by an exchange of potassium hydroxide by rubidium hydroxide, and finally a maximum shift (about 140 mV) occurred when cesium hydroxide was used instead of rubidium hydroxide [593]. These effects, which were not correctly interpreted by Ashworth [593], can be easily explained on the basis of the theory of slow discharge, with consideration of appearance and increase of excessive cation adsorption in the Li^+ to Cs^+ series.

Fedorenko [615] found a shift of the ketone reduction wave for phenylacetyl carbinol and methylbenzoyl carbinol in alkaline solution toward positive potentials at increasing supporting electrolyte concentrations. The smallest effect was observed for Na^+, a larger for K^+, and then the effect increased for Ca^{2+} and finally for $CH_3NH_3^+$ and NH_4^+. Kuta [616] observed an increased reduction rate for oxalic acid esters on increasing ionic strength of the solution, particularly if polyvalent cations were introduced into the solution. A similar effect of cations on the reduction of tetrose was described by Zuman and Zinner [214].

Fischerová and Fischer [458, 617] observed an increase in the kinetic reduction wave for the chromium hexathiocyanate complex anion when the alkali cation of the background was changed from lithium to cesium. Schmid and Reilley [618] found that the kinetic wave of cadmium ethylenediamine tetracetate increases at pH > 5 if foreign electrolytes are added to the solution. The nature of the anion does not affect the wave and the cations can be placed in the following series on the basis of their effect: $Li^+ \simeq Na^+ < K^+ < NH_4^+ < Mg^{2+} < Ca^{2+} < Sr^{2+} < Ba^{2+}$. Zhdanov and co-workers [134, 619, 620] investigated in detail the increase of the anion-reduction rate on addition of different cations, especially polyvalent cations to the solution.

Alkaline earth cations in a tetraethyl ammonium perchlorate supporting electrolyte permit the reduction of nitrate ions even in dimethylsulfoxide medium [621]. The easier reduction of nitrates in this solution also was observed in presence of lanthanum and cerium cations [621]. Frumkin and co-workers investigated in quantitative terms the relationship between the double-layer structure in the presence of various cations of alkali metals and the effect of these cations on the kinetics of hydrogen ion and anion discharge [589, 622].

The cations of tetrasubstituted ammonium salts have even a larger effect than cesium cations on the decrease of the absolute value of the ψ_1 potential. These cations have a larger surface activity (see, for example, [623]) and they show the greatest effect in the potential range corresponding to their adsorption. Due to the relatively larger volume of the tetrasubstituted ammonium salt cations, their adsorption on the electrode is controlled by the same relationships as the adsorption of uncharged organic molecules. Van der Waals attraction forces act between the adsorbed particles which cause an S-shaped adsorption isotherm (see page 58); therefore, the surface coverage changes almost in a discontinuous increase of the adsorbed compound concentration or of the potential change [273]. The adsorptivity of tetrasubstituted ammonium cations, as was mentioned earlier, increases with the hydrocarbon chain length of the substituents at the nitrogen [303, 623, 624]. Therefore, the effect of the cations on the kinetics of electrode processes increases with increasing size.

Levin and Fodiman [575] first investigated the effect of adsorption of tetrasubstituted ammonium salts on the reduction of organic compounds from the viewpoint of the change of the double-layer structure. According to their data, half-wave potentials of the first reduction wave of 2-iodonaphthalene (−1.66 V in a potassium salt solution) becomes −1.51 V in the presence of tetramethyl ammonium salt. Feoktistov and Zhdanov investigated quantitatively the shift of the reduction wave for several halide derivatives of organic compounds when the lithium salt was replaced by salts containing cations having a larger radius. An even larger shift was observed for the half-wave potentials on adsorption of tetrasubstituted ammonium salts in the reduction of anions. Thus, the effect of these salts on the last reduction wave for fluorenone in alkaline medium is much larger than the effect of cesium salts [593]. The effect of the tetramethyl ammonium cation on the reduction wave for chromate anions is also somewhat larger than that of cesium ions [624]. The rate-increasing effect of tetra-alkyl-substituted ammonium salts at low concentrations on the electroreduction of chromate increases substantially if tetraethyl ammonium salts are used [624], and even more with tetra-n-propyl- and tetra-n-butyl ammonium salts.

Bezuglyi, Mel'nik, and Dmitrieva [579] found an increase of the reduction wave for β-acetyltetraline in a neutral 0.05 M halide solution in 60% ethyl alcohol as the supporting electrolyte if the cation was changed successively in the following series: Li^+, Na^+, K^+, $(CH_3)_4N^+$, $(C_2H_5)_4N^+$, Cs^+. The effect of cations in this case can most likely be explained by the increased rate of the second step of the process — the reduction of the anion radical formed as an intermediate product. This second step apparently includes a chemical reaction at the electrode.

The increase of the anion-discharge rate by adsorption of cations is caused not only by the decrease of the negative ψ_1 potential and the easier transport of anions to the negatively charged mercury surface but also by the cations adsorbed on the cathode pulling anions into the double layer. The anions under these conditions can increase the mutual attraction of adsorbed cations in the double layer and thus increase the adsorptivity of cations [303, 623].

A most interesting case of electrochemical inhibition is connected with this pulling-in effect by adsorbed ions for ions of opposite charge into the double layer. This was observed in the reduction of persulfate anions in the presence of tetraamyl ammonium salts by Sat'yanarayan and Nikolaeva-Fedorovich [627]. The inhibition appears at not too high positive and not too high negative charges of the mercury electrode surface. This effect is caused by the tetraamyl ammonium cations adsorbed on the electrode pulling the anionic components of the supporting electrolyte into the double layer [627]. A similar phenomenon, the decrease of the effect of adsorbed tetraalkyl ammonium salts on the kinetics of electrode processes in presence of certain anions, was called the specific-anion effect [624]. The specific-anion effect was studied in detail by Gierst and co-workers [624]. They found that the inhibiting effect of anions (exerted in different electrochemical reaction types) depends both on the nature of surface-active cations and on the nature of other anions present in the solution. Thus, in the presence of tetramethyl ammonium salts, the anion effect can hardly be observed, but in the presence of tetraethyl ammonium in the solution, the effect can be clearly recognized. The reduction wave of chromate ions in these solutions is significantly displaced toward negative potentials if the indifferent electrolyte anions (1-M solutions) change in the following sequence: PO_4^{3-}, CO_3^{2-}, SO_4^{2-}, OH^- < F^-, CH_3COO^- < Cl^-, CN^- < BrO_3^-, NO_3^-, Br^-, ClO_3^- < I^- < SCN^-, ClO_4^- [624]. This series does not correspond to the sequence of increasing adsorptivity of anions on mercury. It must be mentioned that the largest anion effect is shown by ClO_4^- ions; the effect of OH^-, CO_3^{2-}, SO_4^{2-}, and PO_4^{3-} ions is relatively small and almost equal for each. In solutions of tetra-n-propyl and tetra-n-butyl ammonium salts and for the trimethyl phenyl ammonium salt, the specific-anion effect is even more pronounced. The sequence of anions according to their effect is exactly the same as for tetraethyl ammonium salt solutions, and the absolute magnitude of the anion effect increases with the adsorptivity and size of the tetra-substituted ammonium cation [624]. Similarly to the anion effect a cation effect can be observed in other cases which also is caused by the simultaneous adsorption of cations and anions [420, 628]. In the investigation of the double-layer effect on the kinetics of electrode processes in solutions of electrolyte mixtures with cations of different charges, the deviation of the ψ_1 potential in the Helmholtz surface (especially close to ions of high charge) from its average value must be considered [629]. Such a "local field effect" was investigated on the Ga^{3+} discharge in the presence of Na^+, Mg^{2+}, and Al^{3+} cations [629].

According to the opinion of Gierst and co-workers [624], the simultaneous adsorption of cations and anions is not the only cause of appearance of the anion (and cation) effect. One of the probable causes of this effect, according to Gierst, is the formation of ion pairs or associations between anions and cations in the bulk of the solution which affects the transport of anions to the cathode. This viewpoint is not shared by Frumkin [630].

The ion-pair formation effect appears clearly in nonaqueous solvents for reversible redox systems. The cations of light metals show a greater tendency toward ion-pair formation [631]. This means that in this case the effect of the nature of cations is opposite to their effect on the ψ_1 potential. Thus, in the polarography of benzoquinone and anthraquinone in dimethyl formamide in the presence of lithium salts, the half-wave potential of the reduction wave is shifted to positive potentials due to association of semiquinone anions with lithium cations. Other cations do not form ion pairs with the semiquinone anions of these compounds and therefore the half-wave potentials of their reduction wave are the same in solution of sodium and potassium ions and tetraethyl and tetrabutyl ammonium salts [631]. For the second wave on the polarograms of quinones in dimethyl formamide (this wave corresponds to the addition of the second electron to the semiquinone anions and formation of divalent anion) the shift of the half-wave potentials to positive potentials is larger when the cation radius is smaller [631]. Sodium and potassium cations also form associates with the dianion to a significant degree. In the reduction of p-dimethyl quinone in acetonitrile medium, the semiquinone anion forms ion pairs not only with lithium ions, but also with sodium ions [631]. Formation of ion pairs, affecting the half-wave potentials of waves, also can be observed in the reduction of nitro compounds [632].

The high surface activity of many organic cations results in certain characteristic effects on electrochemical processes. One of these is that these cations show a substantial effect even at very low concentrations. Thus, for example, the kinetic reduction wave for sulfite anions at pH 6.0 increases significantly on addition of methylene blue to the solution, even if the concentration of the latter compound in the solution is only $9 \cdot 10^{-7}$ M [633]. However, at higher concentrations the high surface activity of organic cations in aqueous solutions often results in inhibition of electrochemical processes by a layer of adsorbed compound (see page 143). The inhibition results since it is more difficult for the depolarizer to pass through the barrier formed by the adsorbed material (Loshkarev effect) and also since the concentration of adsorbed depolarizer is decreased due to its displacement from the electrode surface by the surface-active compound particles that were added to the solution [624]. The latter effect is strongly exhibited in processes with antecedent surface reaction. Figure 11, for example, shows the effect of tetraethyl ammonium salt (benzenesulfonate) addition on the limiting kinetic current for reduction of undissociated maleic acid. The latter is formed by protonation of monoanions of this acid adsorbed on the electrode. It can be seen from this figure that at very low tetraethyl ammonium salt concentrations the kinetic current increases due to lower negative ψ_1 potential and easier adsorption of maleic acid anions. At higher tetraethyl ammonium salt concentrations, the surface coverage by cations becomes apparent and results in the displacement of adsorbed monoanions from the electrode surface. Therefore, their protonation rate decreases and the kinetic current also decreases (see Fig. 11). For purely volume currents as, for example, in the reduction of maleic acid monoanions that are formed in the electrode layer, introduction of tetrasubstituted ammonium salts results only in an increase of the kinetic current (see Fig. 12). In this case only the effect of a smaller negative ψ_1 potential is expressed due to cation adsorption, and the inhibition caused by a layer of surface-active compound does not appear. Figure 12 also shows the height of this kinetic wave as a function of concentration of surface-active isobutyl alcohol. This compound does not affect the ψ_1 potential. It can be seen from this figure that addition of an uncharged surface-active compound does not affect the value of a purely volume-kinetic current.

Gierst and co-workers [624, 634] described certain interesting phenomena, caused by the effect of surface-active cations and anions on the first positive waves of chromate ion reduction. This peaklike positive wave has a kinetic character and its height is limited by the rate of surface protonation of chromate dianions to form monoanions [634]

$$CrO_4^{2-} + AH \rightleftarrows HCrO_4^- + A^-.$$

The monoanions are reduced at less negative potentials than the chromate anions. In this case the change of ψ_1 potential caused by the change of concentration of salts and nature of anions and cations of these salts affects not only the concentration of chromate anions at the electrode, but also the concentration of proton donors. In the presence of uncharged (H_2O) and negatively charged (HCO_3^-) proton donors and also in the presence of added salts whose ions do not have a significant surface activity, the height of the positive wave increases with the ionic strength of the solution [634]. An addition of surface-active anions to the solution results in a decrease of the positive wave due to increasingly negative ψ_1 potentials (the positive wave is close to the potential of the electrocapillary maximum). This effect becomes more apparent if the surface activity of added anions increases. Thiocyanate and bromide ions had the largest effect [634]. Addition of cesium ions to the solution in-

creases the height of the positive wave. An unusually strong increase of the height of the kinetic wave of chromate anions was
found on addition of very small concentrations of tetraalkyl ammonium cations to the solution [624]. Their effect increases with
the length of alkyl substituents at the nitrogen. When their concentration is increased, their desorbing effect becomes noticeable,
and reduces the height of the kinetic wave. The desorption effect
increases sharply with increasing radius of these cations, which
means that it significantly increases in the series: tetramethyl,
tetraethyl, tetra-n-propyl, and tetra-n-butyl ammonium salts [624].
On addition of tetrabutyl ammonium salt, a strong inhibition is already observed at concentrations of 10^{-3} M.

The coverage of the electrode surface by the surface-active
compound increases with adsorption time, which means the time
passed from the beginning of drop formation. Therefore, in the
lifetime of a drop the effective ψ_1-potential value also changes with
time. Volková [635, 636] first observed the unusual dependence of
the limiting kinetic current, limited by the rate of recombination
of acid anions, with time over the time interval corresponding to
the life of a single drop. She found that the limiting current determined by the rate of recombination of phenylglyoxalic acid
anions increases sharply on addition of small amounts of a surface-
active alkaloid (atropine) to the solution. The beginning section of
the limiting current—time curves (i vs. t curves) in this case is a
parabola; the exponent is larger than unity, and in certain cases
reaches the value 1.6. At the same time, for purely volume-kinetic currents, in the absence of double-layer effects, this exponent does not exceed the value $^2/_3$. The limiting current increase
and its fast rise with time during the life of the drop on addition of
atropine to the solution can be explained by accumulation of adsorbed atropine on the electrode surface. Thus atropine diminishes
the negative ψ_1 potential increasingly with increasing time and leads
to increased adsorptivity of phenylglyoxalic acid anions. The high
exponent found in this case for the i vs. t curves is most probably
caused by the S-shaped adsorption isotherm for phenylglyoxalic
acid anions. The rate of recombination of these anions in the adsorbed state limits the observed kinetic current. Furthermore,
the adsorption isotherm of atropine is also S-shaped.

Kůta and Weber [637] gave equations for the form of the i

vs. t curves in the absence of concentration polarization for the case of adsorption of surface-active ions. These ions not only change the ψ_1 potential but also can inhibit the process by blocking the electrode surface (Loshkarev effect). In the development of these equations it was assumed that the effective ψ_1-potential changes linearly with the fraction of the electrode surface covered by adsorbed ions, and that the electron transfer is the sum of two components: the rate of the process on the free and occupied portions of the surface. In many cases the i vs. t curves show a hump. The drop of the current on these curves is characteristic of the blocking of the surface and of the inhibition of the electrode process. Similar curves were observed [637], for example, in the irreversible reduction of VO^{2+} in 0.1-M sulfuric acid in presence of surface-active dodecylsulfonate anions. If their concentration is low at the electrode surface, the ψ_1 potential becomes more negative, and this increases the rate of electroreductions of VO^{2+} cations. At higher concentrations inhibition caused by blocking of the surface can be observed.

It is interesting to note that an addition of dodecyltrimethyl ammonium chloride (which is a strongly surface-active material) to acid solutions of nitrobenzene and certain of its derivatives, significantly shifts their second wave to negative potentials. If the concentration of this salt is $5 \cdot 10^{-3}$ M, the shift of the half-wave potential of the second wave for nitrobenzene is as much as 600 mV [638]. Such a large shift is caused by the surface character of the process determining this wave. The second wave of nitrobenzene in acid media corresponds to the reduction of phenylhydroxylamine cations, which are generated at the electrode by protonation of adsorbed phenylhydroxylamine formed in the first step of the electrode process. The considerable shift of the half-wave potential in this case is caused by displacement of phenylhydroxylamine from the electrode surface, and also by the pH increase in the reaction layer due to the change of ψ_1 potential by adsorption of dodecyltrimethyl ammonium cations. A similar effect also can be observed on increase of the ionic strength of the solution.

The effect of gelatin, which causes a shift of the second wave toward negative potentials, in this case can be explained only by its displacing action, resulting in the desorption of hydroxylamine derivatives from the electrode surface [638].

The unusual effect of supporting electrolytes in the presence of surface-active agents observed by Losev [398] must be mentioned. The inhibiting effect of tetrabutyl ammonium and tribenzylamine on electrode processes taking place on amalgam electrodes of zinc and cadmium increases significantly on addition of magnesium sulfate to the solution investigated. The effect of magnesium sulfate can be explained by the salting-out of surface-active agents from the solution, thus increasing their concentration at the electrode.

Finally, the peculiar effect of supporting electrolytes containing proton-donor cations on electrode processes with antecedent protonations must be mentioned. Among such electrolytes are ammonium salts, amines (except tetrasubstituted amines), pyridine salts, and salts of other onium compounds, which can easily lose their proton. We will compare protonation reactions in borate and ammonium buffer solutions.

Although the ammonium ions have a somewhat smaller dissociation constant than boric acid (see Table 1) they often show, according to experimental data (see, for example, [518, 546, 547]) a higher protonation activity. This is caused mainly by a strong increase of the ammonium ion concentration in the cathode layer, where they take part in the structure of the outer sphere of the double layer. Apparently, the highest concentration of ammonium ions is realized at the electrode surface. Therefore, the increase of the apparent proton-donor activity is larger if the antecedent protonation takes place in a thin reaction layer. The maximum proton-donor effect of ammonium ions is exhibited in surface processes (see, for example, the different effect of NH_4^+ and H_3BO_3 on the surface protonation of weakly alkaline solutions of thienoyl trifluoro acetone [518]). The other fact that can cause a relative increase of the proton-donor activity for NH_4^+ ions is their positive charge, which facilitates the interaction of NH_4^+ with bases having a negative charge.

H. CHANGE OF REACTIVITY OF PARTICLES
POLARIZED IN THE ELECTRODE FIELD

Besides the already discussed effects of the electrical double-layer structure (change of effective potential drop and the concentration of the reacting ions at the electrode), the change of the reactivity of particles caused by their polarization in the electrical field of the electrode must be considered, as Grabowski [564, 639, 640] and Bezuglyi [641, 642] pointed out. Due to the very high field intensity in the immediate neighborhood of the electrode surface $(10^6-10^7$ V/cm [639]), a redistribution of the electron density can be expected when the molecule is oriented in a certain way at the electrode. This effect is similar to the Wien effect [643] and leads to a change in the reactivity of the depolarizer molecule, partly shown as an increase of the acid or base dissociation constant.

Timan [644] showed in several investigations a displacement of the equilibrium of reversible chemical reactions (taking place in the gas or liquid phase) by the action of an external electrical field. The equilibrium was shifted toward the reaction products when the products had a higher dipole moment that the starting materials.

Koryta [218] mentioned among other effects the possibility of the field effect on electrode reactions. Hoijtink [645] discussed in detail the effect of field on the charge distribution in radical anions that were the products of electron uptake by aromatic hydrocarbons. The field affected the structure of the compounds formed by protonation of these anions.

Grabowski and Kemula [639] first discussed quantitatively the polarization of benzaldehyde derivatives that contained dimethylamino, amino, and oxy groups in the para position in the electrode field. The molecules of benzaldehyde derivatives are oriented at the electrode (cathode) in such a way that the negative dipole end (the aldehyde group) is directed toward the solution. In this case the induced dipole coincides with the permanent dipole. The increased polarization of benzaldehyde derivative molecules in the electrode field results in a sharp increase of the basicity of the carbonyl group. The basicity of the carbonyl group becomes even larger than that of the amino group, and the antecedent protonation taking place at the carbonyl oxygen proceeds at a much higher rate, exceeding the rate of protonation in the bulk solution in the absence of additional polarization of the molecule [639].

Independently of Grabowski, and about at the same time, Bezuglyi [641, 642] discussed the behavior of aldehyde and ketone molecules in the field of the electrode. He related qualitatively the change of the electrochemical reduction rate to the polarization of the molecule.

A very interesting example of the polarizing effect of the electrode field on a chemical reaction accompanying the electrode process was investigated by Vincenz-Chodkowska and Grabowski [640]. They showed that the trans form of stilbenediol formed as a result of a reduction of benzil with two electrons is converted into the cis isomer in the electrode field. The polarization of the trans isomer in the field of the electrode, which leads to the formation of a quinoidal structure with opposite charges on the two phenyl rings and a single bond between the central carbon atoms, makes possible the free rotation of the parts of the molecule around the central $C-C$ bond, and therefore facilitates the trans–cis transformation. It is interesting to note that the ratio of cis to trans isomer in the reaction products increases at higher negative potentials, with increasing field intensity. After reaching a maximum value at $E = -0.95$ V (S.C.E.), if the negative potential further increases, the ratio drops. It is possible that this drop is caused by the desorption of stilbenediol from the electrode surface at increasingly negative potentials [640]. An increase of the cis isomer yield was also observed at increased temperature and solution pH. The effect of higher pH values can be explained by considering that isomerization of the anion containing a single central bond is easier than that of the undissociated diol molecule. An addition of cesium salt to the solution, which decreases the negative ψ_1 potential, decreases the cis isomer yield.

Apparently the electrode field also shows an effect on the dehydration rate of the nickel aquocomplex [646], which precedes the discharge of Ni^{2+} on the electrode [221, 646].

It must be emphasized that a polarization of particles and the change in the kinetics of chemical reactions accompanying the electron transfer thus caused can occur only for adsorbed particles or for particles that are in the immediate neighborhood of the electrode surface. Chemical reactions taking place at a certain distance from the electrode, even when they occur in the diffusion part of the double layer, are not affected by the polarizing action of the electrode. The field of the electrode does not affect volume reactions at all.

I. CONSIDERATION OF THE EFFECTS OF THE DOUBLE-LAYER STRUCTURE IN DETERMINING THE RELATIONSHIP BETWEEN THE STRUCTURE OF ORGANIC MOLECULES AND THE HALF-WAVE POTENTIAL OF THEIR REDUCTION WAVE

The examples given in Chapter V show the effect of the double-layer structure on the kinetics of electrode processes that establish a relationship between the half-wave potentials of irreversible polarographic waves and the structure of a given series of organic compounds, or the character of substituents in these compounds, not only must the pH and ionic strength of the medium be considered, but also the nature and concentration of buffer components and supporting electrolytes [480, 606].

Furthermore, the charge of discharging particles must be considered, since it has a significant influence on half-wave potentials. Thus, in the reduction of isomeric pyridine aldehyde [647] and p-iodoanilinium cations [648] the observed value of the half-wave potential is more positive by about 300 mV than the calculated value determined from their structure. On the other hand, p-iodophenolate anions are reduced at a potential more negative by about 300 mV than that predicted by theory without taking the double-layer effect into consideration [647, 648]. Another example is the cleavage of the lactone ring of narcotine by alkaline treatment and the appearance of an anionic carboxyl group shifts the reduction wave of narcotine N-oxide by 300 mV to negative potentials [372] compared with the free lactonic form of narcotine N-oxide under identical conditions.

Besides the factors already mentioned, reactions at the electrodes, including the dimerization of electrode products, can affect the half-wave potentials even for one-electron reversible

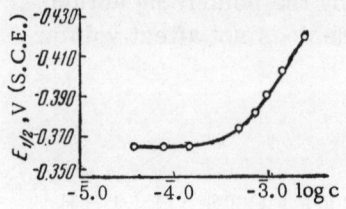

Fig. 32. The half-wave potential of
the nitrobenzene wave as a function
of its concentration in the solution.

waves. Thus, in the reduction of benzaldehyde and acetophenone the products of the reversible electrode reaction dimerize. In this process, with increasing depolarizer concentration, the half-wave potential of the first waves becomes more positive in accord with the theory (see Chapter VIII). It is interesting that the half-wave potential of the more negative waves on these polarograms, which are characteristic of the further reduction of the free radicals formed in the first stage, become more negative with increasing depolarizer concentration. Thus the first and second waves are better separated at higher depolarizer concentrations. This will be discussed in detail in Chapter VIII.

In other cases, the shift of half-wave potentials with increasing depolarizer concentration is caused by a change of the ψ_1 potential, owing to adsorption of the electrode reaction products which have an ionic (or dipole) nature [606]. This happens in the reduction of certain nitro compounds [649-652]. The reduction products of these compounds in acid solution are the corresponding hydroxylamine derivatives. Figure 32 shows the dependence of half-wave potentials on nitrobenzene concentration in an acetate buffer solution at pH 4.80, at sufficiently high buffer capacities (the concentration of the undissociated acetic acid is about 0.1 M). At low nitrobenzene concentrations, when the surface coverage by electrode products is low (particularly for the electrode used in these experiments which had a short drop-time due to forced drop removal, t = 0.22 sec), the half-wave potential is independent of depolarizer concentration. The effect of nitrobenzene concentration on the half-wave potential becomes apparent only at concentrations above $2.5 \cdot 10^{-4}$ M. For 1-nitropropane (the adsorptivity of the propylhydroxylamine formed in its reduction is smaller than that of phenylhydroxylamine) the depolarizer concentration under the same conditions does not affect the half-wave potential of the wave [606].

The half-wave potential is very strongly affected by the adsorption of the depolarizer. This will be discussed in the following chapter.

Chapter VI

Electrode Processes with Participation of Adsorbed Compounds

A. CHARACTERISTICS OF ELECTRODE PROCESSES WITH ADSORPTION OF THE COMPONENTS TAKING PART IN CHEMICAL AND ELECTROCHEMICAL REACTIONS

As was already mentioned (see page 128), the adsorption of an inactive component that participates in an antecedent chemical reaction leads to a sharp increase of the concentration of this component at the electrode. This usually results in a significant increase in the rate of the antecedent reaction taking place at the electrode. The phenomenon is very similar to the adsorption of the depolarizer in the absence of antecedent chemical reactions. Thus, the concentration increase of a reversibly reduced depolarizer at the electrode – as a result of its adsorption on the electrode – results in a significant increase in the peaks measured by the Breyer method with superimposition of alternating voltage. Such peaks can be observed at extremely low depolarizer concentrations in the solutions – as low as 10^{-7} M [653, 654]. It must be mentioned that Breyer, Bauer, and Hacobian [653] first pointed out the rate increase of electrochemical reactions – at conditions corresponding to the appearance of Breyer-type waves – caused by depolarizer adsorption. They qualitatively discussed the concentration change of the depolarizer in several solution layers at the electrode surface [655].

Antropov [656-658] mentioned many times how important it is to consider depolarizer adsorption in electrode processes of organic compounds.

The depolarizer adsorption often changes the character of the electrochemical process. Thus on the current intensity vs. potential curves recorded with an oscilloscope (which means curves of the Ševčik–Randles type) a much sharper current drop can be observed after the maximum on adsorption of the electrochemically active compound, than on such i vs. E curves that correspond to diffusion-limited processes [659]. The diffusion-limited current decreases relatively slowly with time, as the thickness of the diffusion layer increases (proportionally with $t^{-1/2}$). If a compound is reduced in the adsorbed state, usually much higher peaks are observed, and the decrease in current following these peaks is more steep due to consumption of the depolarizer present on the electrode surface in the adsorbed state. The current decreases to the value corresponding to the diffusion transport of the depolarizer from the bulk of the solution. The effect of depolarizer adsorption on chronopotentiometric curves was discussed by Reinmuth [660]. Matsuda and Delahay [661] showed that depolarizer adsorption must also be considered in the determination of exchange currents by galvanostatic and potentiostatic methods. Laitinen and Randles [662] showed that in the investigation of fast electrochemical reactions by the method of measurement of faradaic impedance, the adsorption must be included in the consideration.

Depolarizer adsorption can result in a large increase in its concentration at the electrode and can significantly increase the rate of the electrochemical process. This rate increase can be so large that in certain cases two separate waves appear on the polarogram. The first wave has a kinetic character and corresponds to the discharge of strongly adsorbed (or energetically favorably oriented) particles. The second wave, the normal wave, corresponds to the discharge of nonadsorbed (or weakly adsorbed, less favorably oriented) depolarizer molecules [663]. The doubling of waves for certain organic halide derivatives on their polarograms observed by several investigators [664-665] can be explained by the effect of depolarizer adsorption (see also Fig. 209 in [370]). Adsorption also explains the unusual character of benzyl chloride reduction at the dropping electrode [666, 667].

A characteristic property of electrode processes with adsorbed depolarizer or a component of an antecedent chemical reaction is that the electrode potential shows two effects. First, the electrode potential determines the rate constant of electron transfer, and, second, it affects the adsorption of the components taking part in the electrode process. Thus, finally, it also affects the rate of the electrochemical reaction. An increase of negative potential increases the rate of electron transfer on the particles being reduced. At the same time, if the process proceeds at potentials that are more negative than the potential corresponding to maximum adsorption, the concentration of the compound on the electrode decreases with increasing negative potential. This double action of the potential sometimes leads to decreases in the polarographic wave that are not connected with the decrease of tangential motion of the mercury surface. It can be concluded (if there is no electrostatic repulsion present as, for example, for the reduction of anions, see page 163) that a decrease in polarographic waves indicates either the surface nature of the antecedent chemical reaction [480] or a strong adsorption of the depolarizer itself [666, 667]. We will now discuss quantitatively this double effect of the potential on the shape of a surface kinetic wave.

B. EQUATIONS OF SURFACE KINETIC WAVES

An exact derivation of equations for surface kinetic currents meets serious mathematical difficulties when one of the limiting stages of the electrode process is diffusion. But with the introduction of several simplifying assumptions, equations can be derived for kinetic waves with antecedent surface reaction.

We will discuss the kinetic current, limited by the rate of a first-order (or pseudo-first-order) chemical reaction, taking place on the electrode surface with participation of adsorbed material. Let us consider a protolytic interaction in a buffer solution [see, for example, scheme (XVIII)]. We will limit this discussion to the case where the protolytic equilibrium (XVIII) is shifted to the left at the electrode surface and also in the solution volume (which means that the depolarizer is present in its unprotonated form). Furthermore, the concentration of the buffer components AH (or BH^+) and A^- (or B) in the solution is much higher than the concentration of R (and even higher than RH^+) and therefore reaction (XVIII) can be considered to be a pseudo-first-order reaction.

At sufficiently negative electrode potentials (for cathodic processes) the rate of the electrochemical stage (electron transfer) becomes so high that all the RH^+ particles formed at the electrode surface immediately take part in the electrode reaction. The observed limiting kinetic current (i_{lim}) is determined by the rate of RH^+ particle formation [according to the forward direction of reaction (XVIII)] on the electrode

$$i_{lim} = nsF\rho_s\Gamma, \tag{103}$$

where n is the number of electrons taking part in the electrode process, s is the surface area of the electrode, F is the Faraday number, ρ_s is the rate constant of adsorbed R particle and acid interaction, and Γ is the quantity of adsorbed R particles on $1\,cm^2$ of electrode surface.

If the adsorption rate is determined by diffusion and if the adsorption equilibrium follows the Langmuir adsorption isotherm, then the approach to equilibrium $\Gamma_t / \Gamma_e = y$ (see page 62) at not too high y values, is proportional with sufficient accuracy to the square root of the adsorption time for many practical cases

$$y \simeq kt^{1/2}. \tag{104}$$

Based on the data given by Delahay and Fike [290], it can be shown [668] that for $y \leq 0.55$, for the same adsorption time, the approach to the equilibrium can be more exactly expressed by equation

$$y = k'(1 + \beta c). \tag{105}$$

From Eqs. (104) and (105), and also considering that the y value (at not too high values) is proportional to D/Γ_∞ [290] (where D is the diffusion coefficient of the adsorbed compound), $y \leq 0.55$ can be given by

$$y \simeq 1.5 \cdot 10^{-5}(1 + \beta c)\frac{Dt^{1/2}}{\Gamma_\infty}. \tag{106}$$

The numerical coefficient was calculated from Delahay and Fike's data [290] under the condition that D is expressed in cm^2/sec and Γ_∞ in $moles/cm^2$. From Eqs. (37) and (106) it follows that

$$\Gamma_t \simeq 1.5 \cdot 10^{-5}\beta cDt^{1/2}. \tag{107}$$

It was already mentioned, on page 63, that in the work of Delahay and Fike [290] the independent variable was improperly selected, and therefore Eq. (107) is incorrect. However, in certain particular cases, the plots of Delahay and Fike [290] correctly gave the character of the dependence of y on certain factors, primarily on $t^{1/2}$ and β. Therefore, the expressions derived in several investigations [668, 669] on the basis of Eq. (107) are correct, as will be shown in the following discussion (although the physical meaning of certain constants in these equations is not correct).

Levich, Khaikin, and Belokolos [670] proposed an approximate method to solve the approach to equilibrium problem. Their method can be used for all adsorption isotherms on planar and dropping electrodes. Based on this method they derived analytical expressions for y. The y values calculated by these expressions deviate by less than 8% from the exact values determined by the same authors with the help of computers; the latter were in the form of plots of values as functions of the dimensionless τ parameter, for different Γ_e and γ values (γ is the attraction factor in the Frumkin equation, see page 58). The precision of the approximate solution given by Levich, Khaikin, and Belokolos is sufficient for many practical purposes.

If we limit this discussion to processes taking part at low equilibrium saturations by the adsorbed compound on the electrode (which means that $\beta c \ll 1$), then for the mercury dropping electrode the equation of Levich, Khaikin, and Belokolos has the form [670]:

$$y = \frac{\Gamma_t}{\Gamma_e} = \frac{\tau^{1/2}}{1 + \tau^{1/2}}, \tag{108}$$

where

$$\tau = \frac{12}{7\pi} \frac{Dt}{\beta^2 \Gamma_\infty^2}. \tag{109}$$

On the basis of Eqs. (37), (108), and (109),

$$\Gamma_t = 0.74\beta c \Gamma_\infty \frac{\dfrac{D^{1/2} t^{1/2}}{\beta \Gamma_\infty}}{1 + \dfrac{0.74 D^{1/2} t^{1/2}}{\beta \Gamma_\infty}}. \tag{110}$$

The expression (110) can be simplified for special cases.

If the compound has a very high adsorptivity at low elapsed adsorption times (which means that $\tau^{1/2} \ll 1$), when all material reaching the electrode by diffusion is adsorbed at the surface, Eq. (110) can be used in the form

$$\Gamma_t = 0.74 c D^{1/2} t^{1/2}. \tag{111}$$

Otherwise — when the material has low adsorptivity and the adsorption time is sufficient (i.e., at $\tau^{1/2} \gg 1$), when adsorption equilibrium is almost achieved — on the basis of (110), we obtain

$$\Gamma_t = \Gamma_\infty \beta c. \tag{112}$$

It follows from Eqs. (111) and (112) that the Γ_t value changes either proportionally to $\tau^{1/2}$ or proportionally to β, depending on the conditions. This means that in special particular cases both relationships (107) are fulfilled. These equations were used [668, 669] for the interpretation of observed phenomena. Generally, as predicted by Eq. (110), Γ_t depends on both $\tau^{1/2}$ and on β, but the character of this dependence is somewhat different than that given by expression (107), derived on the basis of the data measured by Delahay and Fike.

The Γ_t value from Eqs. (107) or (110)-(112) can be used directly only for the calculation of those surface catalytic currents (see Chapter IX) in which a fast and complete regeneration of the adsorbed catalyst takes place. For kinetic currents the removal of the adsorbed compound by chemical reaction (and subsequent electrochemical process) must be considered. To a first approximation, Γ (considering the effect of chemical reaction) can be given by

$$\Gamma = \Gamma_t \left(1 - \frac{i_{\lim}}{i_D}\right), \tag{113}$$

where i_D is the diffusion current [to which $i_{\lim}$ tends if the rate of the forward reaction (XVIII) increases].

If we consider that the adsorptivity of a compound on the electrode depends on the potential, according to Frumkin's equation (42), then for conditions when Eq. (112) is valid, we obtain from Eqs. (42), (103), (112), and (113)

$$\log \frac{i_{\lim}}{i_D - i_{\lim}} = A - 0.43 a \varphi^2, \tag{114}$$

where

$$A = \log nsc\, F\rho_s \beta^0 \Gamma_\infty / i_D. \qquad (115)$$

The expression for A differs somewhat for the corresponding value given in [669], which was obtained from Eq. (107), and not from Eq. (112).

Expressing the electrode surface area by the flow rate of mercury (m) in mg/sec and the drop period (t), and also applying the Ilkovič equation (see page 3), the average A value during the drop period (t) is

$$\overline{A} = \log 0.81 \rho_s \beta^0 \Gamma_\infty\, t^{1/3}/D^{1/2}, \qquad (116)$$

where β^0 is expressed in $cm^3/mole$.

Equation (114) shows that the limiting current for surface kinetic waves decreases with increasing electrode potential (φ). This explains the characteristic shape of these waves, which are limited in height by the rate of the surface reaction. If $\tau^{1/2} > 1$, a current decrease is observed instead of a flat limiting current.

At potentials more negative than those where the limiting current is already seen, both R particles and RH^+ particles are adsorbed at the electrode surface [see scheme (XVIII)]. The current then can be given by

$$i = nsF(\rho_s \Gamma - \rho_s \sigma_s \Gamma'), \qquad (117)$$

where Γ' is the quantity of electrochemically active RH^+ particles adsorbed on 1 cm^2 of surface, and $\rho_s \sigma_s$ is the rate constant of their interaction with bases on the electrode surface, according to the reverse direction of reaction (XVIII).

The Γ value in Eq. (117) can be given similarly to Eq. (113) as

$$\Gamma = \Gamma_t (1 - i/i_D). \qquad (113a)$$

This equation together with Eqs. (103) and (113) yields:

$$i_{\lim} - i = nsF\rho_s \left(\sigma_s \Gamma' - \Gamma_t \frac{i_{\lim} - i}{i_D} \right). \qquad (118)$$

A few words must be said about the $i_{\lim}$ value. This is the limiting current, determined only by the rate of antecedent chemical reaction. Therefore, at potentials at which the electron transfer is not sufficiently high and when the measured current does not reach the i_{lim} value, such a fictitious i_{lim} value must be selected, which would be measured at the given potential if the rate of the electrochemical stage were infinitely fast.

From Eqs. (103), (113), and (118) we obtain

$$(i_{lim} - i)\frac{i_D}{i_D - i_{lim}} = nsF\rho_s\sigma_s\Gamma'. \tag{119}$$

Equation (69) of the theory of slow discharges is valid for the case of adsorbed depolarizer when the surface coverage by adsorbed particles is not too high. In this equation $[RH^+]_s$ must be replaced by Γ', E by φ, and the k^0_{el} value, which was related to the volume concentration at the electrode, must be replaced by k'_{el}, which is related to $\varphi = 0$. The dimension of the latter value corresponds to the adsorbed depolarizer concentration, which is expressed in moles/cm². Equation (69) has the following form for this case:

$$i = nsF\Gamma'k'_{el}e^{-\frac{\alpha n_a \varphi F}{RT}}. \tag{69a}$$

The equation of the wave can be obtained from Eqs. (119) and (69a)

$$\log\frac{i_D - i_{lim}}{i_D} + \frac{0.43\alpha n_a F\varphi}{RT} = \log\frac{k'_{el}}{\rho_s\sigma_s} - \log\frac{i}{i_{lim} - i}. \tag{120}$$

This equation can be used for both average and instantaneous currents.

C. THE SHAPE OF SURFACE KINETIC WAVES

At low surface coverages by adsorbed particles, and at high rates of approach to the adsorption equilibrium (which can be measured in the absence of the electrode process), which means when Eq. (112) is valid, the form of surface kinetic waves can be expressed by Eqs. (114) and (120) together. We will compare these equations with experimental data for several examples of surface kinetic waves.

In alkaline solution 5-bromo-2-acetyl thiophene gives polarographic reduction waves of which the first shows a decrease characteristic for surface kinetic waves. The circles in Fig. 33 give the experimental average current values corrected for the

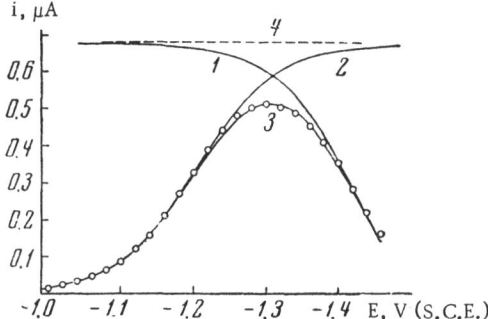

Fig. 33. The first reduction wave of 5-bromo-2-acetyl thiophene
(0.215 mM 5-bromo-2 acetyl thiophene in 0.1 M KCl + 0.1 M KOH
at 25°C; electrode characteristics: m = 1.07 mg/sec, t = 0.26 sec).
Circles indicated experimental data. (1) i_{lim} as a function of potential [according to Eq. (114)]; (2) i vs. E function according to
(123) for i_{lim} = const $\approx i_D$; (3) theoretical wave constructed using
Eq. (114) and Eq. (123); (4) diffusion current level.

background. The electrochemical cleavage of the C − Br bond, corresponding to the first wave, is preceded by surface protonation of
the carbonyl group by water [671]. At sufficiently negative potentials (in this case E < −1.4 V) the current (i) can be considered to
have attained its limiting value (i_{lim}).

In this case Eq. (114) can be used to describe the dependence
of the observed current on potential. Figure 34 shows a plot of
$\log [i_{lim}/(i_D - i_{lim})]$ vs. φ^2 for the decreasing portion of the
5-bromo-2-acetyl thiophene wave. In plotting the curves in Fig.
34 it was assumed that the potential of maximum adsorption is
E_M = −0.45 V (S.C.E.) and i_D = 0.68 μA. As can be seen from
Fig. 34, the experimental points over a certain potential range
(limited from the positive side by decrease of the current compared with i_{lim} and from the negative side by the current increase
caused by appearance of the second wave − reduction of the keto
group) lie on a straight line. From this line the a and A values of
Eq. (114) can be determined. Thus, for bromoacetyl thiophene,
under the condition given in Fig. 33, a = 11.3 V^{-2} and $\overline{A}$ = 4.45. If
we consider the a value to be constant, the i_{lim} values, can be determined from Eq. (114) at any potential. These values then can
be used to construct the theoretical i vs. E curve or the i vs. φ
curve according to Eq. (120).

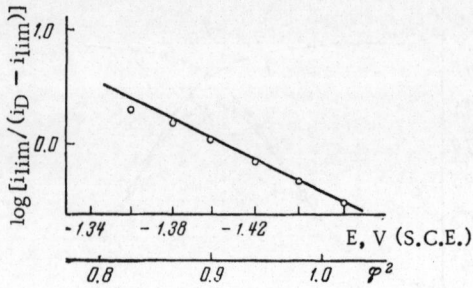

Fig. 34. The quantity $\log[i_{lim}/(i_D - i_{lim})]$ as a function of φ^2 for the decreasing portion of the wave for 0.215 mM 5-bromo-2-acetyl thiophene in 0.1 M KOH + 0.1 M KCl.

Equation (120) can be given in a somewhat simpler form. The i_{lim} values increase at more positive potentials and tend toward i_D (curve 1 in Fig. 33). Therefore, the $\log[(i_D - i_{lim})/i_D]$ function at $i_{lim} \rightarrow i_D$ becomes equal to $\log[(i_D - i_{lim})/i_{lim}]$ which is given by Eq. (114). When the values of $|\varphi| = |E - E_M|$ are not too small, and over a relatively small potential range $\pm\triangle\varphi = \varphi - \varphi_{1/2}$ (around the $\varphi_{1/2}$ value), from Eq. (114) for $i_{lim} \rightarrow i_D$ the following equation can be given:

$$\log\frac{i_D - i_{lim}}{i_D} = -\overline{A} + 0.43a\,[\varphi_{1/2}^2 + 2\varphi_{1/2}(\varphi - \varphi_{1/2})] = \text{const} - \frac{1}{b'}\varphi, (121)$$

where

$$\left.\begin{array}{l} \text{const} = -\overline{A} - 0.43a\,\varphi_{1/2}^2 \\[2mm] b' = -\dfrac{2.3}{2a\varphi_{1/2}}, \end{array}\right\} \qquad (122)$$

and at potentials more negative than E_M, $b' > 0$, since $\varphi_{1/2} = E_{1/2} - E_M < 0$.

Noting $b = 2.3RT/\alpha n_\alpha F$, we obtain from Eqs. (120) and (121)

$$\varphi = \varphi_{1/2} - \frac{bb'}{b' - b}\log\frac{i}{i_{lim} - i}, \qquad (123)$$

where $\varphi_{1/2} = E_{1/2} - E_M$, i.e., the potential at which $i = i_{lim}/2$. This potential is

$$\varphi_{1/2} = \frac{bb'}{b' - b} \left(\log \frac{k'_{el}}{\rho_s \sigma_s} + \bar{A} + 0.43 a \varphi_{1/2}^2 \right).$$ (124)

If we now consider the area of more negative potentials, where the limiting current is significantly lower than the diffusion current $(i_D > i_{lim} > 0.5 i_D)$, the $\log[(i_D - i_{lim})/i_D]$ value becomes somewhat lower than $\log[(i_D - i_{lim})/i_{lim}]$ and, therefore, the $\log[(i_D - i_{lim})/i_{lim}]$ vs. φ^2 plot at higher $|\varphi|$ values begins to deviate from the straight line, corresponding to Eq. (114), and tends toward smaller $\log[(i_D - i_{lim})/i_D]$ values. On the other hand, if we construct a plot of $\log[(i_D - i_{lim})/i_D]$ as a function of φ (or E), and not as a function of φ^2, the curve becomes more linear. Thus, for $i_D > i_{lim} > 0.5 i_D$ values, at potentials in the -0.6 to -1.4 V range, the function $(i_D - i_{lim})/i_D$ vs. φ can be expressed for many purposes by a straight line. The slope of this line is nearly identical with b', given in Eq. (122). The curve in Fig. 35 was constructed for the i_{lim} data for 5-bromo-2-acetyl thiophene in a solution containing 0.1 N KCl + 0.1 N KOH. The i_{lim} values for constructing this curve were determined by Eq. (114) with the previously given a and $\bar{A}$ constants. The same Fig. 35 shows a straight line corresponding to Eq. (121) with b' = 0.135 V and const = -1.433 V. As can be seen from Fig. 35, curve 1 practically coincides with the line 2 up to $\log[(i_D - i_{lim})/i_D] = -0.65$.

Therefore, Eq. (123) is correct for surface waves in the $i_D > i_{lim} \geq 0.55 i_D$ potential range. At more negative potentials, the function $\log[(i_D - i_{lim})/i_D]$ tends toward zero (see Fig. 35) and, as it follows from expression (120), the wave can be expressed by Eq. (123). This equation contains a constant in front of the logarithm, which gradually changes from bb'/(b' - b) to b over a relatively narrow potential range. In this transition zone both Eqs. (114) and (120) must be used for the theoretical construction of the wave. Thus the form of a surface kinetic wave (except in the range of transitional potentials) under conditions when expression (112) is valid can be described by Eqs. (114) and (123). For potentials at the foot of the wave, i_{lim} often is very close to i_D, and therefore the i_D value can be used instead of i_{lim} for the determination of the bb'/(b' - b) value from Eq. (123). From the slope of the beginning of the bromoacetyl thiophene wave plotted in semilogarithmic coordinates, it was calculated that bb'/(b' - b) = 0.128 V, using this method. In Fig. 33, curve 2 was constructed

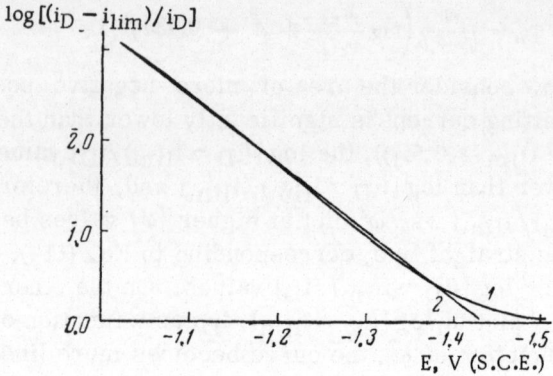

Fig. 35. The quantity $\log\left[(i_D - i_{lim})/i_D\right]$ as a function of E for
5-bromo-2-acetyl thiophene (1) [the straight line (2) was constructed by using Eq. (121)].

using Eq. (123) assuming that $i_{lim} \approx i_D$. At the very beginning of
the wave this curve coincides with the experimental curve. Curve 3
of Fig. 33 was constructed using Eqs. (114) and (123), the previously given constant values. This curve, as can be seen from the
figure, agrees very well with the observed experimental kinetic
wave.

We will deal further with cases of processes with antecedent
protonation. Figure 36 shows the kinetic surface currents for the
reduction of a 1.46-mM phenylacetaldehyde oxime solution in
citrate-phosphate buffer at pH 6.4 in the presence of 10% ethyl alcohol (the electrode characteristics were: m = 1.07 mg/sec, t =
0.26 sec). The same figure gives a continuous curve constructed
using Eqs. (114) and (123). This curve was calculated for i_D =
5.78 μA, bb'/(b' - b) = 0.240 V, $E_{1/2}$ = −1.64 V, a = 5.5 V^{-2}, and
const = 2.57 [in Eq. (122)].

In Fig. 37 is given the wave of 4-bromopyridine at pH 8.3 in
50% ethyl alcohol, constructed from the data from Fig. 1 (curve 6)
of Holubek and Volke [672]. The same figure gives the theoretically calculated curve corresponding to Eqs. (114) and (123) for a =
7.3 V^{-2}, const = 3.58 [in (122)], $E_{1/2}$ = −1.32 V, and bb'/(b' - b) =
0.113 V.

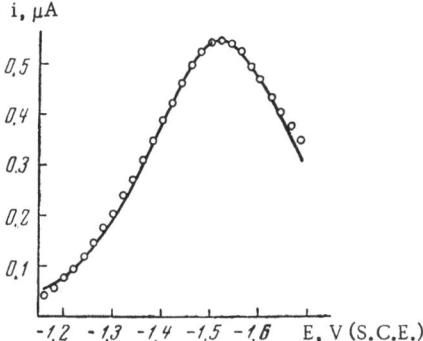

Fig. 36. Surface kinetic wave for phenylacetaldehyde
oxime reduction at pH 6.4. Circles indicate experimental
current data; curve was calculated using Eqs. (114) and
(123).

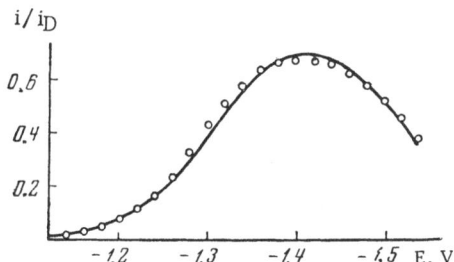

Fig. 37. Experimental current data (from [672]) and the
theoretically calculated wave for reduction of 4-bromo-
pyridine in 50% ethyl alcohol (pH 8.3).

It follows from Figs. 33, 36, and 37 that Eqs. (114) and (123)
correctly represent the shape of surface kinetic waves.

It must be mentioned that if the adsorption of the electrochemically inactive form (evaluated under the same conditions but
in the absence of the electrode reaction) is substantially different
from the equilibrium adsorption value, then instead of Eq. (112) the
more general expression (110) must be applied. It is not hard to
show that the current decrease on the waves will be less deep in
this case. If the approach to the adsorption equilibrium is very
small (evaluated under conditions of absence of electrochemical

reaction), i.e., when Eq. (111) is valid, the adsorption is independent of the electrode potential. In this case the decrease cannot be observed on surface kinetic current waves. If the electrode process takes place only at higher surface coverages of the adsorbed compound, which means that the inequality $\beta c \ll 1$ is no longer satisfied, more complicated equations must be used [670] instead of Eq. (110).

The appearance of decreases on the limiting kinetic current lines is a quite frequent phenomenon in the polarography of organic compounds. Such decreases can be found, for example, on the first and second kinetic waves for maleic acid [137], phthalic acid [145], and phthalimide [673], on polarograms of azomethine derivatives [160, 161], phenolphthalein [674], certain quinones [675], and for many more compounds which are reduced after antecedent protonation. Savéant observed decreases on the reduction waves of ketones containing α-sulfonium [676, 677] and ammonium [676] groups and explained them by antecedent surface protonation.

Kinetic waves show surface character not only for antecedent protonation but also for other types of antecedent reactions. Thus Holleck and Lehmann [678, 679] observed this surface character in the polarography of ninhydrin and certain other diketo compounds (phenylglyoxal, indanedion-2,3-ol-1). A detailed investigation of the character of the "prewave" on the ninhydrin polarograms, which in certain cases shows the form of a peak [679], led Holleck and Lehmann to the conclusion that this prewave is a kinetic wave, limited by the rate of dehydration of the carbonyl group both in the electrode layer volume and at the electrode surface. If camphor is added to the solutions, the height of the prewave decreases and a normal wave form is obtained, with a flat portion of the limiting current. The effect of camphor was explained [679] by displacement of the adsorbed (hydrated) ninhydrin from the electrode surface. Then, only the volume component of the kinetic current remains on the polarograms. Most probably, a mixed volume–surface character plays the determining role for the first kinetic wave on diethanolamino benzoquinone polarograms [675]. Addition to the solution of dextrane which has a high surface activity reduces the height of the first wave and straightens out the hump on its limiting current.

In concluding this section, the theory proposed by Brainina and Stromberg [680, 681] must be mentioned. This theory deals with the formation of decreases on waves for processes with depolarizer adsorption. Brainina and Stromberg discussed the adsorption of simple or complex ions close to the potential of zero charge and assumed that, due to the dipole character of complicated ions, they can be oriented in a different way (in the adsorbed state) on negatively and positively charged electrodes. Therefore, their reduction requires different activation energies. It was further assumed that the ions oriented with their negative end toward the electrode (which takes place at potentials on the positive branch of the electrocapillary curve) are reduced reversibly, and the reduction of oppositely oriented ions goes on with a certain overpotential. Taking into consideration the gradual transition from one orientation to the other as the potential changes around the electrocapillary zero, a region of decreased current can be expected. This happens because the current of the reversible reduction decreases due to the decrease of the fraction of "favorably" oriented ions, and because, at the same time, the current of the irreversible process (at a constant overpotential value) has not yet reached its limiting value. The equations derived by this theory correctly reproduce the observed waves with a decrease [681]. It must be mentioned that the theory given by Brainina and Stromberg explains only the decrease observed close to the electrocapillary zero. The same phenomenon, however, can be quantitatively explained by a more general theory, taking into consideration the chemical reactions at the electrodes (particularly those with participation of adsorbed particles) and the effect of electrode potential and double-layer structure on these reactions. It is well known that the effect of an antecedent reaction also appears in the reduction of aquo complexes of simple ions (for example, Ni^{+2} ions) [221, 646]. For this reason it can be assumed that they also play a role in the reduction of other complex ions.

D. THE EFFECT OF DOUBLE-LAYER STRUCTURE ON THE ADSORPTION OF COMPONENTS OF ANTECEDENT REACTIONS

The effect of supporting electrolyte concentration on surface kinetic waves is twofold. First, it is expressed in the concentra-

tion change of the charged components that take part in the antecedent reactions, in the layer at the electrode [for example, in the
change of $(pH)_S$]. Second, it affects the change of the adsorptivity
of the electrochemically inactive depolarizer form [R in scheme
(XVIII)].

It follows from the Frumkin equation (42) that the effect of
supporting electrolytes on the adsorption of a given compound is
determined by the capacity change of the double layer in the absence of adsorbed compounds (C). An increase in salt concentration results in a higher C value. Therefore, the adsorption decreases and the rate of desorption becomes more rapid with increasingly negative electrode potential.

The change of adsorptivity with potential is most pronounced
for waves whose height is limited by the rate of an antecedent
chemical reaction that takes place with participation of adsorbed
components. Under these conditions, Eq. (112) can be applied. In
Fig. 38, polarographic curves of 2-acetylthiophene semicarbazone
obtained in two experimental series are given: (a) at constant
ionic strength and pH of the solution at different buffer capacities
of the solution ($c_{CH_3COOH} = 0.1$ and 0.4 M, respectively), and, (b)
at constant solution pH and buffer capacity, but at different ionic
strength. On all polarographic curves shown in Fig. 38 a decrease
can be observed on the limiting current portion, which corresponds
to the desorption of semicarbazone. The semicarbazone is desorbed to such a degree that the kinetic current, limited by the
rate of protonation, becomes smaller than the diffusion current.
With increasing buffer capacity, which also means an increased
rate of protonation, the decrease becomes less deep. When the
ionic strength increases, which results in an increased pH in the
electrode layer and suppresses the adsorptivity of semicarbazone,
the decrease becomes deeper [480].

It must be mentioned that the decrease on surface waves is
not always more pronounced with increasing solution ionic strength.
Thus, for example, in strongly alkaline medium, where water seems
to be the only proton donor and pH does not affect the half-wave potential, a change of solution ionic strength affects mainly the effective potential drop and the depolarizer adsorption. Therefore, at
increasing ionic strength the dip can even disappear. Such a behavior can be observed, for example, for the first reduction wave

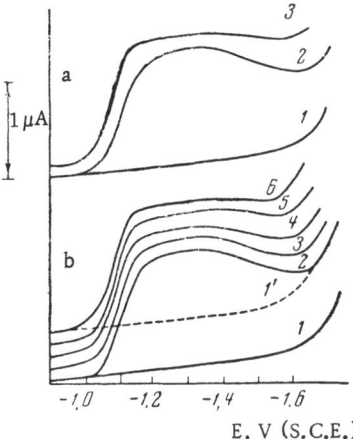

Fig. 38. Polarographic reduction waves for 2-acetyl-
thiophene semicarbazone (4.0 · 10⁻⁴ M) in acetate buffer solutions (CH₃COOH + CH₃COOK + KCl) at pH 4.65.
(a) Ionic strength μ = const = 0.5 N: (1) supporting
electrolyte; (2) 0.1 M CH₃COOH; (3) 0.4 M CH₃COOH.
(b) Ionic strength constant CH₃COOH concentration
(0.1 M). (1, 1') Supporting electrolyte at 1.6 N and
0.1 N ionic strength, respectively; (2) 1.6; (3) 0.8;
(4) 0.4; (5) 0.2; (6) 0.1.

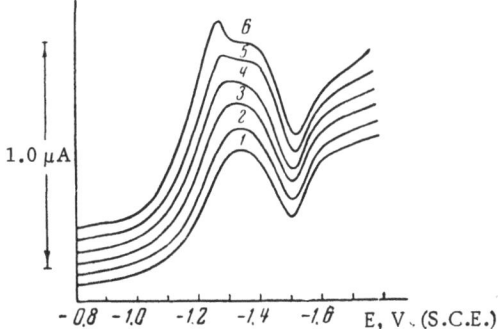

Fig. 39. The effect of the ionic strength on the first
reduction wave for 5-bromo-2-acetyl thiophene in alkaline medium (0.1 M KOH + x KCl). Ionic strength
(in N): (1) 0.1; (2) 0.2; (3) 0.4; (4) 0.8; (5) 1.6; (6) 2.85.

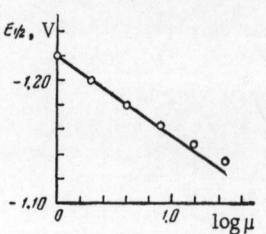

Fig. 40. Half-wave potential for the first reduction wave of 5-bromo-2-acetyl thiophene as a function of the logarithm of the ionic strength (KOH + KCl).

of 5-bromo-2-acetyl thiophene in alkaline medium (electrochemical cleavage of the C−Br bond in the molecule protonated at the CO group). Figure 39 shows polarographic curves for bromoacetyl thiophene at a series of ionic strengths. It can be seen in the figure that the height of the surface kinetic wave increases with the solution ionic strength and the depth of the decrease becomes smaller. The same can be observed if cesium salts are added to the solution. This behavior is similar to the polarographic behavior of anions, and it could be assumed that nonreducible anions are formed in alkaline solutions of bromoacetyl thiophene (for example, by cleaving of a hydrogen ion from its enol form), and that these anions are protonated in adsorbed state before reduction. Figure 40 shows the dependence of the half-wave potentials of the first wave [which was determined from the intercept of a plot in E vs. $\log i/(i_D - i)$ coordinates or the extension of its linear section at the ordinate with the $\log i/(i_D - i) = 0$ line] on the ionic strength of the solution (on addition of KCl to a 0.1 N KOH solution). It follows from Fig. 40 that upon a tenfold increase in potassium ion concentration (c_K) the half-wave potential shifts by about 65 mV to more positive values [671]. Since in alkaline solutions protonation takes place only under the influence of water and $\Delta E_{1/2}/\Delta pH = 0$, the observed shift of the half-wave potential with $\log c_K$ is close to the value that is characteristic for particles which do not carry charge in the solution volume (see page 156). Therefore, uncharged particles arrive at the electrode and they are protonated at the electrode in the adsorbed state. The $\Delta E_{1/2}/\Delta \log c_K$ value is somewhat larger than the change of the ψ_1 potential in these solutions. This indicates that the transport of particles to the electrode and their adsorption is somewhat facilitated by the increased ionic strength. The explanation of this phenomena is that in the adsorption of 5-bromo-2-acetyl thiophene the negative pole of the dipole molecule (at the sulfur atom) approaches the negatively charged mercury surface.

$\log\left[i_{lim}/(i_D - i_{lim})\right]$

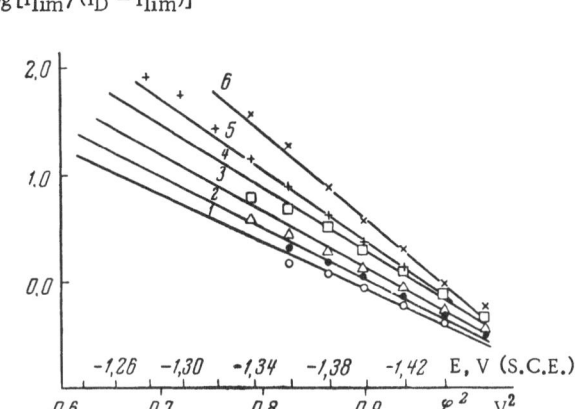

Fig. 41. The effect of solution ionic strength on the decrease of
the first reduction wave for bromoacetyl thiophene in alkaline solution (KOH + KCl). For ionic strength see Fig. 39.

A small, but still significant increase in adsorption with increasing ionic strength of the solution is indicated by a certain
shift toward negative potentials of the lines (Fig. 41) in φ^2 vs.
$\log[i_{lim}/(i_D - i_{lim})]$ coordinates. These lines express the shape
of the decrease on the waves. As would be expected from the
Frumkin equation (42), these lines become more curved with increasing KCl concentration (see Fig. 41). This means that the desorption becomes more pronounced with increasingly negative potentials in solutions of higher ionic strength.

An increase of the first wave on the polarographic curves of
5-bromo-2-acetyl thiophene at increasing ionic strength of the alkaline solution (see Fig. 39), as the experimental data show, is
caused mainly by the shift of the wave into a region of more positive potentials. Under these conditions the adsorptivity of the unprotonated 5-bromo-2-acetyl thiophene form is significantly higher
and the limiting kinetic current reaches the diffusion current value.
The increase of the adsorptivity for 5-bromo-2-acetyl thiophene
with increasing ionic strength is relatively small. The result of
this increased adsorptivity is only a small shift of the waves at the
point where the decrease begins toward negative potentials. It is
interesting to note that when the height of the first wave approaches

the diffusion current values, a first-order maximum develops on the wave. This maximum is absent on polarograms if the kinetic current is significantly smaller than the diffusion current.

The kinetic current of the first wave in alkaline bromoacetyl thiophene solutions increases and the current approaches the diffusion current value [671] with increasing drop period (t) of the electrode [with an electrode with an adjustable glass disk which made it possible to change t at almost constant mercury flow rate (m)]. This is not related in any way to the degree of the approach to equilibrium (which can be evaluated under identical conditions, but in absence of the electrode process), which in this case is obviously close to unity. This is indicated by the absence of a significant effect of t on the straight line, which expressed the decrease of the first wave in φ^2 vs. $\log{[i_{lim}/(i_D - i_{lim})]}$ coordinates. The increase of the wave is caused by its shift to less negative potentials, where the adsorptivity of 5-bromo-2-acetyl thiophene is larger. The $\Delta E_{1/2}/\Delta \log t \approx$ 72 mV value is close to one half of the reciprocal slope of the logarithmic plot of the initial portion of the first wave. This value corresponds to a theoretical value for irreversible and quasidiffusion waves with almost completely established adsorption equilibrium (see the following section).

With increasing t the slope of the wave increases slightly (the reciprocal slope value is 144 mV^{-1} at t = 0.3 sec and 120 mV^{-1} at t = 2.4 sec) due to the shift to more positive potentials (see the following section).

It is interesting to note the unusual polarographic behavior of 4,5-dibromo-2-acetyl thiophene in alkaline medium. The first wave for this compound corresponds to the electrochemical cleavage of the bromine in the α position, which results in formation of 4-bromo-2-acetyl thiophene. The bromine in this compound cannot be removed electrochemically [671]. The presence of bromine in the β position shifts the half-wave potential of the first wave for 4,5-dibromo-2-acetyl thiophene by about +110 mV compared to that for 5-bromo-2-acetyl thiophene. Furthermore, the half-wave potential for 4,5-dibromo-2-acetyl thiophene ceases to be as pH-dependent at lower pH than the wave for 5-bromo-2-acetyl thiophene [671]. These two factors result in a peculiar shape for the first wave for 4,5-dibromo-2-acetyl thiophene in alkaline

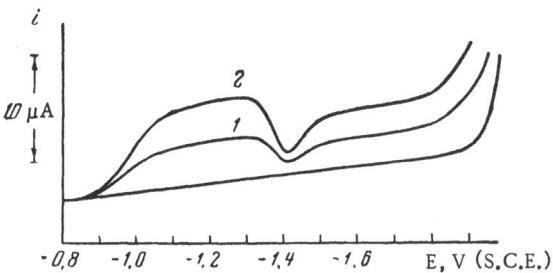

Fig. 42. Polarographic curves of (1) 0.12 mM and (2) 0.23 mM
4,5-dibromo-2-acetyl thiophene in a buffer solution at pH 11.0.

medium (Fig. 42). The wave has a major plateau corresponding to
the pseudodiffusion (with n = 2) limiting current, after which a decrease is caused by desorption and diminished protonation rate of
the carbonyl. Then a new wave can be observed, corresponding to
the single-electron reduction of the keto group. This wave does
not begin at the diffusion-current level of the first wave, but rather
at the limiting current of the wave that drops at sufficiently negative potentials almost to the level of the background current. The
second wave in this case corresponds to the reduction of the 4,5-
dibromo-2-acetyl thiophene carbonyl, and not 4-bromo-2-acetyl
thiophene, because it occurs in less alkaline solutions where no decrease can be observed on the first wave, and the reduction of the
keto group precedes the split of the C − Br bond in the α position
[671].

Decreases on kinetic surface waves become less pronounced
and even disappear entirely with increasing ionic strength of the
solution if anions take part in the antecedent reaction. Thus, for
example, on addition of sodium chloride to a sodium hydroxide solution, the decrease on the kinetic "prewave" for ninhydrin (anions
take part in the antecedent reaction) becomes less deep, and for a
4 M NaCl solution it disappears entirely (see Fig. 6 in [679]). In
exactly the same way, the decreases on the first wave for phthalimide solutions becomes less deep with increasing ionic strength in
the pH interval from 5.6 to 8.0; the kinetic wave itself becomes
slightly higher [673]. The higher adsorptivity of phthalimide anions
with increasing ionic strength of the solutions is indicated by the
shift of the apparent ("polarographic") dissociation curve toward

higher pH values: if the ionic strength increases from 0.2 to 2.0 M the pK' value increases by 0.5 units [673]. A decrease on the kinetic surface wave (limited by the protonation rate of anions) can be observed in the polarography of phenolphthalein in weakly alkaline solutions [674]. It is interesting to note that this decrease disappears if a buffer solution consisting of pyridine and its chloro-hydrate is used; but if potassium chloride is added to this solution the decrease again appears on the phenolphthalein wave [674]. The disappearance of the current decrease in pyridine-containing buffer solutions is most probably connected with a significant decrease of the absolute ψ_1-potential value in the presence of pyridinium ions and also with the presence of a proton donor (pyridinium ions) in the outer shell of the double layer. The reappearance of the decrease on the phenolphthalein wave on addition of potassium chloride to the solution is caused by displacement of pyridinium from the electrode layer by potassium ions. It must be mentioned that the decrease on the phenolphthalein wave becomes less deep at higher temperatures [674]. This is most probably caused by a larger temperature effect of the antecedent reaction rather than the adsorption of the electrochemically inactive phenolphthalein form.

An unusual phenomenon was observed by O. Manoušek and P. Zuman in the polarography of a complicated compound, a pyridylthiazolidine derivative (a product of condensation of pyridoxal with cystein), in acid solution. The first wave of this compound decreases with increasing pH and a decrease was formed on the wave. The half-wave potential of the first wave for this compound became more negative with increasing pH ($\Delta E_{1/2}/\Delta pH \approx$ −60 mV). Consequently, the electrochemical stage was preceded by a surface protonation of the particles (the wave is close to the zero charge potential). It can be assumed that the decrease of the kinetic current at higher pH values is connected with smaller adsorptivity of particles if they acquire positive charges by protonation on one of the nitrogen atoms. Furthermore, since the protonated nitrogen does not form a part of the electrochemically active center (the first wave corresponds to a C−S bond cleavage), the addition of a positive charge to the particle also can make the protonation at the electrochemically active group more difficult.

E. QUASIDIFFUSION SURFACE KINETIC WAVES AND ELECTRODE PROCESSES WITH ADSORBED DEPOLARIZER WITHOUT ANTECEDENT REACTIONS

If the rate of the antecedent surface reaction is sufficiently high, the i_{lim} value of the kinetic wave is close to i_D. Under these conditions the wave has a well-defined limiting current plateau, and the wave height is determined by the Ilkovič equation. However, the shape of such a quasidiffusion surface wave and the dependence of its half-wave potential on various factors does not follow the relationships that are valid for normal irreversible polarographic waves or for volume quasidiffusion waves (see section A-3 in Chapter IV). We will discuss the specific characteristics of quasidiffusion surface waves.

If the adsorption of the inactive depolarizer form follows the Langmuir isotherm, then at $\beta c \ll 1$ and $y \to 1$ the shape of the quasidiffusion surface wave can be described by Eq. (123), in which the i_{lim} value must be replaced by the nearly equal i_D value. The half-wave potential can be calculated using Eq. (124). From Eq. (123), it follows that the slope of quasidiffusion surface waves in semilogarithmic coordinates is not $b = 2.3RT/\alpha n_a F$, which value is characteristic for the usual irreversible processes or for quasidiffusion volume processes, but is significantly higher. As follows from Eq. (122), the b' value decreases when the half-wave potential is different from the potential of maximum adsorption (E_M). Therefore, the slope of quasidiffusion waves $bb'/(b'-b) = b/(1 + 0.87ab\varphi_{1/2})$ must increase if the half-wave potential shifts from E_M to more negative potentials (for example, on increase of the pH). This was often observed experimentally (for example, in [484, 492, 672]) but no explanation was given [669].

We will discuss the dependence of $E_{1/2}$ $(= \varphi_{1/2} + E_M)$ for surface quasidiffusion waves on various factors. It follows from Eqs. (115) and (124) that if expression (112) can be applied

$$\varphi_{1/2} = \frac{bb'}{b'-b}\left(\log\frac{0.81k'_{el}\,t^{1/2}}{\sigma_s D^{1/2}} + \log\beta^0\Gamma_\infty + 0.43a\varphi_{1/2}^2\right). \tag{125}$$

If we compare Eqs. (125) and (71), we find that besides a different multiplication factor preceding the logarithm, Eq. (125) contains two more terms in addition to Eq. (71), which is correct for the half-wave potentials of quasidiffusion volume waves. These additional terms take into consideration the change of depolarizer adsorptivity with the electrode potential. They result in certain characteristic properties of surface quasidiffusion waves. At increasing pH, due to increased σ_S values [see Eq. (26)], the first term in the parenthesis of Eq. (125) decreases. The decrease of this term shifts the $\varphi_{1/2}$ value to negative potentials by $(bb'/b' - b)$ · ΔpH. But the total shift of $\varphi_{1/2}$ toward negative potentials is somewhat smaller because, at the same time, the third term in the parentheses of Eq. (125) increases. The relative shift of the $\varphi_{1/2}$ value of the wave with increasing pH, caused only by the change of the third term, increases for higher $\varphi_{1/2}$ values. This means that the shift becomes larger if the potential at which this wave occurs differs more from the maximum adsorption potential. Thus, the $\Delta\varphi_{1/2}/\Delta pH$ value of quasidiffusion surface waves usually is smaller than the value of the slope of the logarithmic plot of the wave, which is $bb'/(b' - b)$, and the difference between the $\Delta\varphi_{1/2}/\Delta pH$ and $bb'/(b' - b)$ values increases as the potential region in which the wave can be observed becomes more distant from the E_M value.*

As an example, the deviation of $\Delta E_{1/2}/\Delta pH$ from the slope value on logarithmic wave plots for several nitro compounds observed by Stradyn', Giller, and Yur'ev [683, 684] can be cited. The reduction of these compounds is preceded by surface protonation. The correctness of this conclusion is supported by the greatest deviations being observed for strongly adsorbed nitro compounds, such as thiophene and selenophene derivatives [684]. Holleck and Kastening [119] pointed out the surface character of the antecedent protonation in the reduction of certain nitro compounds.

Equation (125) can be given in a slightly different form, which permits a comparison with the corresponding equation for volume quasidiffusion waves [Eq. (71)]

---

*All this is valid only as long as on the electrode surface a sufficient amount of adsorbed depolarizer remains and the process is entirely a surface process.

$$\varphi_{1/_{2}} = b \left(\log \frac{0.81 k'_{e1} t^{1/_{2}}}{\sigma_{s} D^{1/_{2}}} + \log \beta^{0} \Gamma_{\infty} - 0.43 a \varphi^{2}_{1/_{2}} \right).$$ (125a)

To obtain Eq. (125a), the bb'/(b' − b) value in Eq. (125) was replaced by the equivalent $b/(1 + 2 \cdot 0.43 ab\varphi_{1/2})$ value, and a few simple algebraic operations were made. It must be mentioned that the b factor before the parenthesis on the right-hand side of Eq. (125a) does not represent the observed but rather the true slope of the wave, which means the slope that can be observed for the given electrochemical process in absence of the adsorbed depolarizer effect (on its concentration at the electrode) at a constant α value. A comparison of Eqs. (71) and (125a) shows that at equal b values with increasing pH, $\varphi_{1/2}$ shifts more for surface quasidiffusion waves than the half-wave potential for volume waves [the absolute value of the third term in the parentheses of Eq. (125a) increases, which results in an additional shift of $\varphi_{1/2}$ toward negative potentials]. The additional shift of $\varphi_{1/2}$ of surface waves compared with the half-wave potential shift of volume waves with increasing pH can be explained by the smaller depolarizer adsorptivity, caused by the shift of the wave from the potential corresponding to maximum adsorptivity. This also results in a smaller overall rate of the electrode process. Thus, although the $\Delta\varphi_{1/2}/\Delta pH$ value for surface waves is smaller than the slope of their plot in semilogarithmic coordinates, the absolute value of $\Delta\varphi_{1/2}/\Delta pH$ for surface quasidiffusion waves is larger than the $\Delta E_{1/2}/\Delta pH$ value for volume waves.

Holleck [601] pointed out, as early as 1956, the adsorption of the depolarizer or reaction product as a potential cause for the deviation of the $\Delta\varphi_{1/2}/\Delta pH$ value from −60 mV/pH unit.

It must be mentioned that, due to the difference of pH values in the bulk of the solution $[(pH)_{0}]$ and at the electrode surface $[(pH)_{S}]$, caused by the effect of the electrode field, if the pH increases and the half-wave potential shifts to negative potentials, the inequality $\Delta(pH)_{S} < \Delta(pH)_{0}$ is valid. This can be the reason for the smaller half-wave potential change for surface waves with $(pH)_{0}$ than is predicted by the theory. The $\Delta(pH)_{S}/\Delta(pH)_{0}$ value is determined by the ψ_{1}-potential change caused by increasing cathodic polarization of the electrode. Due to a certain change of ψ_{1} [and also a change of $(pH)_{S}$] in the potential limits of a polarographic wave, the steepness of the curve can change (usually only to a very small extent) for waves corresponding to processes with antecedent surface protonation.

If the pH decreases and approaches the pK_S value of the depolarizer, the shift of half-wave potentials with pH decreases for surface waves. To describe the dependence of half-wave potentials (or $\varphi_{1/2}$) on pH under these conditions, the σ_S value in Eq. (125) must be replaced by the expression for the concentration of the inactive depolarizer form from Eq. (73). In the latter expression, K must be replaced by the dissociation constant K_S of the same depolarizer present in the adsorbed state (the K_S value is affected by the double-layer structure − see section H, Chapter V). Finally, $[H^+]$ must be replaced by $[H^+]_S$. Furthermore, if $(pH)_S \rightarrow pK_S$ the proportion of the protonated depolarizer increases sharply. Due to this change the overall adsorptivity of the depolarizer changes, since the adsorptivities of the protonated and unprotonated forms are different. From the dependence of the half-wave potential for surface quasidiffusion waves on pH, the K_S value cannot be determined directly (without considering the change of adsorptivity and K_S), as was possible for volume quasidiffusion waves (see page 123).

If $(pH)_S \ll pK_S$, the quasidiffusion surface wave changes into a purely diffusion wave, but with depolarizer adsorption. The pH in this case no longer affects the potential of the half wave, and Eq. (125) can be given as

$$\varphi_{1/2} = \frac{bb'}{b'-b} \left(\log \frac{0.81 k_{el}' t^{1/2}}{D^{1/2}} + \log \beta^0 \Gamma_\infty + 0.43 a \varphi_{1/2}^2 \right), \qquad (126)$$

in which a, β^0, and Γ_∞ correspond to the adsorption characteristics of the depolarizer itself. Equation (126), and also the expression for the wave form, which is identical to Eq. (123), can be obtained directly from Eqs. (6a), (42), (69a), (112), and (113a).

If the degree of the approach to adsorption equilibrium is small for the depolarizer,* then, as was already pointed out earlier in the derivation of equations for the wave form and for

─────────────

* During an electrochemical process in which the adsorbed depolarizer is consumed, the degree to which adsorption equilibrium is established is estimated from its value in the absence of consumption of the adsorbed material.

the half-wave potential, the more general Eq. (110) must be used instead of Eq. (112). It can be easily shown that the equation for the half-wave potential will contain, in this case, an additional term which is dependent on the drop time. This term will take into consideration the change of adsorbed material quantity with time. With increasing drop time of the capillary, the half-wave potential of such surface quasidiffusion waves must become more positive. The same also occurs for volume quasidiffusion waves and for surface waves at y → 1 [see Eqs. (10), (71), and (126)]. But the $\Delta\varphi_{1/2}/\Delta\log t$ value must be larger for surface waves if the adsorption equilibrium is not established. The value obtained must be larger than $\frac{1}{2}bb'/(b'-b)$. This is the case for the first reduction waves on 5-bromo-2-acetyl thiophene polarograms at low concentrations of this compound. In a borate buffer solution at pH 9.3 the half-wave potentials of these waves becomes more positive by 73 mV, if the t value is increased tenfold, as long as the slope (average) of the semilogarithmic wave plot, which is equal to $bb'/(b'-b)$ is 135 mV [671].

If the approach to the adsorption equilibrium is very small, and if the quantity of adsorbed material is independent of potential, Eq. (111) can be applied. The wave form can be expressed by the usual equation, valid for irreversible, diffusion-limited processes [Eq. (7)], and the half-wave potential is expressed, as it can easily be shown from Eqs. (69a), (111), and (113a), by

$$E_{1/2} = b\log 0.60k'_{el}t. \tag{126a}$$

It follows from Eq. (126a) that the half-wave potentials of waves of systems, where the approach to adsorption equilibrium does not favor the depolarizer adsorption is strongly dependent on the drop time of the electrode. For such, waves, $\Delta E_{1/2}/\Delta\log t = b$. carbon–halogen

Until now we have discussed only those cases where the adsorption followed the Langmuir adsorption isotherm (more exactly, Henry's law) and the surface coverages were very low. But under conditions when quasidiffusion waves can be observed, the coverages can be substantial, and the S-shape of the Frumkin isotherm begins to show up. This isotherm is characteristic for the adsorption of the majority of compounds (at very low coverages when

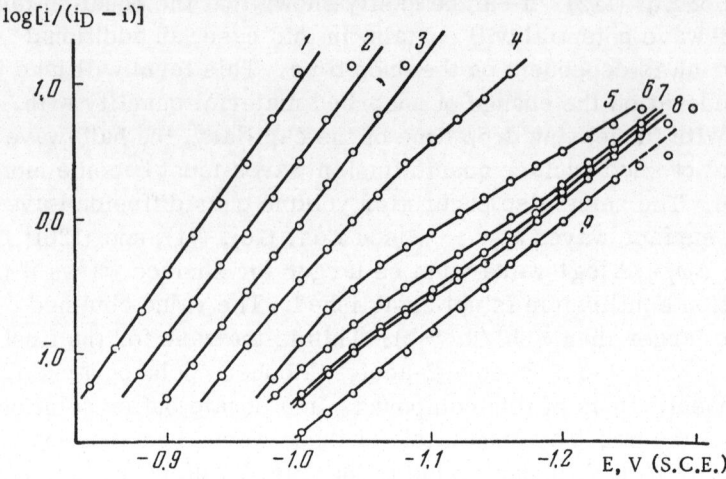

Fig. 43. The quantity $\log [i/(i_D - i)]$ as a function of E for the wave in 0.102 mM 5-bromo-2-acetyl thiophene solution for the following pH values: (1) 5.70; (2) 6.16; (3) 6.43; (4) 7.08; (5) 7.59; (6) 8.25; (7) 8.62; (8) 9.07; (9) (0.1 M KOH + 0.1 M KCl).

kinetic waves with decreases can be observed, the adsorption follows the linear Henry's law and the S-shape of the isotherm does not appear). If the adsorption isotherm has an S-shape, Eqs. (121) to (124) cannot be used for quantitative calculations, but the general character predicted by these equations for quasidiffusion waves remains valid. Figure 43 shows, as an example, the function $\log [i/(i_D - i)]$ vs. E for 5-bromo-2-acetyl thiophene in a 0.102-mM solution at several pH values in acetate−borate buffer solutions at constant ionic strength $\mu = 0.2$ (the concentration in the final solution was: 0.1 M CH_3COOK, 0.1 M KH_2BO_3; HCl was added to obtain the given pH). Up to a pH value of 7.0 the semilogarithmic graphs are linear. The lines shift with increasing pH to negative potentials and their slopes increase significantly [in qualitative accord with Eqs. (121)-(124)]. At pH $\geq$ 7.0 an inflection appears. The upper portion of the curve has a larger slope than the lower portion. At pH $\geq$ 8.3, a second inflection can be observed. The slope of the top portion of the curve increases in comparison with that of the central section. In strongly alkaline solutions the graph becomes linear with a slope close to the slope of the upper portion

of the wave at pH = 8.3. This character of the change of the logarithmic graphs of the wave can be explained by the S-shape of the
adsorption isotherm for the unprotonated depolarizer form. In
acid solutions, where the potential is close to E_M, the adsorptivity
is very high and the coverage also is high. In this case almost all
particles transported to the electrode by diffusion are adsorbed.
In alkaline solutions, as the cathodic potential increases, the adsorptivity is diminished. The maximum change in adsorptivity
with potential corresponds to the rise of the S-shaped isotherm,
which can be found on the middle portion of lines 4 to 8 (with
maximum slope) in Fig. 43. At even higher negative potentials
(and high pH) the coverage becomes so small that the adsorption
corresponds to the linear portion of the isotherm [which means
that the adsorption can be described by Henry's equation (41)], and
the slope of the wave can be given by Eq. (123).

Inflections on logarithmic wave plots are encountered quite
frequently. Thus, for example, P. Elving and E. Olson observed
relationships very similar to those given in Fig. 43 in the reduction of aromatic nitrosohydroxylamines (see Fig. 3 in [486]).

It can be assumed [480] that the observed increase of the
slope for nitromethane at increasing pH of the solutions in the
work reported by Suzuki and Elving [685] (and consequently the
shift of the potential region, in which the wave can be observed, to
negative values) also is connected with the desorption of nitromethane. At pH > 5.0, the slope reaches its maximum value [685],
and it can be considered that, in this case, due to the nitromethane
desorption, the antecedent protonation (which seems to take place
up to pH 8.0) actually takes place only in the solution and the observed αn_a = 0.97 value [685] corresponds to the true value. In
acid solutions, where the reduction occurs at less negative potentials, some adsorption of nitromethane is possible, and as the
negative potential increases (in the limits of the wave section corresponding to the current change) the quantity of adsorbed material decreases. This results in a reduced protonation rate, an increase in the proportion of the unprotonated form of the reduced
material, and a shift of the upper portion of the wave toward more
negative potentials. The slope of the wave becomes smaller and
the apparent value of αn_a also becomes smaller.

In strongly acid solutions, an inflection can be observed at
$E \approx -0.65$ V (S.C.E.) on the logarithmic graphs of nitromethane
waves. The upper part of the wave has a larger slope than the
lower part [685]. This again can be explained by the effect of
nitromethane desorption, which begins to be significant only at
0.65 V.

That the nitromethane adsorption indeed plays a role in its
reduction is shown by the unusual change of the wave with temperature. The slope of the wave for irreversible processes, which
is equal to $b = 2.3RT/\alpha n_a$, must change very little with temperature and the $\Delta E_{1/2}/\Delta t°$ value must be positive. The experimental
data show that for the nitromethane wave at pH 4.3, $\Delta E_{1/2}/\Delta t°$ is
negative. The slope of the wave increases at higher temperatures
[685]. These observations can be explained [480] by the desorption of nitromethane at higher temperatures and by the decrease
in its protonation rate which causes, as already pointed out, a
shift of the half-wave potential to the more negative potentials and
an increase of the steepness of the wave or an increase of the observed αn_a.

It was already mentioned that electrode processes with depolarizer adsorption but without antecedent reaction essentially
follow the same relationships as quasidiffusion waves at $pH_S \ll pK_S$.
Thus, the change of half-wave potentials (or $\varphi_{1/2}$ for waves with
depolarizer adsorption but without antecedent protonation follows
Eq. (126). If the adsorption proceeds in accord with the Frumkin
isotherm, the increase of the effective β^0 value with increasing
coverage of the electrode surface by adsorbed depolarizer must
be taken into consideration [by Eq. (40)]. It follows from Eq. (126)
that at increasing adsorptivity of the compound [with increasing
β^0 value), the half-wave potential of the reduction becomes more
positive. Indeed, a significant shift of the half-wave potential of
the reduction wave has been observed for α-bromocarboxylic acids
(under such conditions, when the antecedent protonation cannot
take place) [686], for esters [686, 687], and also for α-bromoalde-
hydes [688] with increasing hydrocarbon chain length and branching. Elving and co-workers [686] first called attention to this effect and correctly assumed that the process is connected with increased depolarizer adsorption.

Also related to the depolarizer adsorption, as was mentioned by Elving and co-workers [689, 690], are deviations from correlations between the half-wave potentials of waves for certain organic compounds and other characteristics of these same compounds. Examples are the relationship to the Hammett and Taft constants and also to the chemical shift of nuclear magnetic resonance spectra.

With increasing chain length, the slope of the waves of α-bromocarboxylic acid esters increases significantly [687]. The reduction of α-bromocarboxylic acid esters takes place at potentials of -0.7 to -1.0 V, which are potentials favorable for their adsorption. With increasing chain length the adsorptivity of the compounds increases under these conditions, and the half-wave potential becomes less negative (or the absolute value of $\varphi_{1/2}$ becomes smaller); the b' value, according to Eq. (122), increases sharply and thus the $bb'/(b' - b)$ value decreases, tending toward the b value, which means that the slope of the wave increases.

For electrode processes taking place under conditions such that depolarizer adsorption practically does not occur, an increase of the hydrocarbon chain length and its branching results in a shift of the half-wave potential of the wave to negative potentials due to the steric factor and induction effect. Such a shift can be observed, for example, in the reduction of iodomethyl trialkyl silanes [40] and bromoalkanes [691]. The reduction of these compounds takes place at such negative potentials that the depolarizer adsorption does not appear. The half-wave potential in this case can be expressed by Eq. (10), which is applicable to irreversible processes without antecedent chemical reaction and depolarizer adsorption.

Usually the half-wave potential is independent of the depolarizer concentration. But if the depolarizer adsorption follows Frumkin's S-shaped isotherm, then the effective β^0 value increases with increasing depolarizer concentration and the half-wave potential (or $\varphi_{1/2}$) must shift to positive potentials, according to Eq. (126). Such a dependence of the half-wave potential on depolarizer concentration was observed, for example, by Feoktistov and Zhdanov [692, 693] for the first wave of β-iodopropionitrile. This wave appears at a potential of about -1.0 V, in the potential region favorable for adsorption. Therefore, the shift of the wave toward positive potentials with increasing β-iodopropionitrile concentration is most probably caused by the S-shape of its adsorption isotherm. The β-iodopropionitrile adsorption also is indicated by a significant shift of the half-wave potential for its first reduction wave with increasing drop time. The $\Delta E_{1/2}/\Delta \log t$ value, according to data given by Feoktistov [693], is significantly higher than one half the slope on the logarithmic plot of the wave. The S-shaped adsorption isotherm for Δ-iodopropionitrile explains, most probably, the marked increase in current intensity on the beginning portion of the first wave, plotted as a function of time during the life of the drop. Its value is proportional to the 0.75 power of the time [693], while usually this exponent for currents at the base of the wave is equal to 0.67.

If the adsorption follows the Langmuir adsorption isotherm, and if the establishment of the adsorption equilibrium is very fast (which can be proved in the absence of the electrode process) but yet the surface coverage by adsorbed material is not too high ($\theta < 1$), Eq. (126) becomes

$$\varphi_{1/2} = \frac{bb'}{b'-b}\left(\log\frac{0.81 k' e1^{t^{1/2}}}{D^{1/2}} + \log\frac{\beta^0\Gamma_\infty}{1 + \beta^0 c e^{-a\varphi_{1/2}^2}} + 0.43 a\varphi_{1/2}^2\right). \tag{127}$$

Equation (127) shows that the dependence of $\varphi_{1/2}$ on $\log t$ is given by a straight line, and the slope of the lines is equal to one half of the $bb'/(b'-b)$ value. If the adsorption follows the S-shaped Frumkin adsorption isotherm, then additional multipliers must be introduced into Eq. (127); these multipliers determined the dependence of β on θ and γ [see Eq. (40)].

F. DIFFERENCE IN THE PROPERTIES OF DEPOLARIZER FORMED AT THE ELECTRODE AND TRANSPORTED FROM THE SOLUTION (EFFECT OF SUCCESSION IN MULTISTAGE ELECTRODE PROCESSES)

Stable intermediate products of multistage electrode processes usually have electrochemical properties identical to those of the same materials if they are directly added to the solution. Thus, for example, the second wave on the reduction polarograms

for carbon tetrachloride is practically identical with the chloroform wave measured under the same conditions.

But in several cases a different electrochemical behavior
was observed among these depolarizer types. This is caused by
certain differences that manifest themselves during a given time
interval (usually very short) between the compounds formed on
the electrode and the same compounds present in the solution
[694]. One of the causes of these differences is the so-called
"carry-over effect," in which the electrode product retains to a
certain degree ("carries") the properties of the starting compound
which entered in the electrochemical reaction. Thus, the product
of the electrode reaction can retain the orientation of the starting
compound, on the electrode surface, its adsorptivity, its degree of
solvation, and some other properties. Apparently, taking over the
orientation of the starting material causes a shift of the half-wave
potentials (by several tens of millivolts) of the subsequent wave of
certain polybromothiophenes compared with the half-wave potential
of the first waves for the corresponding bromothiophenes with a
smaller number of bromine atoms [695]. Stackelberg and Stracke
[468] report that methyl iodide gives a wave with $E_{1/2}$ = −1.63 V,
whereas the second wave of methylene iodide and the third wave
of iodoform show a half-wave potential of −1.53 and −1.50 V, respectively. Several other examples are given in this work [468].

The data of Levin and Fodiman [575] show that the half-wave
potentials of the second waves of isomeric trichlorobenzene polarograms (corresponding to the reduction of dichlorobenzenes) are
close to each other and can be found between −2.52 and −2.53 V.
At the same time, dichlorobenzene isomers give waves with half-
wave potentials ranging from −2.55 to −2.59 V.

In the polarography of α-halide aldehydes two waves can be
observed: the first wave corresponds to the carbon−halogen bond
cleavage and the second to the reduction of the carbonyl group of
the aldehyde formed [208]. If we compare the second waves on the
polarograms of α-halide aldehydes with the waves of the aldehydes
(for example, the second waves of α-bromo or α-chloroacetalde-
hydes with the reduction of acetaldehyde itself), we see that the
half-wave potentials of these waves are very close to each other,
but the limiting currents are significantly different. The wave on
the aldehyde polarogram is much higher than the second waves for

the α-halide aldehydes [208]. The explanation of this phenomenon is that the carbonyl group of the α-halide aldehyde is much more hydrated than that of the aldehyde itself. Therefore, the aldehyde formed on the electrode from its halide derivative retains for a certain time period the hydration envelop of the α-halide aldehyde [208]. This diminishes the height of the $C = O$ group reduction wave since this wave is limited by the dehydration rate (see the section on page 34 and following). With an increasing number of halide atoms in the α-position of the aldehyde molecule, the hydration of the carbonyl increases [209], and the retention of the degree of hydration of the starting material by the product reduces even further the kinetic reduction wave of the carbonyl group [209].

The differences between the products of electrode reactions and of the same compounds present in the solution also can be caused by a decreased protonation rate of anions formed by proton transfer to the starting compound. This phenomenon is frequently observed in media with small proton-donor activity. The reduction of subsequent waves on naphthalene polysulfonic acid polarograms investigated by Levin [172] may be mentioned as an example (see page 33). Thus the carry-over effect can be affected both by the electron transfer rate and by the course of the chemical reactions at the electrode.

The carry-over effect must be kept in mind in comparing waves on polarograms of compounds investigated with those of model compounds. This method is often used to determine the mechanism of complicated electrochemical processes.

Chapter VII

Mixed Volume—Surface Waves

In several instances the antecedent chemical reaction takes place at comparable rates both at the surface of the electrode (with participation of adsorbed particles) and in the reaction-layer volume. The observed kinetic current is therefore a mixed, surface—volume current. Usually, the condition for appearance of mixed kinetic currents is a relatively low adsorptivity of the electrochemically inactive depolarizer form (at the given potential). Rigorous development of equations giving the dependence for mixed volume—surface currents from the determining factors raises serious mathematical problems.

An investigation of the mixed kinetic limiting current was first reported in [696]. It was shown that the apparent protonation rate constant for maleic acid monoanions increases with increasing drop period if in the calculation using Koutecký's equation (see page 16), the surface current component was disregarded. This effect was explained by the increasing quantity of electrochemically inactive depolarizer adsorbed on the electrode (in this case [696], maleic acid monoanions) with increasing drop period and, consequently, also with the time of adsorption. In this work [696], the effect of adsorption of the inactive depolarizer form on its concentration in the reaction-layer volume was not considered. Furthermore, it was incorrectly assumed that the observed effective rate is simply the sum of two independent components — the volume and surface rates.

On the basis of an approximate theory of depolarizer adsorption on the electrode under conditions of the electrode process which was developed by Levich, Khaikin, and Belokolos [670], equations were derived for the limiting volume—surface kinetic current. These equations can be given for the current averaged over the drop period and for the instantaneous current in the following form:

$$\frac{i_{\lim}}{i_D - i_{\lim}} \frac{1}{t^{1/2}} = \frac{0.46\,(\rho_s\beta\Gamma_\infty + \rho^{1/2}\,D^{1/2}\,\sigma^{-1/2})\,t^{1/2}}{\beta\Gamma_\infty + 0.60 D^{1/2} t^{1/2}}, \tag{128}$$

$$\frac{i_{\lim}}{i_D - i_{\lim}} \frac{1}{t^{1/2}} = \frac{0.86\,(\rho_s\beta\Gamma_\infty + \rho^{1/2}\,D^{1/2}\,\sigma^{-1/2})\,t^{1/2}}{\beta\Gamma_\infty + 0.74\,D^{1/2} t^{1/2}}. \tag{129}$$

In these equations, ρ_S is the total protonation rate constant of the compound present in the adsorbed state; the t value in Eq. (128) is the drop period and in Eq. (129) it is the time elapsed from the beginning of the drop formation.

If we assume that in Eq. (128) $\beta\Gamma_\infty = 0$, which means that the electrochemically inactive compound is not adsorbed on the electrode, Eq. (128) simplifies exactly, with the exception of the numerical coefficient, to Eq. (25), which was derived for kinetic volume currents. The numerical coefficients in Eqs. (25) and (128) at $\beta\Gamma_\infty = 0$ deviate only by a little more than 10%, and in Eqs. (24) and (128) by only about 4%. This is due to the approximate character of Eq. (128). But the deviation is so small that Eq. (128) can be used for many practical calculations.

A very important conclusion can be drawn from Eqs. (128) and (129): if the $[i_{\lim}/(i_D - i_{\lim})](1/t^{1/2})$ function depends on the drop period of the electrode or depends on the time elapsed from the beginning of drop formation, respectively, one can assume that the electrode process is complicated by adsorption phenomena. Furthermore, if this function increases with time, the adsorption of the electrochemically inactive depolarizer form on the electrode is indicated.

But, if the $[i_{\lim}/(i_D - i_{\lim})](1/t^{1/2})$ function is independent of time, as follows from Eq. (128), it does not mean that the inactive form of the compound is not adsorbed on the electrode. This independence can also indicate purely volume character for the antecedent reaction, where Eq. (25) is valid and also a purely surface or volume surface character for the chemical reaction that takes place under such conditions that the adsorption process is close to its equilibrium state and the compound is not consumed in the reaction.

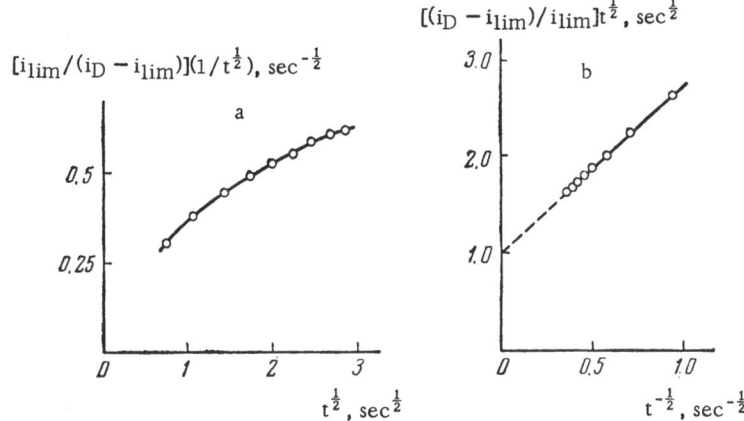

Fig. 44. (a) The dependence of $[i_{lim}/(i_D - i_{lim})](1/t^{\frac{1}{2}})$ on $t^{\frac{1}{2}}$ and (b) the dependence of $[(i_D - i_{lim})/i_{lim}]t^{\frac{1}{2}}$ on $1/t^{\frac{1}{2}}$ for instantaneous currents for the second wave of maleic acid in acetate buffer solution at pH 5.9.

Thus, the curve corresponding to the $[i_{lim}/(i_D - i_{lim})](1/t^{1/2})$ vs. $t^{1/2}$ dependence must pass through the origin of the coordinates. At low $t^{1/2}$ values $[i_{lim}/(i_D - i_{lim})](1/t^{1/2})$ increases proportionally to $t^{1/2}$; at higher $t^{1/2}$ values the increase of $[i_{lim}/(i_D - i_{lim})](1/t^{1/2})$ slows down, and at sufficiently high $t^{1/2}$ values this function tends to a limit, which is equal to $(0.77/D^{1/2})(\rho_s\beta\Gamma_\infty + \rho^{1/2}D^{1/2}/\sigma^{1/2})$.

In the most thoroughly investigated case of mixed currents, namely for the second wave on maleic acid polarograms in acetate buffer solutions (the wave height is limited by the protonation rate of the maleic acid monoanion), only the middle portion of the curve can usually be observed. This section can be considered to be linear to a first approximation. Figure 44a shows as an example the dependence of the $[i_{lim}/(i_D - i_{lim})](1/t^{1/2})$ function on $t^{1/2}$ for the instantaneous current. The data were determined oscillopolarographically on the "first drop" of a special electrode with a long drop period. This figure clearly shows the bending of the curve at large and small $t^{1/2}$ values.

If we use Eq. (128) [or, in the same way, Eq. (129)] in the form

$$\frac{i_D - i_{lim}}{i_{lim}} t^{1/2} = \frac{\beta\Gamma_\infty/t^{1/2} + 0.60D^{1/2}}{0.46(\rho_s\beta\Gamma_\infty + \rho^{1/2}D^{1/2}\sigma^{-1/2})} \tag{128a}$$

and we plot the $[(i_D - i_{lim})/i_{lim}]t^{1/2}$ value as a function of $1/t^{1/2}$, we obtain a straight line (see Fig. 44b) from which the $\beta\Gamma_\infty$ value can be determined. To obtain the $\beta\Gamma_\infty$ value, the slope of the line given in Fig. 44b must be divided by the ordinate intercept, and the result must be multiplied by $0.60D^{1/2}$. From the ordinate intercept of the line constructed by Eq. (128a), the $\rho_S\beta\Gamma_\infty + (\rho D/\sigma)^{1/2}$ value can be calculated.

Variables that affect the surface and the volume components of the current in a different way change the character of the $[i_{lim}/(i_D - i_{lim})](1/t^{1/2})$ vs. $t^{1/2}$ relationship. Thus, with increasing temperature, while the D and ρ values are becoming larger, the adsorptivity may decrease significantly, which means that the β value decreases. Under these conditions the value of the first term in the denominator of Eq. (128) decreases. Therefore, the $[i_{lim}/(i_D - i_{lim})](1/t^{1/2})$ value as a function of $t^{1/2}$ tends to its limit at small $t^{1/2}$ values. This was clearly observed in experimental data (see, for example, [697]). If at higher temperature the ρ_S value increases less than the β value decreases, then the first term in the numerator on the right-hand side of Eq. (128) decreases in comparison with the second term. This means that the proportion of the volume component of the limiting kinetic current increases.

On the basis of Eq. (128a) and graphs of the type given in Fig. 44b for average currents over the life of the drop, the $\beta\Gamma_\infty$ values were calculated [697] for maleic acid monoanions at a potential of -1.20 V (S.C.E.). The solution was an acetate buffer with pH 5.6, and the total concentration of CH_3COOH and CH_3COOK was 0.1 M. The $\beta\Gamma_\infty$ values at 25, 35, and 45°C were $4.1 \cdot 10^{-4}$, $2.8 \cdot 10^{-4}$, and $1.6 \cdot 10^{-4}$ cm, respectively. At 55°C and at higher temperatures, due to a larger scatter of data caused by experimental errors, the $\beta\Gamma_\infty$ value could not be determined with sufficient precision in this work [697].

Figure 45 shows the $\log(\rho_S\beta\Gamma_\infty + \rho^{1/2}D^{1/2}\sigma^{-1/2})$ values as a function of the reciprocal absolute temperature. As can be seen from the figure, the data fall on two straight-line sections. At low temperatures, the change of the $\rho_S\beta\Gamma_\infty + (\rho D/\sigma)^{1/2}$ sum is evidently determined by the change of the first term, because its value is larger at low temperatures than the second term. At high tem-

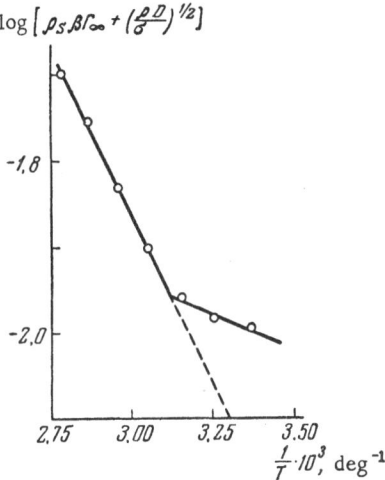

Fig. 45.   The quantity $\log(\rho_s \beta \Gamma_\infty + \rho^{\frac{1}{2}} D^{\frac{1}{2}} \sigma^{-\frac{1}{2}})$ as a function of $1/T$ for the protonation of maleic acid mono-anions in acetate buffer solution.

peratures, due to a decrease of β, the first term of the sum becomes much smaller than the second term, and the total activation energy corresponds to the second term and determines the slope of the line in Fig. 45 at high temperatures.

Thus, from the data obtained at high temperatures, the values of ρ or $(D\rho/\sigma)^{1/2}$ can be determined, and by extrapolating these values to low temperatures (dotted line in Fig. 45) the $\rho_s \beta \Gamma_\infty$ and, consequently, ρ_s, can be determined. In [697] the values determined by this method were given for maleic acid monoanions in acetate buffer solutions with an ionic strength of 0.5 N, for a total concentration of $CH_3COOH + CH_3COOK$, 0.1 M at pH 5.6 (25°C).

It must be mentioned that at pH > 5 the kinetic current begins to be affected by the second dissociation step of maleic acid: a part of this acid is present in the solution in the form of dianions. Koutecký [46] developed equations for the volume recombination limiting current of the acid under conditions where the effect of the second dissociation stage of the acid is already apparent. The equations were given for the case when the diffusion coefficients of the monoanions and undissociated acid molecule (D_1) are close to each other and the diffusion coefficient of the dianions (D_2) is different from D_1. The equations for the limiting kinetic current of the dibasic acid differs from Eq. (25) in that, in the expression for i_D, the D value is replaced by an averaged diffusion coefficient

$$D_{\mathrm{av}} = \frac{D_2 K_2 + D_1 [\mathrm{H^+}]}{K_2 + [\mathrm{H^+}]} .$$

The ρ_2 value calculated considering the second dissociation stage and the ρ_1 value calculated using Eq. (25), can be related to each other by the following equation:

$$\rho_1^{1/2} = \rho_2^{1/2} \frac{[\mathrm{H^+}]}{[\mathrm{H^+}] + K_2 \dfrac{D_2}{D_1}} \sqrt{\frac{[\mathrm{H^+}] + K_2}{[\mathrm{H^+}] + \dfrac{D_2}{D_1} K_2}} ,$$

where K_2 is the dissociation constant of the second stage for the acid (K_1 is included in the σ value: $\sigma = K_1 / [\mathrm{H^+}]$).

The D_{av} value does not have to be calculated for the wave limited in height by the rate of recombination of maleic acid monoanions if in Eq. (25) an i_D value corresponding to the total height of both waves on the polarograms of the solution investigated is used (the first wave corresponds to the kinetic current and the second to the reduction of maleic acid monoanions).

It should be mentioned that in the determination of ρ the effect of the double-layer structure on $[\mathrm{H^+}]_S$ and σ must be taken into consideration.

By constructing graphs of the type given in Fig. 44b on the basis of Eq. (128a), the $\beta \Gamma_\infty$ values were determined for maleic acid monoanions at several ethyl alcohol concentrations [697]. At 25°C, in an acetate buffer solution with a pH 5.15 (in water) containing 0.048 M CH_3COOH, 0.152 M CH_3COOK, and 0.348 M KCl, the $\beta \Gamma_\infty$ values for ethyl alcohol concentrations of 0, 10, 20, and 30 weight percent were $3.2 \cdot 10^{-3}$, $1.3 \cdot 10^{-3}$, $0.4 \cdot 10^{-3}$, and $0.2 \cdot 10^{-3}$ cm, respectively.

Chapter VIII

The Effect of Dimerization of Electrode Products on Processes with a Reversible Electrochemical Stage

A. THE EFFECT OF SUBSEQUENT CHEMICAL REACTION ON ELECTRODE PROCESSES

The effect of a chemical reaction following an electron transfer is apparent only when the rate of the electron transfer itself (in the forward and reverse direction) is very fast, which means processes with a reversible electrochemical stage. For processes with a moderate electrochemical reaction rate, the overall rate (in the absence of diffusion limitations) is determined by the rate of electron transfer, and the subsequent chemical reaction does not affect the kinetics of the electrode process at all.

Let us discuss the process

$$A \underset{\leftarrow}{\overset{el}{\rightleftharpoons}} B \overset{k}{\to} C, \qquad \text{(XXIV)}$$

in which the product of a reversible electrochemical reaction (B) is converted irreversibly into an electrochemically inactive compound (C).

The electrode potential (E) in the system $A \rightleftharpoons B$ is determined by the ratio of the concentrations of the A and B compounds at the surface. If the rate of the $B \to C$ reaction is of the same order as the rate of removal of B from the electrode by diffusion or exceeds this rate, the effect of the subsequent chemical reaction becomes apparent in a shift of E toward potentials at which the starting material A is present in excess. But the subsequent reaction will not affect the limiting current of the wave.

Koutecký [59], Kern [48, 698, 699], and Kivalo [700] discussed electrode processes with subsequent first-order chemical
reaction theoretically. Equations were given describing the polarographic wave [59, 698] and the effect of various factors on the half-
wave potentials considered. It is interesting to note that a change
in the rate constant (k) of the subsequent reaction, and a change of
the drop time (t) of the electrode only shift the wave along the potential axis and do not affect the shape of the wave.

Kern [698] showed that the anodic oxidation of ascorbic acid
follows scheme (XXIV). A chemically unstable intermediate product is formed in a reversible electrochemical reaction and is further converted chemically into dehydroascorbic acid. The shape
of the oxidation wave of ascorbic acid is well described by the
theoretically derived equations, and its half-wave potentials change
with the drop time of the electrode, as is predicted by the theory.

If the rate of the subsequent reaction (chemical conversion
of the product) is sufficiently high, then the process ceases to be
reversible. Kivalo [700] gave the criteria for reversibility of such
electrode processes. Process (XXIV) remains reversible if the
rate of the reverse electrochemical reaction (B → A) is higher
than the sum of rates of removal of reaction product B from the
electrode by diffusion and by its conversion into C.

An example of processes with a fast chemical deactivation of
the electrode product is the anodic oxidation of cadmium from
amalgam, when the cadmium ion formed is complexed with ethylenediamine tetraacetic acid [701]. Other such processes are the
reduction of manganese ions [702] and the reduction of mercury
ion complexes with ethylenediamine tetraacetic acid in presence
of magnesium ions [703].

Most interesting from the viewpoint of organic electrochemistry are electrode processes with second-order consecutive
chemical reaction, which will be discussed in the following section.

B.   SECOND-ORDER INTERACTION OF

THE PRODUCTS OF A REVERSIBLE

ELECTROCHEMICAL REACTION

1.   General Relationships and Equations

The solution of depolarization problems for electrochemical

processes including second-order chemical reactions encounters serious mathematical difficulties. Hanuš [49] solved first these problems by the approximate method. At about the same time, Koutecký [54] gave the general principles for the solution of certain of these problems.

Koutecký and Hanuš [704] gave the exact and approximate solution of the depolarization scheme for the case of a reversible electrochemical reaction which is followed by a fast dimerization of the electrode products. This process can be given by the scheme

$$O \underset{}{\overset{el}{\rightleftharpoons}} R; \quad 2R \overset{k_d}{\rightarrow} C. \tag{XXV}$$

The electrochemical stage itself of the process is reversible. Therefore, according to the Nernst equation, (5), the concentration ratio of the oxidized and reduced form of the depolarizer $[O]_S$ and $[R]_S$ at the electrode surface corresponds to each electrode potential.

The compound O is transported to the electrode surface by diffusion. Therefore its average concentration (during the drop life) at the electrode surface can be expressed approximately by the Ilkovič equation: $[O]_S = (i_D - i)/\varkappa$. The $[R]_S$ value can be determined from the balance of the concentration change of the compound at the electrode. For sufficiently high values of the dimerization rate constant k_d on the basis of the reaction-layer concept, can be given

$$i = nsF k_d \mu_d [R]_s^2, \tag{130}$$

and the thickness of the reaction layer (μ_d) in which the second-order reaction between the B particles occurs, is

$$\mu_d = \sqrt{\frac{D}{k_d [R]_s}}. \tag{131}$$

The diffusion coefficient (D) is assumed to be equal for the O and R compounds. From Eqs. (130) and (131) follows:

$$[R]_s = \left(\frac{i}{ns F D^{1/2} k_d^{1/2}} \right)^{2/3}. \tag{132}$$

Substituting the $[R]_S$ and $[O]_S$ values into the Nernst equation, (5), we obtain an expression for the wave with reversible electrochemical step and consecutive fast dimerization of the electrode products

$$E = \varepsilon_0 - \frac{RT}{nF} \ln \frac{i^{2/3}}{i_d - i},\qquad(133)$$

in which the characteristic potential ε_0 is independent of the depolarizer concentration and is given by

$$\varepsilon_0 = E_0 + \frac{RT}{nF} \ln (nsF)^{2/3} (Dk_d)^{1/3} \varkappa^{-1}.\qquad(134)$$

Equation (133) can be easily transformed into an equation derived by Hanuš [704] in a slightly different way.

The rigorous solution of the same depolarization scheme, given in the form of tabulated functions [704], gives for the shape of the wave a curve that is very close to the shape of the curve described by Eq. (133). But the exact solution gives a curve shifted along the E axis by about 16 mV to the side corresponding to higher reaction product concentration, which means that for a reduction process the curve is shifted toward negative potentials.

If we substitute in Eq. (133) the $i = \frac{1}{2}i_D$ value, we obtain the expression for the half-wave potential

$$E_{1/2} = E_0 + \frac{RT}{3nF} \ln \frac{k_d c t}{\text{const}},\qquad(135)$$

where c is the concentration of compound O in the solution, and the constant value determined by the rigorous solution of the depolarization scheme [704] is equal to 4.0.

Therefore, in contrast to normal reversible waves, the half-wave potential for processes with fast dimerization of electrode products shifts toward positive values both when the drop period of the electrode increases and also when the concentration of the material being reduced increases. It follows from Eq. (135) that this shift must reach 19.7 mV (at 25°C) if either c or t increases by a factor of ten.

Savéant and Vianello [705] developed equations for the peak potential (E_p) during a linear potential change of the electrode (in oscillographic polarography). The peak potential was expressed as a function of the depolarizer concentration (c) and the rate of linear potential change (V). These equations apply to electrode processes in which the reversible electron transfer is followed by a fast chemical deactivation of the electrode products, for example, by a second-order reaction. It follows from these equations that E_p must become more negative with increasing V, and $\Delta E_p / \Delta \log V = -\frac{RT}{nF} \left(\frac{1}{m+1} \right)$. With increasing c, E_p becomes more positive and $\Delta E_p / \Delta \log c = \frac{m-1}{m+1} \frac{RT}{nF}$, where m is the order of the reaction with respect to the electrode products (for a second-order reaction, m = 2).

These authors mentioned [705] that from the character of the dependence of the peak potential on $\log c$, processes with a slow electron transfer (meaning an irreversible electrochemical step) can be distinguished from processes in which the reversible electron transfer is followed by a fast second-order chemical reaction of the electrolysis product.

2.   Polarography of N-Alkyl

Pyridinium Salts

One of the first processes with reversible electron transfer and fast dimerization of the electrode products investigated was the reduction of N-alkyl pyridinium salts. According to Emmert [706] and Tompkins [707], the formation of hydrodipyridyl derivatives takes place in the electrochemical reduction of N-alkyl pyridinium salts. In this process only a single electron is transferred in the electrochemical step and the free radicals are formed [707]

(XXVI)

which are then dimerized in the 2- and 4-position. The so-called Fournier waves [710-712], which represent curves of the dependence of the current component on potential, measured [708, 709] by superimposition of an additional small sinusoidal potential on the electrode, confirmed the assumption of Colichman and O'Donovan [713] about the reversibility of the electrochemical step in the discharge of N-alkyl pyridinium ions.

The reversible electrochemical step and the fast dimerization of the electrode products formed (radicals) determined the polarographic behavior of the N-alkyl pyridinium salts [708, 709]. The shape of the wave at not too high N-alkyl pyridinium concentrations is well described by Eq. (133), and the ε_0 value is within the limits of the experimental error (± 3 mV) independent from the pyridinium salt concentration in the solution. The half-wave potential of the reduction wave for N-methyl pyridinium, in agreement with the theory, becomes more positive by about 20 mV if the concentration of reduced ions is increased tenfold [708]. But the shift of the half-wave potentials observed at increasing drop periods is almost twice as high as is predicted by the theory [708]. This is caused by the incomplete reversibility of the electrode process, due to adsorption of the reaction product (see Chapter X).

Judging from the half-wave potential shift of the wave to positive potentials with increasing concentration and from a larger slope than $\frac{1}{60}$ mV^{-1} on the logarithmic graph in $\log[i/(i_{lim} - i)]$ vs. E coordinates [446], the reduction of quinolinethoiodide is reversible, with transfer of one electron, and is followed by the dimerization of the formed product.

3. The Effect of Dimerization of
Free Radicals on Polarographic
Reduction Waves of Aromatic Aldehydes
and Ketones in Acid Medium

The polarographic waves for aromatic carbonyl compounds in acid medium [715, 524] represent the second thoroughly investigated example of the effect of dimerization of electrochemical reaction products on electrode processes.

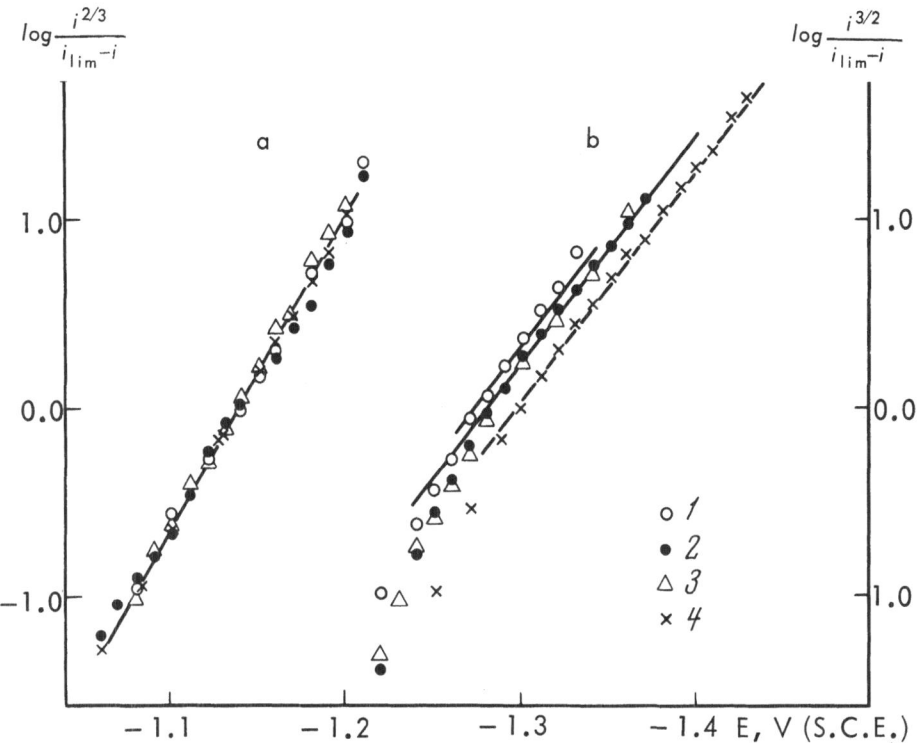

Fig. 46. The quantity $\log[i^{2/3}/(i_{lim} - i)]$ vs. E for the first (a) and $\log[i^{3/2}/(i_{lim} - i)]$ vs. E for the second (b) wave for benzaldehyde polarograms. Benzaldehyde concentration (in M): (1) $6.9 \cdot 10^{-5}$; (2) $19 \cdot 10^{-5}$; (3) $56 \cdot 10^{-5}$; (4) $174 \cdot 10^{-5}$; background — acetate buffer solution at pH 4.65; electrode with forced drop formation: m = 1.02 mg per sec, t = 0.23 sec.

In the reduction of aromatic aldehydes and ketones at the dropping mercury electrode in acid medium, the first step of the process seems to be a reversible addition of an electron and a proton [524, 576, 577, 593, 714-717] with formation of a free radical

$$RR'C{=}O + e^- + H^+ \overset{el}{\rightleftarrows} RR'\overset{\cdot}{C}OH, \qquad (XXVII)$$

where R is an aryl, and R' is H or an alkyl(aryl).

The radicals either dimerize, forming the corresponding pinacols

$$2RR'\overset{\cdot}{C}OH \overset{k_d}{\rightarrow} \underset{\substack{| \quad |\\ OH \; OH}}{RR'C - CRR',} \qquad (XXVIII)$$

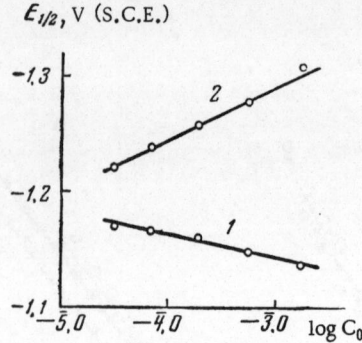

Fig. 47. The half-wave potential of the first (1) and the second (2) wave for the reduction of benzaldehyde as a function of its concentration in the solution.

or if they reach a sufficiently negative electrode potential, they can be further reduced to alcohol

$$RR'\dot{C}OH + e^- + H^+ \xrightarrow{k_{el}} RR'CHOH, \qquad \text{(XXIX)}$$

and this latter process appears on the polarogram as a second wave.

Therefore, the form of the first wave on the polarographic curves of aromatic aldehydes and ketones in acid solution must correspond to Eq. (133) and its half-wave potential must be given by Eq. (135). Figures 46a and 47 (curve 1) show the results of experimental verification [714] of these equations. It follows from these figures that the values of the $\log[i^{2/3}/(i_{lim} - i)]$ function determined from the experimental data (dropping mercury electrode with a glass plate, t = 0.23 sec), plotted as a function of E for several benzaldehyde concentrations, fall close to the line constructed by Eq. (133). In Fig. 47 the points on line 1 show the variation of the half-wave potentials for the first wave of benzaldehyde with its concentration in the solution. The slope of this line, $\Delta E_{1/2}/\Delta \log c \approx 20$ mV, is practically identical with the value (19.7 mV) predicted by Eq. (135).

Suzuki and Elving [718] were not able to determine clearly the effect of radical dimerization on the first wave in benzophenone reduction. They conducted their experiments with a regular (long period) dropping electrode on which the dimeric, strongly adsorbed reaction products accumulated and inhibited the electrochemical stage (see Chapter X).

It was already mentioned that Savéant and Vianello [705] derived equations for the dependence of the potential peak on oscillopolarograms with linear potential change of the electrode on the rate of potential (V) change and on the concentration of the depolarizer (c). This was done for the second-order reaction of products of the electrochemical reaction. The experimentally determined dependence of the E_p value for oscillographic benzaldehyde waves (in acid Britton and Robinson buffers at pH 3.1) on $\log c$ and $\log V$ corresponds exactly to the theoretically predicted curves. These authors [705] could not calculate the second-order rate constant of free radicals, since the redox potential of the benzaldehyde β-hydroxybenzyl radical is not known. This value was only calculated on the basis of the faradaic impedance in the given system [717].

4. Electrode Processes with Second-Order Reaction, Parallel to the Electron Transfer

Such processes correspond to the second wave on polarograms of aldehydes and ketones in weakly acid medium. At potentials where the second wave appears, some of the free radicals formed in the first step of the electrode process (XXVII) dimerize according to (XXVIII); another portion is further reduced according to (XXIX). The pinacol yield depends on the ratio of the rates of reaction (XXVIII) and (XXIX). The rate of the electrochemical reaction (XXIX) depends on the electrode potential. The current of the second wave (i') corresponds to the fraction of free radicals reduced to alcohol; and $i'_{lim} - i'$ corresponds to the fraction of radicals removed from the electrode by dimerization with pinacol formation (the prime here and elsewhere always denotes values of the second wave).

If $[R]_s$ is the concentration of free radicals at the electrode surface, and if we consider the limiting currents of the first and second wave equal, then for the second wave, on the basis of the approximate method given by Brdička, Wiesner, and Hanuš (see Chapter II) we can put

$$i_{lim} - i' = sFD \, [R]_s/\mu_d = sF \, (k_d D)^{1/2} \, [R]_s^{3/2}, \tag{136}$$

since the reaction-layer thickness (μ_d) in which the radical dimerization takes place is given by Eq. (131).

The second wave on the polarographic curves of aldehyde and ketone is irreversible. Therefore, the current (i') is determined by Eq. (69) corresponding to the theory of slow discharge. Substituting in Eq. (69) the concentration of the radicals at the electrode, $[R]_s$, from Eq. (136), we obtain, after changing to logarithms, the expression for the shape of the wave that corresponds to electrode processes with competing electrochemical reaction and second-order reaction of the depolarizer [714]

$$E = \varepsilon_0' - \frac{2}{3} \frac{RT}{\alpha F} \ln \frac{(i')^{3/2}}{i_{lim}' - i'}, \tag{137}$$

where the concentration-independent characteristic potential is

$$\varepsilon_0' = - \frac{2}{3} \frac{RT}{\alpha F} \ln \left(\frac{k_d D}{sF}\right)^{1/2} \frac{10^{-3}}{(k_{el}^0)^{3/2}}, \tag{138}$$

if the current is expressed in μA.

The half-wave potential of such waves is

$$E_{1/2}' = \varepsilon_0' - \frac{1}{3} \frac{RT}{\alpha F} \ln \frac{i_{lim}'}{2} = \varepsilon_0' - \frac{1}{3} \frac{RT}{\alpha F} \ln \frac{\varkappa c}{2}, \tag{139}$$

where $\varkappa$ is the proportionality constant between i_{lim}' and c (according to the Ilkovič equation) (n = 1) in $\mu A \cdot mM^{-1}$.

It follows from Eq. (139) that with increasing depolarizer concentration in the solution, the half-wave potential of the second wave must become more negative. The reason for this is that with increasing c, the rate of the second-order process (XXVIII) increases faster than the rate of reaction (XXIX). Therefore, to equalize the rate of processes (XXVIII) and (XXIX) (which occurs at $E' = E_{1/2}$) the rate of the electrode process must increase, which means that the potential must be more negative.

Experiments prove the validity of expressions (137) and (139). In the right-hand side of Fig. 46 are given the values of $\log[(i')^{3/2}/(i'_{lim} - i')]$ vs. E, according to Eq. (137) for the second benzaldehyde wave. The experimental data obtained with different depolarizer concentrations are quite close to each other (causes for the shift of the lines are discussed in Chapter X). The upper portion of the graph is linear, as expected from Eq. (137). The lower portion is bent due to closeness of the first wave. The i' values for constructing the graph in Fig. 46b were calculated by subtracting the diffusion current of the first wave from the observed current. The first and second waves are very close to each other along the potential axis. Therefore, to determine the true i' value from the observed current, it is not the limiting current of the first wave (i_{lim}) that should be subtracted; that current cannot even be reached at the potentials corresponding to the beginning of the second wave. Rather, the true current value of the first wave at the given potential, which can be calculated by the equation of the first wave, (133), should be subtracted.

The dependence of the value of $E_{1/2}$ for the second wave on the benzaldehyde concentration is given in the upper part of Fig. 47. According to the theory, the value of $E_{1/2}$ of the second wave becomes more negative with increasing benzaldehyde concentration. Since the half-wave potential of the first wave shifts toward more positive potentials at the same time, this leads to a separation of the waves on benzaldehyde polarograms. For the second wave, $\Delta E_{1/2}/\Delta \log c \approx -44$ mV.

A comparison of Eqs. (137) and (139) shows that the $\Delta E_{1/2}/\Delta \log c$ value must be equal to one half the reciprocal slope on the graph in $\log[(i')^{3/2}(i'_{lim} - i')]$ vs. E coordinates. As can be seen from the right-hand side of Fig. 46, the reciprocal value of the slope in the logarithmic plot is equal to about 82 mV, which means that the theoretical relationship is almost met.

From the slope of the logarithmic graph in Fig. 46b the transfer coefficient for the second wave of benzaldehyde was determined according to Eq. (137): $\alpha = 0.48$. The ε_0' value in the investigated benzaldehyde concentration region is in the limits -1.278 to -1.296 V.

Similar results were obtained for the acetophenone wave [714]. The first wave is described satisfactorily by Eq. (133) with $\varepsilon_0 = -1.300$ V (S.C.E.) and the $\Delta E_{1/2}/\Delta \log c$ value for the first wave becomes more negative; it was not possible to determine the $\Delta E_{1/2}'/\Delta \log c$ value for the second wave with sufficient accuracy, since at low acetophenone concentrations the second wave is rather indistinct. The reciprocal value of the slope in the logarithmic plot of the second wave at $c \approx 1$ mM is 88 mV, $\alpha = 0.45$, and $\varepsilon_0' = -1.51$ V.

The assumption mentioned [719], namely that the competition of the electrochemical reaction with the second-order chemical reaction leads to a decrease of the wave height, can be observed in the reduction process of phenylhydroxylamine.

5. Certain Other Reversible Processes with Dimerization of Electrode Products

Numerous examples are now known of processes with a reversible electrochemical step and subsequent second-order reaction. A dimerization of the electrode products occurs in the reduction of tropylium ions [720-722]; the wave shifts toward positive potentials at increasing tropylium concentrations [722]. This indicates that the electron-transfer process is rapid, which means that the electrochemical step of the process is reversible. The same is indicated by the shape of the tropylium wave [721].

According to Zhdanov and Mirkin [497], the reversible reduction of pyridinium cyclopentadienyl, and also that of azulenium cations, is accompanied by subsequent dimerization.

Experiments conducted by Lavrushin, Bezuglyi, and Belous [723] proved that the polarographic behavior of chalcone is similar to that of aromatic aldehydes and ketones; the form of the first and second chalcone wave (mainly for electrodes with a short drop period) are described well by Eqs. (133) and (137), respectively.

Stradyn' found that the same relationships are followed by the reduction waves on fural polarograms in acidic buffer solution [578, 724]. Ivcher [725] showed that the reduction wave of cyclohexene-2-on-1 in acid medium is described by Eq. (133) and the change of the half-wave potential by expression (135). In the reduction of γ-thujaplicin — a tropolone derivative which was investigated by Brdička [726] — an antecedent protonation of the molecule takes place, and judging from the slope of the wave, the reduction of the protonated particle is reversible with participation of one electron and with subsequent fast dimerization of the free radical formed at the electrode. Zhdanov and Pozdeeva showed [440] that tropone and 2,3-dimethylbenztropone behave similarly to aromatic aldehydes and ketones at the dropping electrode in acid and neutral solution.

α-Acetyl thiophene gives two waves at very low concentrations (below 0.3 mM) on a dropping electrode with a regulated drop time in weakly acidic solutions. The properties of these waves can be described satisfactorily by Eqs. (133)-(139). But even at acetyl thiophene concentrations above 1 mM, its waves do not follow these equations, and only a part of the wave is linear in logarithmic coordinates, corresponding to Eqs. (133) and (137) [671]. Linear graphs can, however, be obtained even at concentrations c > 1 mM if, in Eqs. (133) and (137), the current in the numerator after the logarithm is not taken at a $^2/_3$ and $^3/_2$ power, but at $^1/_2$ and 2 power, respectively [671]. Equations with such exponents on the current correspond to electrode processes with dimerization of radicals which do not occur in the reaction–layer volume but rather at the immediate surface of the electrode in the adsorbed state. The same equations also were derived independently by Feoktistov [693]. The surface dimerization of the radicals also is indicated by a larger value of $\Delta E_{1/2}/\Delta \log c$ than that predicted by Eqs. (135) and (139).

Chapter IX

Catalytic Hydrogen Evolution at the Dropping Mercury Electrode Caused by Organic Catalysts

A. INTRODUCTION

1. Certain Properties of Catalytic Hydrogen Waves

Catalytic hydrogen evolution in solutions of organic catalysts, including both antecedent protonation reactions and subsequent second-order reactions, is one of the most complicated electrochemical processes. However, since, in catalytic processes, the catalyst is completely regenerated at the dropping electrode surface, the development of equations for catalytic currents and the explanation of different effects came much earlier than for kinetic waves. Primarily for this reason, several phenomena described in Chapters IV, V, VI, and VIII (such as, for example, the effect of depolarizer adsorption on electrode processes) were first investigated in detail on catalytic hydrogen waves.

Catalytic hydrogen waves can be observed at less negative potentials than the usual hydrogen discharge waves in the same solutions. Thus, the catalysts seem to reduce the hydrogen overvoltage — increase ("catalyze") the rate of hydrogen discharge. This is the reason why these waves are called catalytic hydrogen waves. The catalytic waves are a special case of kinetic waves and have many of the properties of the latter. However, in contrast to the kinetic current, the limiting catalytic current can significantly exceed the limiting current corresponding to the given catalyst concentration.

Herles and Vančura [727] first observed a wave different in its properties from diffusion currents in the polarography of blood serum in a supporting electrolyte of sodium, potassium, and magnesium salts, in 1930. A short description of their results appeared in 1932. Heyrovský and Babička [728] repeated these experiments in 1930, and also observed similar waves in the presence of proteins in an ammoniacal buffer solution. They proved that these waves are not caused by the reduction of proteins, but by catalytic hydrogen evolution. Later, Brdička observed that the catalytic effect of proteins and certain other sulfur-containing compounds increases significantly if salts of di- and trivalent cobalt are added to the ammoniacal buffer solution.

At approximately the same time, Pech [730] found that catalytic hydrogen waves can also be caused by organic compounds not containing sulfur, for example, by alkaloids. The catalytic hydrogen waves caused by proteins in ammoniacal protein solutions containing cobalt salts and the waves caused by alkaloids and other nitrogen-containing derivatives, were the subject of many investigations (see, for example, the literature review in [3-6, 370, 731, 732]). Many experimental investigations were made by Kirkpatrick [733], Stromberg [394, 734], Knobloch [735, 736], Kůta and Drabek [737], Stackelberg and co-workers [738-740], Nürnberg [741], Lamprecht [742], and several other investigators.

In several investigations dealing with the effect of the catalyst concentration (c_{cat}) and the solution composition on the height of catalytic hydrogen waves, it was determined that with increasing c_{cat} the limiting catalytic current increases, first linearly with c_{cat}. Then the increase of the wave height becomes slower. Finally, at high c_{cat} values, the limiting catalytic current becomes independent of c_{cat} and reaches a certain limit. This was first observed by Brdička [743] for catalytic hydrogen waves caused by egg white in unbuffered potassium chloride solutions and also for waves caused by albumin in buffer solutions of different composition and pH. Brdička found that the height of the catalytic wave and the limit that it can reach with increasing catalyst concentration increases for higher buffer capacities of the solution and at lower pH.

Brdička proved [744], using the example of catalytic waves in protein solutions containing cobalt salts, that for catalytic waves the dependence of the limiting current on the catalyst concentration, if all other factors are held constant, follows the form of the Langmuir adsorption isotherm.

The idea developed in the work of Brdička about the regeneration of the catalyst after its participation in the electrode process is very important and forms the basis of the theories of the catalytic hydrogen waves. This regeneration, according to Brdička, consists of hydrogen ion addition to the product of electrochemical reduction of the catalyst by acids, which are the proton donors in the solution.

Stromberg [734] also observed the growth of the catalytic wave caused by increasing protein and cobalt salt concentrations. He first discussed this phenomenon from the viewpoint of the effects of rates of reactions that take place at the electrode and on this basis he developed a relationship for the limiting current of catalytic waves. Klumpar [745] investigated the catalytic waves caused by cysteine in the presence of cobalt salts. He gave an empirical equation relating the limiting catalytic current to the cysteine and cobalt salt concentrations in the solution.

Knobloch [735] first investigated the effect of different factors on catalytic hydrogen waves caused by nitrogen-containing compounds, in part pyridine derivatives, in buffer solutions not containing cobalt salts. He showed that the limiting catalytic current of the hydrogen wave increases when the pH of the solution decreases and also increases if the buffer capacity increases. When the catalyst concentration increases, the current increases, and the graph giving the dependence of the current on the concentration agrees with the Langmuir adsorption isotherm. This led Knobloch to the conclusion [735] that the current is determined by the quantity of catalyst present in adsorbed state, and that the limit of current increase reached at increasing catalyst concentrations is related to the full coverage of the electrode surface by catalyst particles (see also [746]). The pH and buffer-capacity effect was explained by Knobloch by the protolytic regeneration of the catalytically active cationic catalyst form which, according to Knobloch's view, causes the decrease of hydrogen overvoltage according to Heyrovský's scheme as a result of increased H_2^+ formation rate from H atoms and from protons of the cationic catalyst

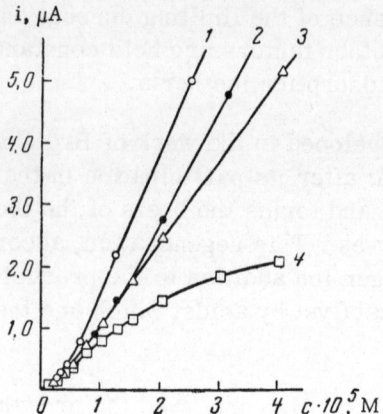

Fig. 48. The dependence of the limiting catalytic hydrogen current on the anabasine concentration in acetate
buffer solution with μ = 0.5 N. (1) pH 4.46, total CH$_3$COOH +
CH$_3$COOK concentration Σc = 0.2 M; (2) pH 5.0, Σc = 0.1 M;
(3) pH 5.0, Σc = 0.2 M; (4) pH 6.0, Σc = 0.1 M.

form. Thus Knobloch first assigned the electrochemical activity
only to the acidic (cationic) form of nitrogen-containing catalysts.

The effect of catalyst and proton-donor concentration on
catalytic hydrogen waves was investigated in detail on the example
of the first quinone wave in solutions containing only a single species of proton donors — hydrochloric acid in a supporting electrolyte consisting of 1 N LiCl [747, 748]. Similarly to other catalytic
hydrogen waves, the catalytic current in quinone solutions increases
with its concentration along a curve that is similar to the Langmuir
adsorption isotherm (see Fig. 1 in [749]). The limit to which the
wave tends and also the quinone concentration at which i$_{lim}$ reaches
the limit depends [748] on the hydrochloric acid concentration in the
solution. If the HCl concentration is low, the limit that the height
of the catalytic hydrogen wave approaches becomes lower and the
quinone concentration required to approach this limit becomes
smaller. Therefore, this limit is not related to the complete surface coverage by adsorbed catalyst. A comparison of the limiting
value for the increase of the catalytic wave with the diffusion current value for discharge of hydrogen ions showed that this limit is
related to the complete exhaustion of proton donors (in the given
case, hydrogen ions) at the electrode surface. Therefore, the value

of this limit corresponds to the diffusion transport of hydrogen ions from the bulk solution to the electrode surface [747-749].

In the presence of an excess of proton donors at very low catalyst concentrations, an unusual relationship can be observed. The height of the catalytic wave increases faster than the catalyst concentration (see, for example, Fig. 48, from [750]). This relationship is apparently frequently encountered. Thus, for example, judging from the polarograms given in the works of Kirkpatrick [733], the height of the catalytic hydrogen wave caused by atropine and pilocarpine in buffer solutions also increases faster than the concentration of these alkaloids.

On the basis of all experimental data, using the views developed by Brdička, Knobloch, and Stromberg, a scheme was proposed [748, 749, 751] for catalytic hydrogen evolution that explains all experimentally observed phenomena. Equations accounting for all observed relationships were developed [749] on the basis of this scheme and by applying the ideas of Brdička and Wiesner about kinetic current limitations and the theoretical conclusions of Frumkin about the effect of double-layer structure on electrode processes and about the character of adsorption of organic compounds on the electrode and also data on the kinetics of adsorption.

2. Mechanism of Catalytic Hydrogen Wave Formation under the Effect of Organic Bases

The indispensible condition for catalytic activity of an organic compound is that the compound be able to exist in two forms — acidic BH^+ and basic B (in the Brönsted sense) — which are in a protolytic equilibrium

$$B + DH^+ \underset{\rho\sigma}{\overset{\rho}{\rightleftarrows}} BH^+ + D,$$

$$(XXX)$$

where DH^+ is the acid, proton-donor present in the solution, and D is the conjugated base.

The ability of certain organic compounds to cause catalytic hydrogen evolution is apparently connected with the presence of unshared electron pairs (at nitrogen, sulfur, phosphor, arsenic, etc., atoms) to which a proton can be added. As the result, an

onium compound is formed, which is able to enter into an electrochemical reaction on the cathode. The electrochemical activity of
onium compounds is not caused by their positive charge, which
facilitates the transport of particles to the cathode and their reduction [752] but rather by the chemical nature of the onium ion.
If the electron transfer on the protonated onium form were caused
only by their positive charge, the reduction of the uncharged, unprotonated catalyst form could also be expected, although at more
negative potentials (by 200-300 mV) as occurs, for example, in the
reduction of uncharged aniline and phenol derivatives [647, 648]
and also with weak acids and their anions (see Chapter V). But
the reduction of the unprotonated catalyst is not observed, and
this leads to the previously given conclusion [749].

The catalytic activity of compounds is not related to the mobility of hydrogen ions: careful investigations showed [739] that
organic compounds containing an extremely mobile hydrogen do
not show any catalytic activity.

As a result of electron transfer on the protonated catalyst
form

$$BH^+ + e^- \underset{<\cdots}{\overset{el}{\longrightarrow}} BH, \qquad (XXXI)$$

an unstable, uncharged particle is formed that has the character
of a free radical. In the reduction of tetrasubstituted ammonium
[753-755] and phosphonium salts [756] or trisubstituted sulfonium
cations [757], the unstable products formed either dimerize, as,
for example, in the discharge of N-alkyl pyridinium salts (see section B-2 in Chapter VIII), or fall apart at the carbon—heteroatom
bond, giving the corresponding tertiary amines, phosphines, or di-
substituted sulfides and hydrocarbon radicals. These later can be
reduced on the cathode, can react with the solvent, dimerize, or
disproportionate.

In the reduction of the protonated onium cation, as, for example, during catalytic hydrogen evolution, it is energetically
more favorable for the BH particles formed [see scheme (XXXI)]
not to release a hydrocarbon radical or hydrogen atom, but to take
part in a second-order interaction

$$2BH \overset{k_d}{\longrightarrow} 2B + H_2. \qquad (XXXII)$$

In this reaction the catalyst is regenerated in its basic form, and hydrogen molecules are evolved. The rate of this reaction, due to its second-order character, decreases significantly at low catalyst concentrations and in certain cases can even limit the rate of the entire process. This occurs partly under the experimental conditions used in the recording of the data given in Fig. 48. By this second-order reaction (in the presence of a catalyst) the faster increase of the limiting catalytic current than the increase of the catalyst concentration, which was observed in certain cases, can be explained [733, 747, 750].

For example, to some extent in the reduction of anabasinium ion [758], intermediate products are formed by the second-order reaction of the radicals according to scheme (XXXII) that are relatively stable HB−BH dimers. They disproportionate relatively slowly into two B particles and a hydrogen molecule. At higher temperature the stability of di-hydrodianabasine HB−BH dimers decreases significantly, and their formation (or more exactly, their decomposition) ceases to affect the overall rate of the (XXXII) process [758].

The basic catalyst (form B) formed according to Eq. (XXXII) is again protonated according to scheme (XXX), discharges according to (XXXI), then is again regenerated, etc., and in each cycle of reactions (XXX)-(XXXII) two protons from DH^+ are converted by electrochemical catalysis into hydrogen molecules. Each catalyst particle can thus transfer a considerable quantity of proton-donor particles.

The catalytic current intensity of hydrogen evolution depends on the rate of the processes (XXX)-(XXXII). Under limiting current conditions, when the rate of the electrochemical step (XXXI) becomes sufficiently high and all arriving BH^+ particles are reduced on the electrode, the height of the catalytic wave is determined by the rate of BH^+ formation at the electrode. If the protolytic equilibrium (XXX) is shifted in the solution toward the electrochemically inactive B form, then, as was shown for the limiting kinetic current (see section B, Chapter II), it is simple to obtain an expression [747, 749] for the catalytic wave height

$$i_{lim} = \frac{k_{DH^+} sFr c_{cat}}{1 + k_{DH^+} sFr c_{cat}} \varkappa_{DH^+} [DH^+], \qquad (140)$$

in which equation r corresponds to the thickness of the reaction layer (μ) for volume catalytic waves or to the adsorption index (β) for surface waves; k_{DH^+} is the interaction rate constant of the catalyst with the proton donor (DH$^+$), and c_{cat} is the catalyst concentration in the volume.

From Eq. (140) it follows, in agreement with experimental data, that the curve representing the dependence of the limiting catalytic current on the catalyst concentration is similar to the Langmuir adsorption isotherm. It is easy to show that with increasing c_{cat}, the height of the catalytic current tends to a limit that is equal to $\varkappa_{DH^+}[DH^+]$, which is the diffusion current of proton donors in the given solution.

The rate of reaction (XXXII) affects the height of the catalytic wave only at very low catalyst concentrations. In the limiting case, when the process is determined by the rate of second-order kinetics, it can be assumed that reaction (XXXII) occurs only in the reaction-layer volume (μ_d), and on the basis of the Brdička−Wiesner method the height of the catalytic wave can be given by

$$i_{lim} = sFk_d\mu_d[BH]_s^2 , \tag{141}$$

where k_d is the rate constant of the second-order reaction (XXXII). Alternatively, applying Eq. (131), the following equation can be given:

$$i_{lim} = sFk_d^{1/2}D^{1/2}[BH]_s^{3/2}, \tag{142}$$

where D is the diffusion coefficient of the catalyst.

When the limiting catalytic current is limited by the rate of the second-order reaction (XXXII), the $[BH]_S$ value is proportional to c_{cat}. Therefore, according to (142), i_{lim} changes proportionally to the $3/2$ power of the catalyst concentration. This was experimentally proven at very low c_{cat} values (see Fig. 48). With increasing c_{cat}, the rate of the process (XXXII) increases fast, and the reaction ceases to be limiting.

It is not possible to develop a general equation for the limiting catalytic current as a function of catalyst concentration in a broad concentration range, on the basis of the Brdička−Wiesner method. However, by this method solutions were obtained [759] for depolarization problems for several partial cases of catalytic hydrogen waves. In a general form, the depolarization scheme for volume chemical reactions (XXX) and (XXXII) and reversible electrochemical stage (XXXI) was solved by Koutecký [760] by the rigorous method.

B. VOLUME CATALYTIC HYDROGEN WAVES WITH REVERSIBLE ELECTROCHEMICAL STEP

1. Reversible Catalytic Hydrogen Waves with Volume Chemical Reactions

Catalytic hydrogen waves with a reversible electrochemical step were first observed in pyridine solutions [541, 751].

The assumption about the reversibility of the electrochemical step in the reduction of pyridinium ions was made [751] on the basis of the analogy with the discharge of N-alkyl pyridinium cations. The reversibility for the latter case was proved in several investigations [708, 713]. The assumption about the reversibility of the discharge of pyridinium ions during the catalytic hydrogen evolution was further confirmed by the presence of reaction pseudocapacity observed in the measurement of the double-layer capacity in solutions of pyridine [761-763] and its homologs [764].

A reversible electron transfer also occurs apparently in electrochemical processes with participation of certain other compounds with onium structure (for example, for ammonium ions [765-768]). The reversible character of the reduction of tetramethyl ammonium ions is in part indicated by the extremely high pseudocapacity, observed by Grahame [769] in the electroreduction of these ions.

In section B of Chapter IV it was shown that due to the significant surface activity of many organic compounds at the mercury−solution boundary, the electrode processes in which they take part have a clearly expressed surface character. The same also applies to catalytic hydrogen waves.

One of the apparently very few exceptions seems to be the wave in solutions of pyridine and its homologs. The height of these waves is determined under specific conditions only by volume chemical reactions. The volume character of the wave in pyridine solutions is explained by the relatively low adsorptivity of pyridine and its homologs at potentials where the catalytic wave appears at about −1.8 V (S.C.E.) [265, 295-297, 401] (see also Figs. 2 and 4, which show the effect of potential on the adsorption of 2,6-lutidine). The investigation of i vs. t curves for purely volume catalytic waves caused by pyridine and its homologs shows that these waves can be obtained only at sufficiently high supporting electrolyte concentrations (for potassium salts >0.2 M [770-772]), when the adsorptivity of pyridine is significantly reduced. The work must be carried out with electrodes having a very short drop period (less than 0.5 sec). For this reason electrodes using glass plates for regulated drop formation were used.

2. Reversible Volume Catalytic

Hydrogen Waves in Buffer Solutions

Usually the limiting catalytic current is limited by the rate of antecedent protonation reaction (XXX) and the rate of the subsequent second-order reaction (XXXII) does not show any effect. In this case the height of the catalytic wave is determined by the quantity of BH^+ particles that are formed in unit time in a layer of thickness μ

$$i_{\lim} = sF\mu \, [B]_{s\rho}. \tag{143}$$

If we assume that $\sigma \gg 1$ (which is usually the case), it means that the catalyst is present in the solution mainly as the basic, unprotonated form. Also, due to complete regeneration of the catalyst, a steady state is rapidly established at the electrode. Thus, the $[B]_s$ value can be taken to be equal to the overall analytical concentration of the catalyst (c_{cat}) in the solution. We include in our consideration Eq. (16) and the Ilkovič equation for the hypothetical average diffusion current that could be observed for an ordinary, diffusion-limited reduction of the catalyst at n = 1. If the Ilkovič equation is used in form (6a), the ratio of the limiting catalytic current and the diffusion current can be given by the following simple equation [759]:

$$\frac{i_{\lim}}{i_D} = 0.81 \sqrt{\frac{\rho t}{\sigma}}. \tag{144}$$

Using Eq. (144) it is very easy to calculate the total rate constant of protonation (ρ) from the experimental i_{lim} data. It must be mentioned that expressions practically identical with Eq. (144) were obtained by the rigorous solution of the depolarization scheme under the given conditions (see [760]).

If the limiting current of the catalytic wave is significantly affected also by the rate of subsequent reaction, then the exact solution [760] of the depolarization scheme yields for ρ the equation

$$\rho = \frac{(i_{lim}/i_D)^2 \, \sigma}{0.81^2 t \left[1 - \left(\frac{3}{2} \frac{(i_{lim}/i_D)^2}{0.81^2 k_d c_{cat} t}\right)^{1/3}\right]^2} . \tag{145}$$

It is easy to show that with increasing rate of second-order reactions of electrode products that results in an increase of c_{cat} or k_d, Eq. (145) reduces to Eq. (144). An equation similar to (145) also can be obtained on the basis of the reaction-layer concept (see [759]); the equation obtained differs from (145) only in the absence of the numerical coefficient $\frac{3}{2}$ in the second term in the denominator.

If a second-order reaction is the rate controlled step for the limiting catalytic current, the exact solution of the depolarization scheme [760] gives for the limiting current

$$i_{lim} = i_D \sqrt{\frac{2}{3}} \, 0.81 \left(\frac{k_d c t}{\sigma}\right)^{1/2} . \tag{146}$$

This equation differs from (142) if we include Eq. (6a) only in the presence of the numerical coefficient equal to $\sqrt{2/3}$. Thus, for this type of catalytic waves the rigorous and the approximate solution of the depolarization scheme lead to practically identical results.

An equation for the shape of the reversible catalytic hydrogen wave occurring in the volume when the limiting step is the antecedent protonation and $\sigma \gg 1$ can be obtained by substituting the $[BH^+]_S$ and $[BH]_S$ values, which means the surface concentrations of the components of the redox system (XXXI) in the Nernst equation [751, 759]. The same was done in deriving Eq. (133)

$$E = \varepsilon_0 - \frac{RT}{F} \ln \frac{i^{2/3}}{i_{lim} - i} . \tag{147}$$

It can be seen that Eqs. (133) and (147) are almost identical. The characteristic potential ε_0 in Eq. (147), which is independent of the catalyst concentration, corresponds to that point of the wave where $i^{2/3} = (i_{lim} - i)$. The characteristic potential is equal to

$$\varepsilon_0 = E_0 - \frac{RT}{3F} \ln\left(\frac{sFD^{1/2}\rho^{3/2}\sigma^{3/2}}{k_d}\right), \qquad (148)$$

where E_0 is the redox potential of system (XXXI).

The half-wave potential in the discussed case is given by the rigorous solution of the depolarization scheme [760] by

$$E_{1/2} = E_0 + \frac{RT}{3F} \ln\frac{c_{cat}k_d}{3\sigma^2\rho} . \qquad (149)$$

A similar expression for the half-wave potential can be derived [751, 759] from Eq. (147) for $i = i_{lim}/2$ and by combination with Eq. (148). This equation differs from (149) only in that the denominator of the fraction after the logarithm contains a constant 2 instead of 3.

Well-defined catalytic waves can be obtained in the presence of pyridine and its homologs in borate and veronal buffer solutions in the pH range of 7.5-10 [773-775]. Under these conditions the limiting catalytic current, in agreement with theory, is proportional to the dropping electrode surface and is independent of the height of the mercury column above the electrode. The limiting step of the process governing the wave in these solutions seems to be the protonation of the catalyst. Therefore, according to the theory, the rate of protonation (ρ) always increases with an increase of the buffer capacity of the solution if the pH and ionic strength remains constant. Consequently, the limiting catalytic current increases too (see, for example, Fig. 49). As was observed in numerous experiments [750, 751, 773-775], a decrease of the catalytic wave is found [see Eq. (145)] with increasing pH of the solution, when $\sigma = K/[H^+]$ increases.

According to Eq. (147), volume catalytic hydrogen waves with a reversible electrochemical step are unsymmetrical. The logarithmic plot of the waves in E vs. $\log[i^{2/3}/(i_{lim} - i)]$ coordinates gives a straight line, as was shown experimentally [750,751] and the reciprocal of the slope of the line is about 59 mV. In agree-

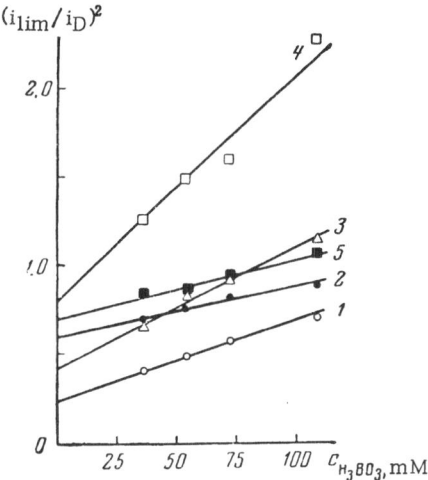

Fig. 49. The quantity $(i_{lim}/i_D)^2$ as a function of undissociated H_3BO_3 concentration in borate buffer solutions
at pH 8.40 (ionic strength 0.5 N) for hydrogen catalytic
waves caused by solutions of ~0.5 mM pyridine and its
homologs. (1) Pyridine; (2) α-pycoline; (3) β-pico-
line; (4) γ-picoline; (5) α,α'-lutidine.

ment with theory, the half-wave potentials of the waves becomes
more positive by 20 mV if c_{cat} increases tenfold [751].

Under such conditions, when the limiting step of the process
is the protonation of the catalyst, the half-wave potential for a reversible volume wave becomes more negative if the pH of the solution increases (which means an increased σ value). In agreement
with the theory, when the pH of the solution is not too high,
$\Delta E_{1/2}/\Delta pH \approx -40$ mV (Fig. 50). Figure 50 shows that at pH > 8.8,
the change of the half-wave potentials with increasing σ becomes
smaller. This can be explained by the effect of the double-layer
structure, which becomes apparent when the reaction-layer thickness decreases.

Most interesting for confirmation of the proposed scheme of
catalytic wave is the effect of the buffer capacity on the half-wave
potentials of reversible waves [775]. For the majority of processes
limited by antecedent protonation, a shift of the half-wave potential
toward less negative potentials should be expected with increasing
buffer capacity. The same conclusion, namely that half-wave potentials must become less negative if the buffer capacity increases,

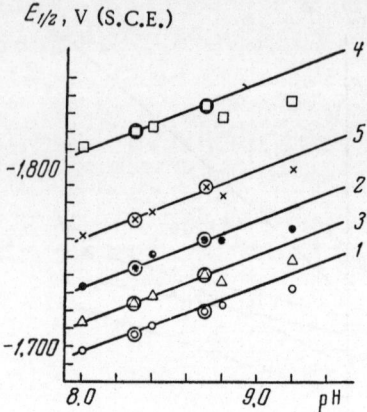

Fig. 50. Half-wave potentials for catalytic waves as a function of pH in borate and veronal buffer solutions (veronal circled points) at a catalyst concentration ~0.5 mM and $\rho \approx 1500$ sec^{-1}. Catalyst: (1) pyridine; (2) α-picoline; (3) β-picoline; (4) γ-picoline; (5) α,α'-lutidine.

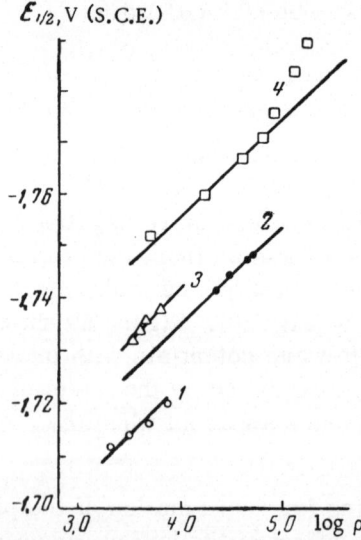

Fig. 51. Half-wave potentials of catalytic waves in pyridine solutions as a function of the observed ρ values. (1) Veronal buffer solution at pH 8.3; (2) ammoniacal solution at pH 9.0; (3) borate buffer solution at pH 9.0; (4) ammoniacal solution at pH 9.0 and 9.5.

can be reached based on the theory of catalytic waves given by Knobloch [735], and Stackelberg and Nürnberg [740]. But Eq. (149) shows that with increasing ρ when the buffer capacity of the solution increases the half-wave potential must become more negative. The experimental data completely confirms this result. Figure 51 shows the dependence of the half-wave potential for waves caused by pyridine [775] on ρ, which means on the concentration of the buffer components of different solutions at a constant pH and ionic strength. In quantitative agreement with Eq. (149), the half-wave potentials of reversible catalytic waves become more negative by about 20 mV if the ρ value increases tenfold.

If pyridine is used as a catalyst, it is not possible to deter-mine precisely the limiting value of the catalytic current governed by a second-order reaction [750]. The catalytic wave caused by pyridine was investigated not only in alkaline but also in neutral solutions up to pH 6.0 (acetate buffer solution). It was shown that under these conditions the current is limited to a significant de-gree by second-order kinetics [750, 774]. From these experiments it was possible to estimate the approximate value of the second-order rate constant for the reaction (XXXII). At 25°C this value is of the order $5 \cdot 10^8$ liters/mole · sec.

A more pronounced limitation of reversible catalytic waves by the rate of the second-order regeneration was found for the wave caused by 2,6-lutidine [750].

3. Determination of the Protonation Rate Constant of the Catalyst; the Effect of Double-Layer Structure on the Protonation Occurring in the Reaction-Layer Volume

From the limiting current values for volume catalytic waves the protonation rate constant (ρ) in the given solution was deter-mined using Eq. (145) [774, 775]. In these calculations the hypo-thetical diffusion current (i_D) for the catalyst was taken to be equal to the diffusion current for the reduction of N-methyl pyridini-um ion, and a k_d value for pyridine of $5 \cdot 10^8$ liters/mole · sec was used.

The specific protonation rate constants for the catalyst, by individual acids present in the solution, can be determined from the dependence of ρ on the concentration of the corresponding proton donor if the concentration of all other components is held constant. Figure 52 shows as an example the dependence of ρ on the concentration of undissociated veronal (barbital), which is one of the components of the buffer solution. The pH and the ionic strength of the solution were kept constant. The ρ value increases linearly with the veronal concentration. The slope for the ρ vs. acid component concentration lines gives directly the specific protonation rate constant of the catalyst by the given proton donor.

The specific protonation rate constants of the catalyst, determined directly from the dependence of ρ on the concentration of undissociated acid, are pH-dependent. Their value increases with increasing pH. The observed apparent increase in the rate constant can be explained by the increasing effect of the electrical double-layer structure with decreasing thickness of the reaction layer, when the solution pH increases [see Eqs. (16) and (26)]. The effect of double-layer structure was investigated by special experiments. In these, changes in volume catalytic waves were studied with increasing ionic strength of the solution, while the pH and buffer capacity of the solution were maintained constant. With increasing ionic strength the thickness of the diffusion part of the double layer decreases, and the pH difference between the electrode layer and the bulk solution becomes smaller. Therefore, the average pH value in the reaction layer increases somewhat and the limiting catalytic current becomes smaller. Calculation showed that the observed [480] decrease of the catalytic pyridine wave height caused by an increase of the ionic strength of the solution can be quantitatively described by equations that take into consideration the effect of the double-layer structure [776].

Particularly significant is the effect of the double-layer structure on catalytic waves in ammoniacal buffer solutions (see page 182). The main proton donor in ammoniacal solutions is the NH_4^+ ion, which also takes part in the formation of the outer shell of the double layer. Therefore, a very large excess of these ions exists at the electrode, and this causes a much larger increase of the ρ value in ammoniacal solutions [775], due to the proton-donor capability of NH_4^+ ions.

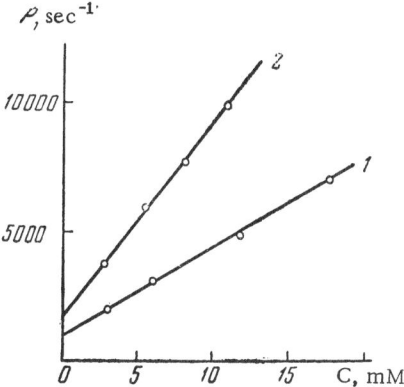

Fig. 52. The dependence of ρ on undissociated veronal (barbital) concentration in solutions at pH 8.30 (1) and 8.70 (2).

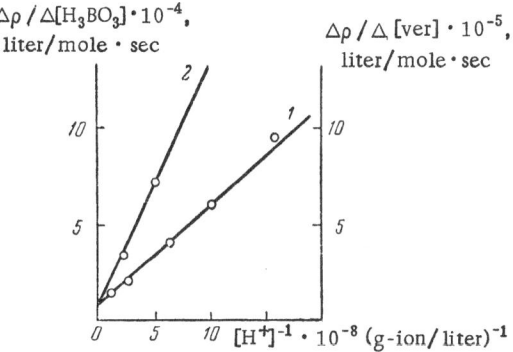

Fig. 53. The observed specific rate constants for the interaction of pyridine with H_3BO_3 (1) and veronal (barbital) (2) as function of $1/[H^+]$.

To determine the true values of protonation rate constants for the catalyst, which are not distorted by the effect of the electrical field of the electrode, the directly determined experimental values must be extrapolated into the region of very large μ values, or low pH values, where the effect of the double-layer structure does not appear [775]. The experiments show that the experimentally determined "apparent" specific protonation rate constants

$k_A^! = \Delta \rho / \Delta [DH^+]_A$ (by graphs of the type given in Fig. 52) change
linearly with the reciprocal of the value of the hydrogen ion concentration. By plotting the $k_A^!$ values vs. $1/[H^+]$ and by extrapolation to $1/[H^+] = 0$, the true values of the protonation rate constants (k_A) can be determined. Figure 53 shows such an extrapolation for the protonation rate constants of pyridine by undissociated boric acid (line 1) and by veronal (line 2) at 25°C and an ionic
strength of 0.5. The true rate constant values determined for pyridine from the ordinate intercept are $k_{H_3BO_3} = 9.0 \cdot 10^3$ and
$k_{ver} = 1.2 \cdot 10^5$ liters/mole · sec.

The intercepts of the lines (of the type given in Fig. 52) with
the ordinate correspond to the "apparent" $\rho_0^!$ values – the sum of
rate constants for interaction with water and hydrogen ions. To
determine the ρ_0 values not distorted by the effect of the double-
layer structure, the $\rho_0^!$ value must be divided by $k_A^!/k_A$ for the
given solution. Using the dependence of the ρ_0 values obtained by
applying this correction on $[H^+]$, the rate constant for pyridine
protonation by hydrogen ions was determined to be $k_{H^+} \approx 2.5 \cdot 10^{10}$
and for protonation by water, $k \approx 7$ liters/mole·sec [775].

4. The Effect of Temperature on

Volume Catalytic Waves in

Buffer Solutions

The height of volume catalytic hydrogen waves increases at
higher temperatures. From the dependence of limiting currents
on the concentration of acid components at different temperatures
and using a correction for the double-layer structure, the specific
rate constants for the protonation of pyridine were determined [774].

It is interesting that the relative change of the apparent rate
constants of protonation caused by the effect of the double-layer
structure at increasing solution pH [which means the value of
$\Delta k_A^! / \Delta (1/[H^+])k_A$ is almost independent of temperature [774].
The explanation of this is that the change of the pH difference at
the electrode surface and in the bulk solution (which causes the apparent change in the rate constant for protonation) is proportional
to the exp $(-\psi_1 F/RT)$ value [see Eq. (99)]. This latter value is almost temperature-independent because the ψ_1 value (if it is not too
small) changes almost proportionally with the absolute temperature

[see Eq. (96)]. If the work is carried out in solutions of higher ionic strength, 1.0 instead of 0.5 N, the apparent increase of rate constants for protonation with increasing solution pH is much less [774], since the thickness of the diffusion part of the double layer is smaller. The activation energies were determined from the dependence of specific protonation rate constants on the reciprocal temperature. The activation energy for the interaction of pyridine with boric acid is 5.1 kcal/mole; with veronal, 6.4 kcal/mole; with hydroxonium ions, 2.4 kcal/mole; and with water, about 6 kcal/mole [774].

It is interesting to note that lines corresponding to the Brönsted equation (92) for pyridine protonation, constructed for different temperatures, have a constant slope ($\alpha = 0.58$) within the experimental error and are only somewhat displaced parallel to each other [774].

5. Reversible Volume Catalytic Hydrogen Waves in Unbuffered Solutions

On the basis of the reaction-layer concept, as was done for kinetic waves observed in solutions of weak acids (see section C-4 in Chapter IV), the equation for the limiting current of a volume catalytic wave was derived in an unbuffered solution [541, 749]

$$i_{\lim} = (Qc_{\text{cat}})^{2/3}, \tag{150}$$

where

$$Q = k_1 s F \,[\text{H}_2\text{O}] \, D^{1/2} \varkappa_{\text{OH}^-}^{1/2}/k_2^{1/2}. \tag{151}$$

It can be seen that the Q value in Eq. (151) is identical to the Q value in Eq. (84) if $n = \nu = 1$, and if the coefficient is 0.81 instead of 0.886. From Eq. (150) it follows that the height of a catalytic wave in an unbuffered solution changes proportionally to the $2/3$ power of the catalyst concentration.

Figure 54 shows [770] the height of the catalytic wave observed in 0.5 N KCl solution as a function of the $2/3$ power of the concentration of pyridine and its homologs, respectively. It can be seen that the agreement with Eq. (150) is satisfactory. From the slopes of graphs constructed in $i_{\lim}$ vs. $c_{\text{cat}}^{2/3}$ coordinates,

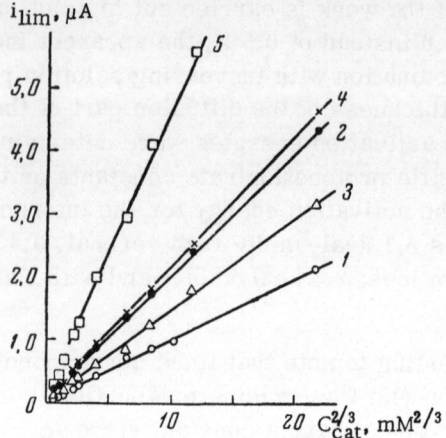

Fig. 54. The dependence of limiting catalytic currents on the $^2/_3$ power of catalyst concentration for waves caused by pyridine and its homologs in unbuffered solutions. (1) Pyridine; (2) α-picoline; (3) β-picoline; (4) γ-picoline; (5) α,α'-lutidine.

which are equal to the Q values, the protonation rate constants of pyridine and its homologs by water were determined by Eq. (151) (Table 2).

The equation for the shape of reversible catalytic waves in unbuffered solutions can be easily derived [541] by substituting the $[BH^+]_S$ value (which is equal, in this case, to $[RH]_S$ from expression (89) at $n = \nu = 1$) and the $[BH]_S$ value [which is equal to $[R]_S$, from expression (132)] into the Nernst equation

$$F = \varepsilon_0'' - \frac{RT}{F} \ln \frac{i^{5/3}}{i_{lim}^{3/2} - i^{3/2}} . \tag{152}$$

If the current intensity in Eq. (152) is expressed in μA

$$\varepsilon_0'' = E_0 + \frac{RT}{F} \ln \frac{k_d^{1/3} \varkappa_{OH^-}^{1/2}}{k_2^{1/2} (sF)^{1/3} D^{1/6}} + 59 \ .mV. \tag{153}$$

Figure 55 shows the dependence of the $\log [i^{5/3}/(i_{lim}^{3/2} - i^{3/2})]$ function of the current on the potential for catalytic waves at a catalyst concentration $c_{cat} = 24$ mM (for waves in pyridine solutions the data are given at several c_{cat} levels). The upper part of the polarographic curves (above a certain current value, at which

TABLE 2. Catalytic Properties of Pyridine and Its Homologs

Catalyst	Pyridine	α-Picoline	β-Picoline	γ-Picoline	α,α'-Lutidine
pK_A (25°C, 0.5 M KCl solution)	5.50	6.15	5.90	6.19	6.83
k_d, liter/mole · sec	$5.0 \cdot 10^8$	$3.8 \cdot 10^7$	$1.4 \cdot 10^8$	$\sim 3 \cdot 10^5$	$1.5 \cdot 10^6$
E_0, V (S.C.E.)	-1.632	-1.670	-1.655	-1.673	-1.710
$k_{H_3BO_3}$, liter/mole · sec	$7 \cdot 10^3$	$(1.5-2) \cdot 10^3$	$(7-10) \cdot 10^3$	$(1-2) \cdot 10^4$	~ 100
k_{ver}, liter/mole · sec	$1.3 \cdot 10^5$	$4 \cdot 10^4$	$(1-2) \cdot 10^5$	$(3-6) \cdot 10^5$	$\sim 8 \cdot 10^3$
k_{H_2O}, liter/mole · sec	5.5	10.1	8.0	12.5	26

the very small but still finite buffer capacity of the potassium chloride solution is exhausted) is described satisfactorily by Eq. (152), which was used for the construction of the lines in Fig. 55. (The buffer capacity is caused by traces of dissolved carbon dioxide and other impurities and partly by leaching of the glass.) The experimental values of the current function corresponding to Eq. (152) for waves in the presence of the same catalyst, independently from the catalyst concentration, should lie on a single line. The data given in Fig. 55 for several pyridine concentrations show that with increasing pyridine concentration the lines shift somewhat to negative potentials. This is most probably due to some increase in the pH in the bulk solution caused by the increased catalyst concentration, which was not taken into consideration in deriving Eq. (152).

C. SURFACE CATALYTIC WAVES OF HYDROGEN

1. The Shape and Properties of Surface Catalytic Waves

For purely surface catalytic waves, the catalyst enters the protonation reaction at the electrode surface in the adsorbed state. The value of the surface catalytic current can be determined using Eq. (117) for $n = 1$. In this equation, Γ and Γ' are the quantities of adsorbed B and BH^+ particles, respectively, on a unit surface area of the electrode. For the limiting catalytic surface current Eq. (103) is correct. The electrode processes corresponding to surface catalytic waves include an irreversible electrochemical step, as was shown by experiments [777]. Therefore, the theory of slow discharge as given by Eq. (69a) can be used to relate the current to the potential.

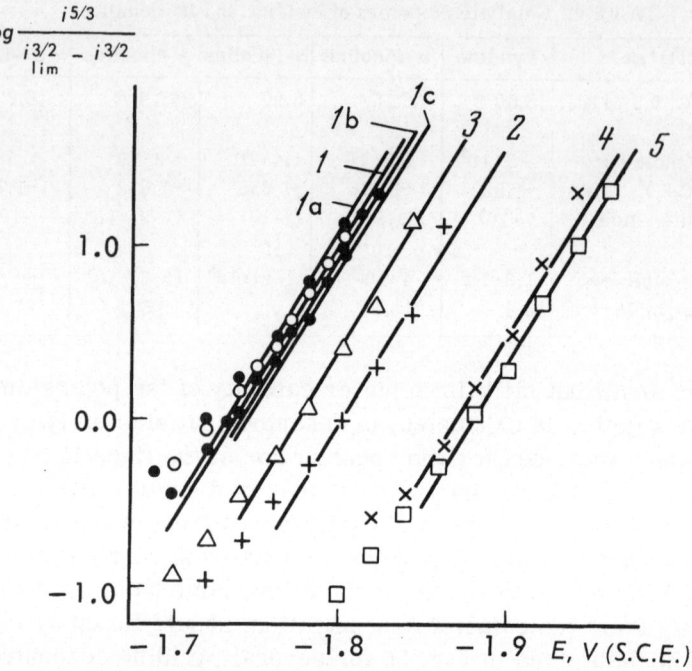

Fig. 55. The quantity $\log[i^{5/3}/(i_{lim}^{3/2} - i^{3/2})]$ as a function of E for catalytic hydrogen waves in a 0.5 M KCl containing 24 mM solutions of pyridine and its homologs. (1) Pyridine (1a, 1b, 1c — pyridine concentration 16, 20, and 100 mM); (2) α-picoline; (3) β-picoline; (4) γ-picoline; (5) α,α'-lutidine.

We will discuss the catalytic processes for the case when an adsorption equilibrium is reached at the electrode surface (or is almost reached). If the adsorption of the catalyst were not affected by the potential of the electrode, and if it were equal to the Γ_0 value which corresponds to the potential of maximum adsorption (E_M), then the wave that would be observed could be given on the basis of Eqs. (69a), (103), and (117) by [778, 670]

$$E = E_{1/2} - \frac{RT}{\alpha F} \ln \frac{i^0}{i_{lim}^0 - i^0} \qquad (154)$$

The zero superscript indicates that the values correspond to the adsorption at E_M.

The total adsorption of the catalyst depends on the adsorption of its basic and acidic forms, which are in a protolytic equilibrium

in the solution. The surface activity of the basic form on mercury
is much higher than that of the acidic form. This conclusion is
based on the data of Gouy and Butler, which pertain to organic bases
in neutral and acid solutions [281, 292]. Furthermore, the adsorption of the cationic form on the cathode changes much less with
potential than that of the basic, uncharged form [281]. For this
reason the adsorption of the catalyst [especially if the equilibrium
(XXX) is shifted to the left] is determined primarily by the change
of the adsorption of its basic form, although the quantity of adsorbed material also depends on the ratio of the forms, which in
turn means a dependence on the pH of the solution.

Due to decreased catalyst adsorption with increasing negative potential, the observed catalytic current (i) is smaller than
current i^0 and $i/i^0 = \Gamma/\Gamma^0$. Taking into consideration that the
catalyst adsorption follows the Henry isotherm, and applying Eqs.
(42) and (112), we obtain for i/i^0 (when $\Gamma_t/\Gamma_e \to 1$)

$$\frac{i}{i^0} = \frac{\beta}{\beta^0} = e^{-a\varphi^2}. \tag{155}$$

Thus, the characteristic peak-shaped form of the observed
surface catalytic waves for hydrogen can be described [777, 778,
670] by Eqs. (154) and (155).

Equations (154) and (155) can also be given in the following
form:

$$E = E_{1/2} - \frac{RT}{an_aF} \ln \frac{i}{i_{\lim} - i} \tag{156}$$

and

$$i_{\lim} = i^0 e^{-a\varphi^2}. \tag{157}$$

These equations can be used for a comparison with the corresponding equations for surface kinetic waves [Eqs. (114) and (123)]. It
can be seen that the equations for catalytic surface waves are much
simpler than the equations for kinetic waves.

For the analysis of catalytic waves it is convenient to construct a logarithmic graph in E vs. $\log [i/(i_{\lim} - i)]$ coordinates,
and it can be assumed that $i_{\lim} \approx$ const and that its value is equal
to the maximum on the peak-shaped current curve. Very often the

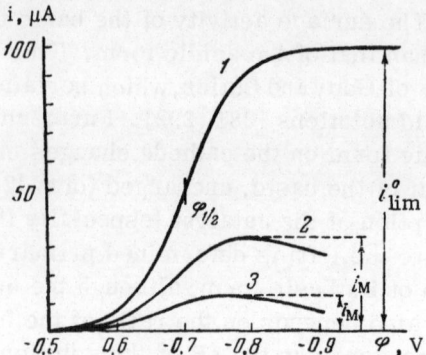

Fig. 56. Scheme for the explanation of maxima on surface catalytic waves. (1) Curve without considering the change of adsorptivity with potential [according to Eq. (154)]; (2, 3) waves with consideration of adsorption changes [$a = 1.63$ and $a = 3.27$ V^{-2}, respectively, in Eq. (155)].

graphs of the waves constructed in this manner are linear, especially their lower section, and the slope of the linear section of such a graph can easily be shown to be equal to $bb'/(b' - b)$ [see Eq. (123)], where $b = RT/\alpha n_a F$, and b' is given by Eq. (122). To determine the a value a graph must be constructed in $\log i_{lim}$ vs. φ^2 coordinates for the descending part of the wave on which the catalytic current attains its limiting value. The slope of the linear section of the graph is equal to $-0.43a$, and by extending to $\varphi^2 = 0$, the i^0_{lim} value can be determined (see Figs. 34 and 61). If we know the a value and $bb'/(b' - b) = b/(1 + 2ab\varphi_{1/2}/2.3)$, it is easy to determine b (see page 211).

To determine the $\varphi_{1/2}$ (or $E_{1/2}$) value, the following equation can be used:

$$\varphi_{1/2} = \varphi_{peak} + b \log\left(-\frac{2.3}{2ab\varphi_{peak}} - 1\right), \tag{158}$$

which was obtained by differentiation of Eq. (156), then putting the result equal to zero, and considering Eq. (157); φ_{peak} in Eq. (158) is the potential of the wave maximum. With the i^0_{lim}, a, b, and $\varphi_{1/2}$ values known, the theoretical curve corresponding to the surface catalytic wave of hydrogen can be constructed using Eqs. (154) and (155) or (156) and (157). An example of the determination of the parameters for a surface catalytic wave and the construction of the theoretical wave are given on page 279.

Figure 56 shows schematically the formation of decreases on surface waves. The maximum height of the observed (i_M) wave is not equal to the limiting current of the wave. In the majority of cases i_M is practically proportional to i^0_{lim}. Therefore, for analytical purposes it is entirely satisfactory to use the directly measurable i_M value of a catalytic wave instead of the i^0_{lim} value.

With increasing buffer capacity of the solution, similarly to other waves that correspond to electrode processes limited by antecedent protonation, the surface catalytic wave increases. The rate constant for protonation was determined from the dependence of the catalytic current in anabasine solutions on the concentration of undissociated acetic acid [778]. For anabasine adsorbed by acetic acid, the rate constant is $2.9 \cdot 10^5$ liters/mole $\cdot$ sec at a potential of -1.84 V (S.C.E.).

An interesting characteristic of surface waves is the unusual dependence on the height of the mercury column (h_{Hg}) observed in several cases. The waves often increase in height with decreasing h_{Hg}. This phenomenon can be observed when the adsorption equilibrium is not attained, and an increase of the drop time (which also means that the time interval for adsorption increases) leads to an increased degree in the approach to the adsorption equilibrium and consequently results in a higher Γ_t value [in such cases Eq. (112) cannot be used for the calculation of catalytic waves; Eqs. (110) or (111) must be used instead]. Thus Vojíř [779, 780] observed an increase of the catalytic wave if the height of the mercury column was lowered.

Similar phenomena also were observed by other investigators (for example, [350, 781]) for the currents caused by cobalt complexes of amino acids and proteins. Thus Stackelberg and Fassbender [738] observed in the measurement of i vs. t curves during the life of a single drop that the catalytic current increases in proportion to t^ν, and in the presence of catalysts with a high molecular weight the ν value is larger than one. Judging from Fig. 2 in [738], in certain cases ν attains a value of about $\frac{7}{6}$, which corresponds to the sum of effects of increasing electrode surface ($\frac{2}{3}$) and increasing quantity of adsorbed catalyst ($\frac{1}{2}$), according to Eq. (111).

As with surface kinetic waves, surface catalytic currents are almost independent of a temperature increase [758]. In this case, an increase of the rate constant of protonation of the catalyst is compensated by a decrease in its adsorptivity on the electrode surface.

It is well known that addition of foreign surface-active compounds to the polarographed solution very often significantly decreases the catalytic wave (see, for example, [530, 736]). The explanation of this is that the foreign adsorbed compounds displace the catalyst from the electrode surface. If the desorption of this added compound occurs earlier than the desorption of the catalyst, and if the potential of desorption of this compound is in the region of catalytic hydrogen evolution, then a sharp increase may be observed on the polarograms when this foreign compound is desorbed [782, 783]. This can be incorrectly regarded as the formation of a new wave.

At very high catalyst concentrations, when the electrode surface coverage by adsorbed particles is almost complete, a decrease of the catalytic current can be observed with increasing catalyst concentration [784]. This phenomenon, observed by Frumkin, Dzhaparidze, and Tedoradze in solutions of diphenylamine containing hydrochloric acid was explained by the hampered formation of the transition complex as $\theta \rightarrow 1$, if the reacting particles occupy a larger surface in the transition state than in the original adsorbed state [784].

2. Mixed Volume – Surface Catalytic Waves

Volume and surface current components usually give one single wave on the polarographic curves although in certain instances the appearance of two separate waves could be expected. Thus, a significant increase in the concentration of irreversibly reduced particles at the electrode (owing to their adsorption) results in formation of an additional reduction wave for these strongly adsorbed particles in certain instances [663, 667, 668]. After this wave, at more negative potentials, the nonadsorbed particles are reduced (more precisely, the weakly adsorbed particles or the energetically less favorably oriented particles). On the other hand, if a large quantity of particles that show a strong adsorptivity is added to a quinine solution, it is evidently possible to change an

irreversible surface wave into a volume catalytic hydrogen wave
[777]. In this process the total wave height decreases by a factor
of about 30, and the half-wave potential becomes somewhat more
positive (in comparison with the half-wave potential of the surface
wave of the same height). This indicates that the volume component of the catalytic wave in a quinine solution is reversible and
can be observed at less negative potentials than the surface wave.
All this shows that under certain favorable conditions a single
catalytic wave can be separated into a volume and a surface step.
Such a separation of the waves was accomplished for catalytic currents in quinine solutions [530].

It can be assumed that the appearance of two humps on catalytic waves in solutions of the leuko form of malachite green and
β-naphthoquinone, observed by Stackelberg and co-workers [739]
was also caused by the separation of the surface and volume components of the current.

At long drop periods and small supporting electrolyte concentrations (especially if the adsorptivity of its cation is small, as
for example for Li^+) it was possible to observe the surface component of the catalytic current in solutions of pyridine and its
homologs [771, 772, 785].

It is possible that under these conditions the surface component is caused in part by the catalytically active products adsorbed at the electrode, which are formed in small quantities as
a result of the electrochemical reduction of pyridine and its
homologs [401].

3. The Effect of Double-Layer Structure

on Surface Catalytic Waves

Effects of the double-layer structure for surface catalytic
hydrogen waves appear as a result of changes in the adsorptivity
of the catalyst, changes of pH in the electrode layer, and, to a certain degree, changes of the dissociation constants of the catalyst,
which are present in adsorbed state. Figure 57 shows the effect
of sodium chloride addition (at constant pH, constant buffer capacity of the solution, and constant catalyst concentration) on the
catalytic wave caused by quinine [786]. Increasing the total concentration of sodium ions in the solution from 0.04 to 0.08 M

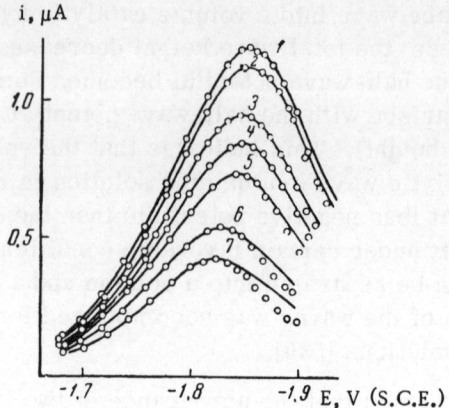

Fig. 57. Catalytic waves of quinine ($c = 3 \cdot 10^{-6}$ M) in
borate buffer solution at pH 9.5. Na^+ concentration
(g-ion/liter): (1) 0.040; (2) 0.045; (3) 0.050; (4) 0.055;
(5) 0.060; (6) 0.070; (7) 0.080.

caused the maximum height of the catalytic wave to decrease almost to one third (see Fig. 57). The potential of the maximum becomes more positive as a result of increased double-layer capacity. Consequently, the a value in Eq. (155) decreased and the absolute ψ_1 value decreased. The curves in Fig. 57 were constructed
using Eqs. (154) and (155). The points correspond to the experimental data.

The change of the nature of the supporting electrolyte shows
a much more significant effect than the increase of the supporting
electrolyte concentration on surface waves. Thus, the catalytic
wave sharply decreases if solutions containing Li^+ are changed to
those containing Cs^+, the over-equivalent adsorption of these cations increases on the mercury electrode, the absolute value of
the ψ_1 potential decreases, and the capacity of the double layer
grows. Consequently, the a value in the Frumkin equation (see
page 60) increases. All these factors result in a decrease of i_{lim}^0
and an increase of the slope of the decrease in surface catalytic
waves if solutions containing cations with larger radii are used
[786].

Zhdanov and Zuman [787] observed a decrease of the catalytic hydrogen wave in solutions of N-troponylamino acids caused
by an increase of the radius of the supporting electrolyte cation.

The very interesting observations made by Pungor and Farsang [789, 790] can be explained from the viewpoint of the double-layer structure. These authors observed that a gradual addition of a tetraethyl ammonium salt to the polarographed solution results first in an increase of the catalytic quinine wave, and then in a decrease. Pungor and Farsang worked with phosphate buffer solutions, in which the main proton donors are the HPO_4^{2-} and $H_2PO_4^-$ ions. Therefore, at very low concentrations of the surface-active tetraethyl ammonium cation the negative ψ_1 potential is decreased, and the transport of proton donors (phosphate anions) to the electrode is made easier. If the concentration of tetraethyl ammonium ion is further increased, the displacement of quinine from the electrode surface becomes apparent, and this leads to a decrease of the catalytic current.

D. CATALYTIC WAVES CAUSED BY SULFUR-CONTAINING COMPOUNDS

1. The Mechanism of the Process

Many organic compounds containing a mercapto group are capable of causing catalytic hydrogen waves to a significant degree. A characteristic property of sulfur-containing catalysts is the sharp increase in their catalytic activity, observed by Brdička [729], on addition of cobalt or nickel salts. Although a great number of investigations have dealt with the catalytic waves in solutions of sulfur-containing compounds (in addition to the already mentioned works see [3-5, 370, 731, 732, 734, 781-783], and see also [791-803]), several aspects of the mechanism of this process have not yet been clarified.

To explain the catalytic effects of compounds containing mercapto groups (thiols), Brdička assumed [3, 729, 732, 743, 744] that the high mobility of the hydrogen in the SH group allows it to be easily cleaved in the electrochemical reaction:

$$RSH + e^- \rightarrow RS^- + H \qquad \text{(XXXIII)}$$

and the RS^- anions formed are rapidly protonated by proton donors in the solution, thus regenerating the electrochemically active RSH form. An important argument supporting this scheme was offered by Brdička [732, 800], who observed that the catalytic activity of thiols is eliminated by addition of salt of monoiodoacetic acid to the solution. The radical of this salt adds to the sulfur and replaces the mobile hydrogen atom.

Mairanovskii and Neiman [804] investigated the reduction of sulfones at the dropping electrode. They found that the reduction product of p,p'-dichlorodiphenyl sulfone (with a transfer of 4 electrons), which is most probably the p,p'-dichlorodiphenyl sulfide, also shows a catalytic wave, although it does not contain an SH group. On the basis of these experiments it was assumed [804] that the mechanism of catalytic wave formation by sulfur-containing catalysts is the same as for nitrogen-containing bases. This means that the sulfonium ion formed by protonation of the sulfide is reduced. A catalytic hydrogen evolution caused by doubly substituted sulfide sulfur was observed quite frequently. For example, it takes place in solutions of ethylene bisthioglycolic acid [739], $HOOCCH_2SCH_2 \cdot SCH_2COOH$, and for sulfides, formed in the electroreduction (with n = 2) of trisubstituted sulfonium ions [757]. The mechanism of the process, assuming a discharge of sulfonium cations as shown by Frumkin and Andreeva [752], seems more probable from the viewpoint of the electrical double-layer structure at the electrode surface. The elimination of catalytic activity of the catalyst by interaction with iodoacetate also can be explained on the same basis: the formed product has a negative charge, which makes its reactions on the cathode more difficult.

Further evidence for the sulfonium scheme for the formation of catalytic waves is that the high activity of cysteine is preserved if mercuribenzoate is exchanged for the hydrogen in its SH group [742] according to the following reaction: $RSH + ClHgC_6H_4COOH \rightarrow RSHgC_6H_4COOH + HCl$. Schwabe and Bär [805] assumed that the reduction products of trithione cause a formation of a catalytic wave on polarograms of its solutions according to the sulfonium scheme. However, the available experimental evidence is not sufficient to draw final conclusions with respect to one or another scheme for the hydrogen-discharge process that is catalyzed by compounds containing mercapto groups.

As was mentioned earlier, a very important factor causing a sharp increase in the catalytic activity of sulfur-containing compounds is the presence of cobalt ions in the solution. The reduction wave for Co(II) precedes the catalytic hydrogen evolution. There-fore, several investigators [732, 734] assumed that catalytically active centers are formed by adsorption of organic catalyst on metallic cobalt deposited at the mercury surface. Březina [806] disproved this assumption, showing that a preliminary electrolysis of the solution on a stable mercury cathode at a potential at which Co(II) is deposited, but at which catalytic hydrogen evolution does not take place (which means an accumulation of metallic cobalt on the electrode), does not affect the catalytic wave measured after-wards at more negative potentials. Therefore, the catalyst must be a complex of the cobalt ion with the organic molecule.

The formation of complexes between Co(II) and sulfur-containing compounds is indicated by the effect of the concentra-tion of the organic compounds on the half-wave potential of the Co(II) reduction wave, which was observed by several investigators.

It can be assumed that Co(II) can form several complexes of different composition [732] with the same thiol, some of which are so stable that they may not enter into electrochemical reactions leading to cobalt deposition, but instead only cause catalytic hydro-gen evolution. Stable complexes that are not reducible are formed between Co(II) and ethylenediamine tetraacetic acid [807], for ex-ample. The simultaneous formation of such a "nonreducible" com-plex (accompanying the deposition of metallic cobalt) from Co(II) with cysteine was confirmed by the decrease of the Co(II) reduc-tion wave with increasing cysteine concentration. This system was investigated by Sunahara [794].

Apparently, the deposition of the metal ion from the complex formed with the thiol takes place parallel with catalytic hydrogen evolution. The effect of this reduction becomes significant at rela-tively low concentrations of the complexing agent in the solution. Due to the low concentration of these ions in the electrode layer, and also due to their partial liberation from the complex, the cata-lytic current begins to be dependent on the diffusion of these ions from the bulk solution, and the observed catalytic wave acquires certain characteristics of a diffusion current. This apparently ex-plains the effects observed in cysteine solutions at low Ni^{2+} ion concentrations [796]. The discharge of nickel from catalytically active complexes proceeds relatively slowly, and the complexing agent can take part in the catalytic process several times before its nickel is reduced. This is why the catalytic current is much higher than the diffusion current corresponding to the reduction of nickel ions [796].

The increase of catalytic activity of thiols due to the formation of their complexes with Co(II) is related, apparently, to the increase of the catalyst charge by two units. From the viewpoint of the effect of double-layer structure the increased charge should significantly facilitate the transfer of electrons on such particles.

The presence of Co(II) and Ni(II) increase the catalytic activity not only of sulfur-containing compounds but also of pyridine [808]. The catalytic activity of pyridine is considerably higher in the presence of Co(II) than Ni(II), at equal concentrations; the catalytic activity is exceptionally high in the presence of both Ni(II) and Co(II). The catalytic hydrogen wave caused by pyridine complexes with Co(II) and Ni(II) has a pronounced surface character. With increasing temperature the height of the catalytic wave decreases rapidly and above 35°C in a solution containing 0.1 N pyridine, 0.1 N pyridinium hydrochloride, and $5 \cdot 10^{-4}$ M Co(II) and Ni(II) salts, the wave cannot be observed [808].

A great majority of compounds of biological interest contain not only catalytically active sulfur but also nitrogen in the form of amino groups, which also are able to catalyze hydrogen evolution. The observed catalytic effect in solutions of such compounds is determined by the overall effect of all active groups in the molecule. Sunahara [795] showed on the example of the wave caused by cysteine and its derivatives in which either the amino groups (N,N-diacetyl cysteine) or the carboxylic groups (diethyl ester of cysteine) were blocked, that the blocking of amino groups leads to a complete loss of the catalytic activity of cysteine, while the esterification of the carboxylic group does not affect the catalytic wave very much. For this reason it can be assumed that in catalysts containing both amino and mercapto groups the catalytic activity is caused mainly by the amino groups, while the adsorptivity of the catalyst is determined mainly by the SH groups. From this viewpoint there is a parallelity, as observed by Ivanov [809], between the ability of the catalyst to suppress a polarographic maximum which corresponds to their adsorptivity, and the height of the catalytic wave, measured at equal weight concentrations of the catalyst. The adsorptivity of catalysts, which determines under equal conditions the limiting catalytic current, increases if the size of the molecule or the number of SH groups in the molecule increases. At the same time the number of amino groups per unit weight of catalysts having a similar structure is almost independent of the size of the catalyst molecule. The same explanation also pertains to the results of Ito [810], who found an approximate proportionality in the increase of catalytic activity with increasing number of cysteine groups in the protein catalysts.

When the catalyst concentration in the solution increases, the height of the catalytic wave, caused by sulfur-containing compounds, increases. The dependence of the wave height on the catalyst concentration can be given by a curve that is similar in form to the Langmuir isotherm. The deviation from linearity between the wave height and catalyst concentration is caused by exhaustion of the buffer capacity of the solution in the electrode layer or by the decrease of the concentration of the complex-forming ion (Co^{2+}, Ni^{2+}) at the electrode.

The effect of pH on catalytic waves caused by cobalt or nickel ion complexes with sulfur-containing compounds is much more complex than in the case of simple nitrogen-containing catalysts. If complex ions are the catalysts, the maximum wave height occurs at a certain pH of the solution [801]. At higher pH values the wave decreases in height because the rate of the protolytic reaction decreases. If the pH of the solution is lower than the optimum value, the wave decreases because it becomes more difficult to form the catalytically active complex. This is due to the protonation of the functional groups contained in the organic molecule that takes part in the complex formation [801].

2. The Nature of the Doubling

of Protein Waves

The catalytic waves observed in the polarography of protein solutions containing cobalt ion, in contrast to the waves caused by many amino acids, often have a characteristic two-peak form (Fig. 58).

To determine the nature of the catalytic double-wave formation in protein solutions, the effect of the drop time was investigated on the individual steps of the wave [350]. The experiments

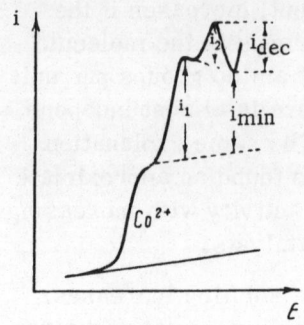

Fig. 58. Polarographic curves of a cobalt(II) salt in the presence of proteins. Co^{2+} indicates the discharge wave of cobalt ion; i_1 is the height of the first catalytic protein step; i_2 is the height of the second step; i_{dec} is the depth of the decrease; i_{min} is the current in the minimum of the wave.

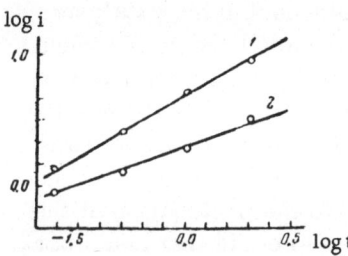

Fig. 59. Dependence of the height of the individual steps of the protein wave in ammonium solutions of crystalline human serum albumin (186 mg/ liter), in presence of Co(III), on the drop time. (1) First protein step; (2) second step.

were conducted on a special dropping electrode equipped with a glass rod for regulated drop separation; this glass rod could be displaced by a microscrew. By this method, the flow rate of mercury was practically constant and the drop time period (t) was adjustable over a wide range.

Figure 59 shows the dependence of the logarithm of the height of the steps as a function of log t on polarographic curves obtained with human serum albumin. For both steps this dependence is linear.

The slope of the line corresponding to the first protein step (line 1) at low protein concentrations and with an excess of cobalt ions is close to $^7/_6$, which means that this step increases proportionally to $t^{7/6}$. A current increase proportional to $t^{7/6}$ is apparently caused by the increased surface concentration of adsorbed catalyst if the catalyst adsorption process occurs before the adsorption equilibrium, which is limited by diffusion, is reached. The slope of the line 2 in Fig. 59, corresponding to the height of the second protein wave, is $^2/_3$. This indicates either a surface character of the wave, after attainment (or close to attainment) of adsorption equilibrium by the catalyst, or it may indicate a volume process. In both cases the current is proportional only to the surface of the electrode. A similar relationship between the heights of the individual steps and t was observed for the wave of sulfosalicylic filtrate of blood serum [350].

When the protein concentration increases (and at the same time the cobalt ion concentration increases) the exponent for t in the relationship for the height of the first protein wave from the drop time gradually decreases from $7/6$ to $2/3$. Under these conditions the catalyst adsorption ceases to follow Eq. (111) and instead follows Eq. (110) or even (112).

The analysis of a great number of polarograms showed [811] that the depth of the current decrease after the second catalytic wave (i_{dec}) (see Fig. 58) is always smaller than the height of the first catalytic wave (i_1) on the same polarographic curve, but is often larger than the height of the second wave (i_2). The factors causing a decrease of i_1 (for example, a decrease of the drop time of the electrode) simultaneously decrease the i_{dec} value (see Fig.1 and Fig. 3 in [350]); therefore, on polarograms on which the first protein wave is nearly absent, no decrease can be observed after the second protein wave. And in exactly the same way, if the concentration of the protein is significantly increased, and the rate of the electrode process becomes independent of its surface concentration, the desorption of the protein with increasingly negative potential does not cause a decrease on the first wave. The decrease does not appear on the second wave either (see, for example, Fig.1 in [812]). These data prove that the second protein wave does not itself have a decrease, and the observed diminishing of the current after the second wave is caused only by the decrease of the first wave at more negative potentials. Therefore, the current intensity at the minimum (i_{min}) (see Fig. 58) is never smaller than i_2. The absence of a decrease for the second wave and the proportionality of i_2 to $t^{2/3}$ indicate that the second wave has a volume character. This is further indicated by the observation that an addition of gelatin to the solution decreases the height of the first wave only (see Fig. 4 in [813]).

An addition of higher fatty acids to the solution (which combine chemically with the catalyst [813]), decreases the height of only the second wave [813]. This indicates that the first and second protein waves are caused by catalysts of different nature. Thus, the solution contains at least two types of catalysts [350] that differ in their adsorptive and catalytic properties as well as in their ability to interact with different reagents (for example, with fatty acids [813], formaldehyde [814], etc.).

The presence of two catalyst types cannot be assigned to the presence of different protein molecules in the solution, since a double wave is observed even for individual proteins in cobalt salt solutions. The appearance of two waves is most probably related to the formation of several types of complexes between the Co(II) and the protein molecule, which are different in their composition or structure and have different catalytic properties (see, for example, [815]). According to the opinion of Ivanov and Rakhleeva [816], the first step is caused by a complex of cobalt with protein in a globular state, the second by the same protein but present in an aglobular state.

The concentration of Co^{2+} ions in the solution affects the heights of each protein wave in a different way, which means that it affects the formation of the corresponding catalytic complexes differently. Numerous data indicate [350] that the cobalt ion is bound more firmly in the catalyst causing the first wave.

If there is a complexing agent present in the solution with which the cobalt forms a more stable complex than that formed with the catalyst causing the second wave, but less stable than that with the catalyst causing the first wave, then only the first wave of the protein remains on the polarograms. This was observed by Ito [810] when he worked with solutions of tris-(ethylenediamine) cobalt chloride.

The different dependence of the heights of the individual protein steps on the drop time makes a comparison of data obtained with dropping electrodes with different characteristics difficult. Such a comparison is very important in the investigation of the Brdička wave for serological diagnostic purposes in medicine (see, for example, [817]). To compare the heights of individual protein waves obtained with different dropping electrodes, it was proposed [818] to make the data relative to a certain standard electrode with the following characteristics: m = 1 mg/sec, t = 1 sec, by using [818]

$$A_1 = i_1/m^{2/3} t^{7/6} \text{ and } A_2 = i_2/m^{2/3} t^{1/6},$$

where A_1 and A_2 are the limiting heights of the first and second protein steps, respectively; i_1 and i_2 are their observed values.

At very high flow rates of mercury from the dropping electrode, when tangential movements occur at the surface of the mercury electrode (maximum of the second kind), a sharp increase can be observed only for the first protein wave (at not too high protein concentrations) [818]. This is caused by increased catalyst transport to the electrode surface. Stirring under conditions of a maximum of the second kind then affects volume catalyst currents only when the reaction-layer thickness is much larger than 20 Å [819]. The appearance of a maximum of the second kind makes it impossible to compare the heights of the first protein waves.

Kalous and Pavliček [820] proposed a simpler (although less accurate) method for the comparison of wave heights. This method is based on relating the observed total wave height to a standard electrode with a constant surface area.

3. The Form of Double Waves on Polarographic Curves of Protein Solutions in the Presence of Cobalt Salts

On the basis of the data given in the previous section, it can be assumed that the observed double wave in solutions of proteins in the presence of cobalt salts is the sum of two waves: (1) a surface, first wave; and (2) a volume, second wave. When the catalyst adsorption has attained the equilibrium state, the first wave can be described by either of two sets of equations − (154) and (155) or (156) and (157). The second wave, a volume catalytic wave which apparently has an irreversible electrochemical step, can be given by

$$E = E'_{1/2} - \frac{RT}{\alpha' n'_a F} \ln \frac{i'}{i'_{lim} - i'}, \tag{159}$$

where the limiting catalytic current (i'_{lim}) is potential-independent.

The set of equations (156), (157), and (159) was compared [821] with the experimentally recorded double wave caused by egg albumin in an ammoniacal buffer solution containing trivalent cobalt salt. In Fig. 60 the points give the observed current values measured from the level of the limiting current for the reduction of cobalt ions. The polarographic curve was recorded with a dropping electrode (with regulated drop formation), with the charac-

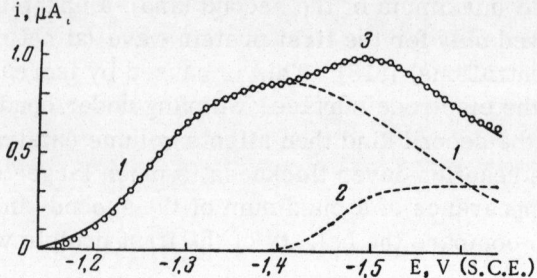

Fig. 60. Catalytic hydrogen double wave in egg albumin
solution in an ammoniacal buffer solution in presence of
$Co(NH_3)_6Cl_3$. Points indicate experimental data measured
from the level of the diffusion current of the cobalt discharge wave. Curve 1 was constructed using Eqs. (156)
and (157); curve 2 using Eq. (159); curve 3 indicates the
total current.

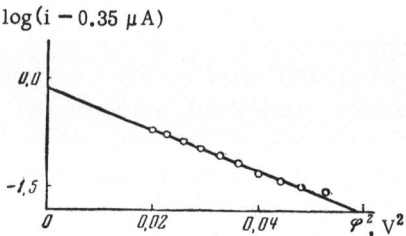

Fig. 61. The dependence of the logarithm of current intensity in the section of the decrease of the first protein
wave on the square of potential, measured from E_M.

teristics m = 0.96 mg/sec, t = 1.2 sec. The solution composition
was: egg albumin, 160 mg/liter (dry albumin was used); 0.25 mM
$Co(NH_3)_6Cl_3$; 50 mM NH_4OH; and 150 mM NH_4Cl. The polarographic curves were recorded using a PE-312 electronic polarograph
with potentiometric potential control using a reference electrode.

It can be assumed that at sufficiently negative potentials
(−1.55 to −1.60 V in Fig. 60), the catalytic current of both steps
practically reaches the corresponding limiting current levels. In
this case, based on the current decrease curve, the constant values
of Eq. (157) that are characteristic of the change of the limiting
current for the first wave with potential can be determined.

Figure 61 shows the dependence of the logarithm of the limiting current for the first wave (determined by subtracting the limiting current of the second wave from the observed total current) on the square of the potential measured from E_M. In constructing this plot the limiting current value for the second wave (i'_{lim}) was assumed to be 0.35 μA and E_M was -1.40 V (S.C.E.). Such a negative E_M value can be visualized by assuming that the catalytically active complex of albumin and cobalt ions has a negative charge. It can be seen from Fig. 61 that the points satisfactorily follow a line corresponding to $i^0_{lim} = 0.88$ μA and $0.43a = 9.0$ V^{-2} ($a = 20.8$ V^{-2}) values. According to the calculations, a deviation in the E_M value of the order of even 100 mV changes the i^0_{lim} and a values only slightly and hardly affects the form of the calculated curve, since in this case the changes of the i^0_{lim} and a values compensate each other. The 20.8 V^{-2} value means [according to Eq. (42)], that each catalytically active center of the adsorbed protein molecule (apparently amino acids containing sulfur) occupies an area of about 200 Å^2 at the electrode surface.

From the determined values of i^0_{lim} and a it is easy to obtain i_{lim} for the first wave at any potential using Eq. (157). But to be able to construct the first wave, the αn_a and $E_{1/2}$ values in Eq. (156) also must be known. The αn_a value can be determined from the slope of the logarithmic graph in E vs. log $[i/(i_{lim} - i)]$ coordinates, assuming that $i_{lim} = $ const. Usually, $E_{1/2}$ corresponds to significantly more negative potentials than E_M and therefore b' > 0 and $[bb'/(b' - b)] > b$. But for the first wave for egg albumin the E_M value (-1.40 V) is at much more negative potentials than the $E_{1/2}$ value [and apparently close to it (see Fig. 60)]. In this case b' < 0 and $[bb'/(b' - b)] < b$, and due to the closeness of the $E_{1/2}$ and E_M values, b' will not be constant but must be determined not only for $E_{1/2}$, but for different E values along the wave. Since the b' values is not constant, the graph of the wave in the usual logarithmic coordinates, assuming that $i_{lim} = $ const, is not linear. Figure 62 shows the graph of the first wave constructed in log $[i/(0.87 - i)]$ vs. E coordinates (for i_{lim} the maximum value of the wave height was taken and not i^0_{lim}). The graph is indeed curved, and as the curve comes closer to E_M its slope increases. It is easy to see that with decreasing $E - E_M$ values, the $bb'/(b' - b)$ value tends toward b. The b value determined by extrapolation to E_M was 84 mV [821].

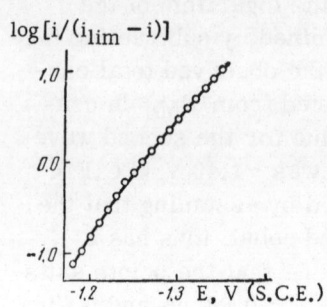

Fig. 62. The $\log[i/(i_{lim} - i)]$ value as a function of E for the first wave assuming that i_{lim} = const.

Equation (158) can be used for the determination of the half-wave potential. Since for cathodic processes the potential of the wave maximum (E_{peak}) is always to the negative side of the adsorption maximum (E_M) (which means that φ_{peak} is always negative), the first term in the parentheses of Eq. (158) is always positive. Usually the maximum appears very clearly on surface waves, but on the curve given in Fig. 60 the maximum is hidden by the rise of the second wave. Still the half-wave potential was determined ($E_{1/2} = -1.22$ V or $\varphi_{1/2} = +0.18$ V) and curve 1 given in Fig. 60 was constructed.

To determine the shape of the second wave, the current values of the first wave calculated by Eqs. (156) and (157) were subtracted from the total observed current. The data obtained by this method were then plotted using Eq. (159) in $\log[i'/(i'_{lim} - i')]$ vs. E coordinates, with an i'_{lim} value of 0.35 μA. This relationship is linear with an $E'_{1/2} = -1.463$ V value and a slope, which is determined by the coefficient before the logarithm in Eq. (159), that is equal to 0.052 V [821]. The dependence of the current intensity on the potential constructed using Eq. (159) is a straight line; it is given in Fig. 60 as line 2. The total catalytic current of both waves is given in Fig. 60 by curve 3, which satisfactorily represents the shape of the experimentally obtained two-step wave. Some deviation of the experimental data and calculated curve at small current values can be expected, since at these potentials the cobalt wave has not yet attained its limiting height (the level of the diffusion current), and at the same time the value of the diffusion current for the reduction of cobalt ions was subtracted from the observed total current values.

E. THE DEPENDENCE OF CATALYTIC HYDROGEN WAVES FROM THE STRUCTURE OF ORGANIC CATALYSTS

1. General Remarks

Knobloch [735] first observed that the overpotential of hydrogen on the dropping mercury electrode decreases more in the presence of more acidic heterocyclic catalysts. The data of Stackelberg and co-workers [739] also indicate a promotion of catalytic hydrogen evolution for decreasing pK_A values in a series of catalysts of similar structure. These data were measured with different classes of organic compounds. Qualitatively this is easy to explain on the basis of the scheme discussed for the formation of the catalytic wave. The reduction of the electrochemically active acidic catalyst form, according to the rule of Shikata and Tachi [585], becomes easier if the molecular frame on which the reducible group is bound is more electronegative [758]. Introduction of a negative group into an acid molecule decreases its pK_A value, as is well known.

The height of the catalytic wave, in contrast, is usually higher for catalysts with larger pK_A [735, 822]. The reason for this is that under strictly identical conditions, with increasing pK_A of the catalyst, the protolytic equilibrium (XXX) between its forms shifts toward the electrochemically active (acidic) form.

2. Reversible Volume Catalytic Waves

The relationship between catalyst structure and the character of the wave caused by them can be most advantageously investigated quantitatively using the example of reversible volume catalytic waves when complicating adsorption phenomena are absent.

Figure 49 shows the dependence of $(i_{lim}/i_D)^2$ on the undissociated boric acid concentration in a solution at pH = const = 8.40. As Fig. 49 indicates, the slope of the lines, which is determined by the rate of protonation, depends on the nature of the catalyst. In exactly the same way, the lines given in Fig. 54, determined with different catalysts in an unbuffered solution, have different slopes. The height of the catalytic wave in the latter case is determined by the rate of protonation by water.

The rate constants for protonation of pyridine and its derivatives by different acids are given in Table 2 (page 261).

For a series of catalysts of similar structure the protonation rate constants with the same acid (k_A) must increase with increasing pK_A of the catalyst in accord with the Brönsted relationship (see page 146). Indeed [749, 770], the $\log k_{H_2O}$ values (the pro-

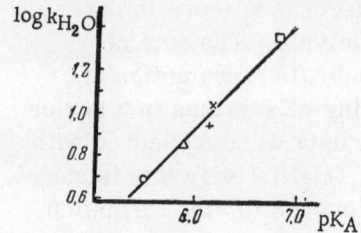

Fig. 63. The logarithms of the protonation rate constants (by interaction with water) for pyridine derivatives as a function of their pK_A
values.

tonation rate constants corresponding to a reaction with water) change
linearly with pK_A (Fig. 63). But for
the protonation effected by H_3BO_3
and veronal, the $k_{H_3BO_3}$ and k_{ver}
values for α-methyl derivatives of
pyridine (see Table 2) are significantly lower than would be expected
on the basis of the pK_A values of
these catalysts.

Brown and co-workers [823]
found that due to steric hindrances
the rate constants for the chemical
reaction according to the Menshutkin reaction for pyridine derivatives having a CH_3 group in α position are significantly lower than
would be expected on the basis of the rule of linear change of the
free energy (see, for example, [824]). For the same α-substituted
pyridine derivative the steric effect is more pronounced if the reagent interacting with the nitrogen of the pyridine ring has a larger
dimension [823].

For the protonation of pyridine homologs with boric acid and
veronal the steric hindrances also become apparent. But for the
protonation of α-methyl derivatives of pyridine by interaction with
water the steric hindrance does not appear: this is most probably
due to the small dimensions of water molecules. Similarly, protonation of these derivatives by hydroxonium ions also occurs without steric inhibition [823].

The dependence of the half-wave potential of catalytic waves
on the catalyst nature can be seen in Fig. 50. In this figure the
half-wave potentials are given for waves in about 0.5 mM solutions
of different catalysts, related to the same $\rho \approx 1500 \ sec^{-1}$ at different solution pH. Furthermore, the dependence of the ε_0'' value in
unbuffered solutions is given in Fig. 55. These figures show that
with increasing pK_A of the catalyst the wave shifts toward negative potentials.

The normal redox potentials E_0 can be determined for reaction (XXXI) using Eq. (149). Substituting k_d values for pyridine
and α, α'-lutidine [750, 774] in Eq. (149) yields for the reduction

of their ions the E_0 values of -1.626 and -1.715 V (S.C.E.), respectively. The E_0 value is related to pK_A by the expression

$$E_0 = \text{const} - \frac{2.3\,RT}{F}\,pK_A. \tag{160}$$

From Eq. (160) the experimental data for pyridine and α, α'-lutidine yield the values -1.302 and -1.312 V, respectively, for the constant. If we consider the low precision of the determination of k_d [750, 774], the deviation in the constant values can be disregarded, and the order of magnitude of k_d can be considered confirmed. The k_d values were determined for picolines (see Table 2) from Eqs. (149) and (160) with const $= -1.307$ V, and these results were very close to the values determined later by independent methods [762, 764]. The set of equations (149) and (160) relates $E_{1/2}$ of catalytic waves with pK_A values of the catalysts.

3. Irreversible Surface Waves

For irreversible surface catalytic hydrogen waves the relationship with the catalyst structure is complicated by adsorption. First, the adsorption increases the catalytic current; second, it can shift the wave to less negative potentials.

For a series of catalysts of about the same adsorptivity the dependence of the potential value (half-wave or the more simply accessible "half-maximum" potential) on the pK_A value of the catalyst can be given by one of the modifications of the Hammett equation (see, for example, [825, 826]). With increasing pK_A the wave must shift toward negative potentials. The limiting current, as was already mentioned, increases with increasing pK_A of the catalyst.

Several examples that illustrate this relationship are given in the studies already mentioned [735, 739, 821].

Zuman and Kuik [826] found that introduction of substituents in the 5-position of pyridine increased its catalytic activity in Co (II) solutions in the sequence OH < NH_2 < SH. The increased activity was explained by these authors by changed acidic properties of these compounds and by their ability to form complexes with Co (II).

Lamprecht and co-workers [742] investigated a great number of sulfur-containing compounds for their activity in catalytic hydrogen evolution. They found that the character of the wave depends on the type of complex formed by the molecule of the sulfur-containing compound with Co(II). Thus, if besides the SH (or SS) group a neighboring carboxylic group takes part also in the complex formation, the catalytic wave is observed immediately after the cobalt reduction wave. The wave shows a characteristic decrease [for example, in solutions of cysteine and Co(II)]. If the complex with Co(II) is formed by the sulfur and by the NH_2 of NH group present in the molecule, the wave resulting is somewhat lower. In this case, instead of a decrease, a plateau similar to that of a limiting current can be observed on the wave. If the organic molecules do not contain other groups besides SH or SS groups, only a gradual rise of the wave is observed before the reduction of the components of the supporting electrolyte without a clearly defined limiting current plateau in ammoniacal cobalt solutions.

The results obtained by Stevančević [827], who investigated the catalytic activity of quinoxalin dithiol-2,3 in cobalt(II) and nickel salt solutions, are very interesting. It follows from his data that the catalytic wave is significantly lower in nickel salt solutions than in cobalt salt solutions, and the nickel wave can be observed at less negative potentials. If it is assumed that the adsorptivity of the complexes formed by quinoxalin dithiol-2,3 with nickel and cobalt ions is about the same, the higher catalytic activity for the waves may apparently be explained by the much higher pK_A value of the complex with Co(II) than with Ni(II). The data of Stevančević further confirm that complexes of organic compounds with cobalt (or nickel) ions cause the appearance of catalytic hydrogen waves in such systems.

Toropova and Elizarova [828] proved that sulfur-containing compounds form catalytically active complexes not only with Co(II) and Ni(II) but also with Fe(II). These authors found [828-830] that introduction of metal ions into the solution that form more stable complexes with the thiols than Ni decreases the height of the catalytic hydrogen wave. A method for the determination of several metal ions at low concentrations [830] was developed based on the displacement of Ni from the catalytically active complex.

The dependence of the catalytic wave on the pK_A value of the catalyst can be used for the approximate determination of unknown pK_A values of catalytically active compounds [831].

For a series of catalysts that show different surface activity, the height of the catalytic wave increases for catalysts of higher adsorptivity [809, 810]. Surface waves are extremely sensitive to minute changes in the catalyst structure. Thus, for example, Zuman [832] found that the second wave on erythrophenyl cysteine polarograms in ammoniacal cobalt chloride solution is almost twice as high as that for the threo-isomer.

4. Catalytically Active Organic Compounds

It was already mentioned that the catalytic activity of organic catalysts is due to the presence of unpaired electrons (at the nitrogen, sulfur, phosphorus, oxygen, arsenic atoms) and their ability to accept protons and form onium compounds. Therefore, catalytic waves are observed in solutions of amines (but not tetra-substituted amines), thiols and sulfides, phosphines [780, 833], and arsines [834-836]. Recently, Knobloch [736, 837] investigated in detail the catalytic hydrogen waves caused by oxonium compounds—chromone derivatives. Zhdanov, Polievktov, and Pozdeeva [838, 839] described "oxonium" catalytic waves caused by diphenyl cyclopropenone in acidic solutions.

Catalytic hydrogen evolution also can be caused by sulfoxides in strongly acid media [840]. The sulfoxides can accept hydrogen ions and form onium compounds, which are reduced more readily than the hydroxonium ions in the same solution [840].

In concluding this chapter, it must be mentioned that besides the discussed theory of formation of catalytic hydrogen waves, other theories have also been proposed; their critical evaluation is given in the last section of the review [749].

Chapter X

The Effect of Composition of Aqueous Organic Solvents on the Polarographic Behavior of Organic Compounds

A. GENERAL REMARKS

The electrochemical behavior of organic compounds differs in several characteristic properties from the behavior of inorganic compounds. These characteristic properties are [841]: 1) significant adsorptivity of organic depolarizers on the electrode surface, leading usually to a significant increase in the rate of the processes taking place on the electrode and at the electrode; 2) participation of hydrogen ions in the potential determining step; 3) the inhibiting effect of the electrode reaction products if their surface activity is higher than that of the initial compounds (or the inhibition caused by almost complete coverage of the surface by adsorbed depolarizer [438, 784]); and, 4) the capability of certain organic compounds to form hydrogen bonds or interact in any other way with the solvent. Therefore, if the composition of the solvent is changed, additional effects can be observed [841] simultaneously with phenomena characteristic for reduction waves of inorganic depolarizers, which are caused by changed diffusion coefficients and activities of ions (see, for example, the review by Schwabe [842, 843], studies of Tur'yan and co-workers [844-846], Tachi and Takahashi [847, 848], and also other investigators [849-852]) for reduction waves of organic compounds.

The half-wave potentials of waves corresponding to reversible reduction can be determined using Eq. (8). In this equation, f_A and f_B are "concentration" activity coefficients. Their values tend toward unity if the ionic strength of the solution and the concentration of the components of the redox system decrease to zero in the given solvent.

If the nature of the solvent or the composition of the solution is altered, both terms on the right-hand side of Eq. (8) change. The change in D_A/D_B is usually small and can be easily measured. But the changes in E_0 and f_A/f_B must be considered in detail.

Since the problem of absolute potential values has not been solved, it is impossible to determine exactly the change of E_0 caused by a change of the nature of the solvent or the composition of an aqueous organic solvent mixture. The problem is even more complex since, in the measurement of the potential of the electrode under investigation relative to any reference electrode that is in a solution of constant composition, a liquid junction potential is formed at the boundary of the two solutions. This potential consists of the diffusion potential (caused by unequal diffusion coefficients of the electrolyte anion and cation) and an interphase potential (caused by unequal solvation of the ions in solutions of different composition). The change of these potentials cannot be calculated when the nature or composition of one of the solutions is altered.

Therefore it has been proposed to relate the electrode potential to the potential of a certain standard redox system, present in the same solutions. Pleskov [853] assumed that the Rb/Rb^+ and Cs/Cs^+ systems are most suited to be used as such "normal" systems. The ions in these systems do not show a tendency toward complex formation and have a small charge and large radius. Therefore they also have a small solvation energy compared to other ions. The potentials of other electrodes in different media should be related to their redox potential. Later, other systems were also proposed for the same purpose, including metalloorganic compounds of iron and cobalt [854-857] (for example, the ferrocene/ferricenium and cobaltocene/cobalticenium systems). But because the solvation energy apparently changes if the solvent nature is altered, even for these ions, the application of these systems as a potential standard leads to some error.

Procedures were developed that minimize the change of diffusion potential in work with external reference electrodes, since its value cannot be determined (by applying intermediate salt bridges between the reference electrode and the investigated solution). The change of the interphase potential in work with these methods remains undetermined. The experiments proved that when the composition of the aqueous organic mixture is changed in the presence of the same supporting electrolyte, the half-wave potentials for the waves of various depolarizers shift either toward positive or toward negative potentials. No relationship was found between the values and the directions of the shifts of half-wave potentials for these waves. This observation leads to the conclusion that the change of interphase potential, especially for mixtures of water with lower alcohols, dioxane, and dimethyl formamide, are apparently very small and that with a certain tolerance they can be neglected in the discussion of a number of phenomena.

In the following discussion, all potentials will be related to an aqueous saturated calomel electrode without correction for the diffusion and interphase potential.

The normal potentials of reversible cation reduction can be calculated using the equation given by Brodskii [858] developed on the basis of theoretical concepts of Born

$$E_0 = A + \frac{ez}{2aD} ,$$
(161)

where A is a constant, z and a are the valence and radius of the cation, and D is the dielectric constant of the solution.

From Eq. (161) it follows that if the dielectric constant of the solution is lowered (for example, by increasing the proportion of organic solvent in an aqueous organic mixture), E_0 becomes more positive. This character of the E_0 shift is considered by Markman and Tur'yan [844] to be primarily responsible for the observed shift of Cd^{2+}, Pb^{2+}, and Tl^+ reduction waves (in the absence of complex formation) to positive potentials when the alcohol concentration is increased in the solution. The magnitude of the shift in E_0 is significantly larger than the shift of the half-wave potential in the opposite direction. The latter is caused by a certain decrease of the concentration activity coefficient f_A [844]. The f_B value, the activity coefficient of the metal in the amalgam, is not affected by the content of organic solvent.

Generally, the change in the half-wave potential for a reversible cation reduction when the composition of the solution is altered is apparently affected by the change of f_A and E_0. If the solvation (hydration) energy of the ion is smaller, which means that the dimension of the ion is larger and its charge smaller, the change of E_0 is less when the composition of the aqueous organic mixture is altered.

We will now discuss the effect of changes of the composition of aqueous organic solvents on reduction waves for organic compounds of different character.

B. ELECTRODE PROCESSES WITHOUT PARTICIPATION OF PROTONS IN THE POTENTIAL DETERMINING STAGE

An example of this case can be the process involved in the reduction of different bromothiophene derivatives [695]. The half-wave potentials of these waves in aqueous organic mixtures are given in Fig. 64.

From Fig. 64 it follows that with increasing organic solvent content in the solution, the waves shift to negative potentials, and the greater the magnitude of the shift the more positive the half-wave potential. A similar relationship was observed earlier by Cisak [859] in the investigation of reduction waves for hexachloro-cyclohexane and dibromocyclohexane isomers in 50 and 90% ethyl alcohol; even before this investigation Levin and Fodiman [575] and Lothe and Rogers [582] found that ethyl alcohol affects waves according to the position of these half-wave potentials.

The authors [582] assumed that the rate-determining step of the process is the transfer of the first electron and they considered the transition complex to be an anion (R−Hal) in which the bond strength between the halogen atom and the rest of the molecules is decreased. For this reason, factors leading to decreased stability of the transition complex must shift the half-wave potentials to positive values. Using this theory they explained [582] the more positive half-wave potentials of the first carbon tetrachloride wave in water compared with its value in aqueous organic solvent. But from this viewpoint it is hard to explain the fact that if several almost water-free solvents are used, the half-wave potentials of the same organic depolarizer (γ-hexachlorocyclohexane) are very close to each other (see, for example, Fig. 9 in [843]).

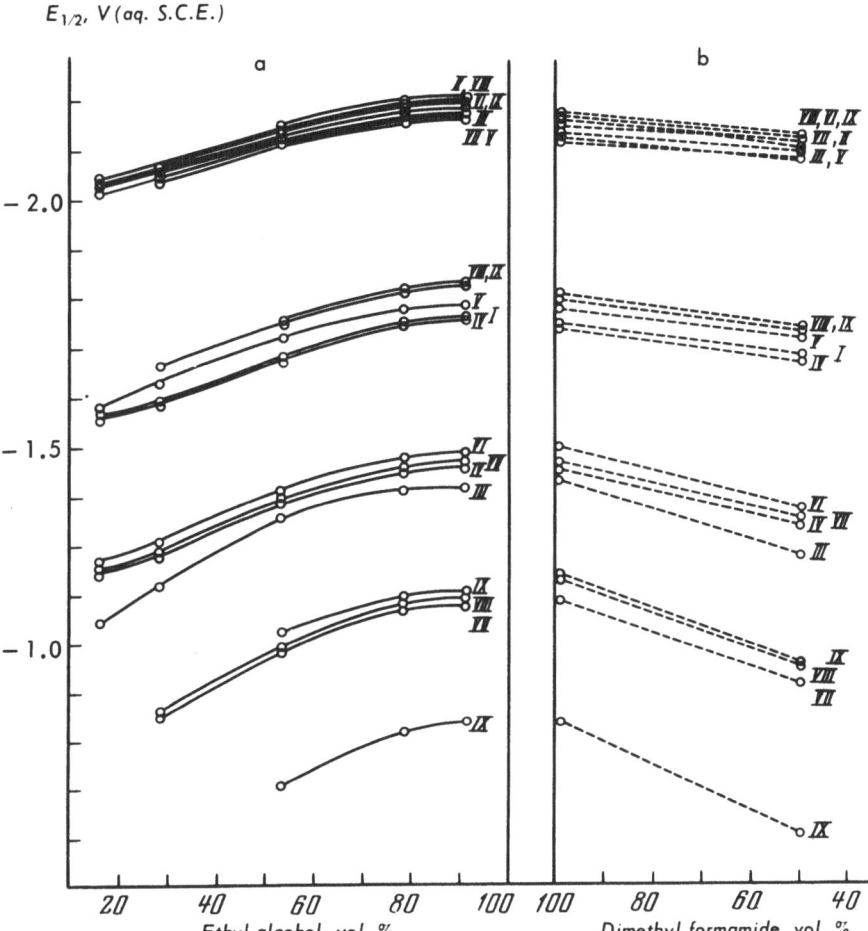

Fig. 64. The half-wave potentials of different bromothiophene waves as functions of the ethyl alcohol (a) and dimethyl formamide (b) content in aqueous organic mixtures. The location of the bromine atom in the thiophene ring for monobromo derivatives: (I) 2; (II) 3; for dibromo derivatives: (III) 2, 3; (IV) 2, 5; (V) 3, 4; (VI) 2, 4; for tribromo derivatives: (VII) 2, 3, 5; (VIII) 2, 3, 4; (IX) tetrabromo derivative 2, 3, 4, 5.

The shift of the wave to negative potentials with increasing alcohol concentration was explained by Cisak [859] by the adsorption of alcohol at the electrode surface, which results in inhibition of the electron transfer. As the potential of the electrode becomes more negative the alcohol adsorption significantly decreases and this, according to Cisak, causes a smaller effect of the alcohol on the wave than can be observed at more negative potentials. Schwabe [843, 860] investigated the effect of the concentration of different organic solvents in aqueous organic mixtures on the polarographic behavior of hexachlorocyclohexane and several other compounds. He also concluded that the cause of the half-wave potential shift for irreversibly reducible depolarizers toward negative potentials with an increase in the proportion of the organic solvent is due to the inhibition of the electron transfer by the solvent particle adsorbed at the electrode.

An increase in the fraction of organic solvent in aqueous organic mixtures usually causes a decrease of adsorptivity for organic compounds at the mercury, and also changes the S-like shape of the adsorption isotherm (see section A in Chapter III, and also [262, 276, 861]). The effect of the organic solvent on electrode processes of the type described (when the depolarizer is adsorbed at the electrode – see section E of Chapter VI) is not, in our opinion, caused by a change (in the limiting case by a significant change) of the rate constant for electron transfer (k'_{el}) but rather by the displacement of the depolarizer from the electrode surface, which means a decrease of the β^0 value [see Eqs. (126) and (127)]. This reduction decreases the overall rate of the process. The effect of organic solvents is stronger if the depolarizer adsorptivity at the electrode is larger. For this reason the maximum effect of a change in the composition of the solution, as pointed out by Cisak [859], can be observed close to the potential of electrocapillary zero. If the possibility of the process proceeding simultaneously with participation of adsorbed and unadsorbed material is considered, it can be explained by an increase in the proportion of the surface electrochemical reaction which is affected more by the increase of organic solvent content.

At low organic solvent concentrations the shift of the half-wave potential is larger if the surface activity of the solvent increases [843, 582]. But as was already mentioned, if solvents with

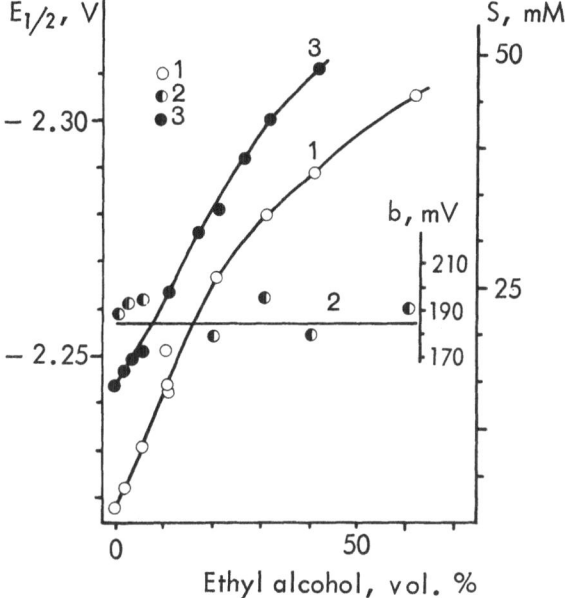

Fig. 65. The dependence of half-wave potentials [1] of the slope of the semilogarithmic graph b = 2.3RT/α n_aF (2) for the butyl bromide reduction wave and also the solubility of C_4H_9Br (3) on the ethyl alcohol concentration in presence of 0.05 N $(CH_3)_4NBr$.

small water content are used, the half-wave potentials of a given depolarizer become almost identical in different solvents (for example, the half-wave potential of bromothiophenes in 91% ethyl alcohol and 99% dimethyl formamide – see Fig. 64). Apparently, at very high organic solvent concentrations, practically complete desorption of the depolarizer occurs, and the nature of the solvent has hardly any effect on the half-wave potential. In this case the half-wave potential is determined not by Eqs. (126) and (127), but by Eq. (10), which is applicable for the half-wave potentials of irreversible waves not affected by depolarizer adsorption. Therefore, only in solutions with high organic solvent content can half-wave potentials that are not influenced by adsorption effects be determined for organic compounds, and only these data can be used to determine the relationship with the structure of these compounds.

However, a shift of the half-wave potentials of irreversible reduction waves for uncharged compounds toward negative potentials with increasing organic solvent content can also be observed in those cases when the compound entering into the electrochemical reaction is not adsorbed at the electrode surface [862]. Figure 65 shows as an example the dependence of reduction half-wave potentials of n-butyl bromide on the ethanol concentration in the solution. The reduction of butyl bromide occurs at potentials that are so negative that its adsorption at the dropping electrode is impossible, and with increasing ethanol concentration the half-wave potential becomes even more negative. It is interesting to note that the slope of the wave remains practically unchanged (line 2 in Fig. 65) which means that the α' value is apparently independent of alcohol concentration. The observed shift in the half-wave potential can be explained by the change of the activity coefficient of the uncharged compound in the solution. The ideas of the absolute reaction rate theory [547] can be used for the semiquantitative explanation of this effect.

The rate constant for electron transfer (k_{el}^0) formulated on the basis of this theory assuming that the concentration of the activated complex is proportional to the depolarizer concentration can be given by

$$k_{el}^0 = k_{el(0)}^0 \frac{f_R}{f^{\ddagger}}, \tag{162}$$

where f_R and $f^{\ddagger}$ are the activity coefficients of the depolarizer and activated complex (transition state), respectively, and $k_{el(0)}^0$ is the rate constant of electron transfer independent of the value of activity coefficients. When a change of organic solvent content does not change the nature and concentration of the supporting electrolyte in the solution, then taking the aqueous solution of the supporting electrolyte as a standard (for which $f_R = 1$) and disregarding the relatively minor effect of the dissolved electrolyte, the change in f_R can be regarded as a change of the zero activity coefficient [863] or as a primary effect of the medium [864].

When the proportion of organic solvent in the solution increases, the solubility of the majority of uncharged organic compounds increases (see, for example, curve 3 in Fig. 65), which means that the activity coefficient f_R decreases. If the solubility of the organic compound is extremely low, the change of its activity coefficient f_R caused by an increase in organic solvent content can be assumed to be inversely proportional to the increase of the solubility (S) of the compound

$$f_R \sim \frac{1}{S}. \tag{163}$$

With respect to the zero activity coefficient of the transition state for the electrochemical reaction $(f^{\ddagger})$ it can be assumed that it changes much less than the corresponding f_R change if the composition of the aqueous organic mixture is altered. This assumption is based on two reasons. In the transition state, the depolarizer is extremely close to the electrode surface (usually it is adsorbed at the electrode) and a certain complex is formed between the depolarizer and the electrode. The properties of this complex and the character of the change of these properties is determined both by the properties of the electrode material and by the properties of the reagent depolarizer. Since the composition of the solution apparently does not affect the activity of the metal electrode, a change in the content of organic component in the solution should affect the $f^{\ddagger}$ value much less than the f_R value.

Furthermore, if the charge of the electrode is sufficiently high, then in solutions containing even small quantities of water, water molecules can be found almost exclusively inside the electrical double layers [865, 866]. These water molecules have smaller dimensions and a larger dipole moment in comparison with organic molecules. Therefore, under the given conditions, an increase of the organic solvent proportion has little effect on the concentration of the water in the layer directly adjacent to the electrode surface, and affects the $f^{\ddagger}$ value hardly at all.

For this reason a decrease of f_R and, at the same time an almost unchanged $f^{\ddagger}$ value caused by an increase of organic solvent proportion in the solution, according to Eq. (162), results in a smaller constant for electron transfer. For the same reason [see Eq. (10)] the half-wave potential shifts to more negative values.

Under conditions where Eq. (163) is valid, the magnitude of the half-wave potential shift, as follows from Eqs. (10), (162), and (163) (if we disregard, as it was mentioned earlier, the changes of diffusion and interphase potentials), is approximately equal to

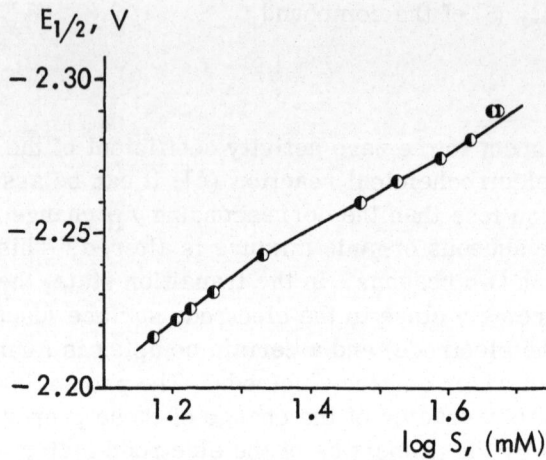

Fig. 66. The dependence of the half-wave potentials of
the reduction wave of n-C$_4$H$_9$Br on the logarithm of its
solubility, changing the ethyl alcohol content in presence of 0.05 N (CH$_3$)$_4$NBr.

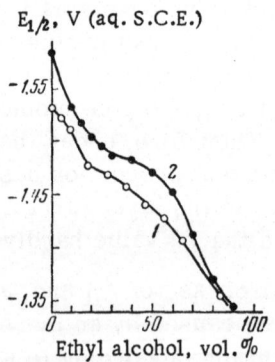

Fig. 67. The dependence of the half-wave potentials of
the reductions of bromoacetic acid (1) and α-bromo-
butyric acid (2) anions from the ethyl alcohol concentration in the solution.

$-(2.3RT/\alpha n_a F)\Delta \log S$. Figure 66 shows the change of the half-wave potentials for the butylbromide reduction wave as a function of its solubility when the ethyl alcohol concentration in the solution changes [862]. At low C_4H_9Br concentrations [which means very low alcohol concentrations in the solution, where Eq. (163) is apparently correct], $\Delta E_{1/2}/\Delta \log S \approx -185$ mV is quite close to $-2.3RT/\alpha n_a F$ (compare with the values for line 2 in Fig. 65).

For the irreversible reduction of charged compounds, an increase in the proportion of organic solvent in the solution causes a sharp change in the half-wave potential to less negative potentials. Figure 67 shows as an example the dependence of the half-wave potential of α-bromocarboxylic acid anions reduction (electrochemical cleavage of the $C-Br$ bond) on the ethyl alcohol concentration in alkaline medium [867]. This character of the organic solvent effect can also be explained [862] on the basis of the theory of absolute reaction rates.

By an increase of the proportion of the organic solvent (such as alcohols, dioxane, and dimethyl formamide) in the mixture with water the solubility of salts usually decreases sharply, which means that the zero activity coefficient of the ions increases significantly [863, 864]. Very often the logarithm of the zero activity coefficients of ions is a linear function of the reciprocal value of the dielectric constant of the solution. Considering a sharp increase of f_R caused by an increase of organic solvent content and an apparently small change of $f\ddagger$ (which was discussed above), the observed shift of the half-wave potentials for reduction of ions toward less negative potentials can be explained using Eqs. (10) and (162).

A change of the zero-activity coefficients caused by the change of the dielectric constant of the medium also affects the half-wave potentials in the reversible reduction of ions. The effect of the f_R change in this case is taken into consideration by a change of the E_0 value in Eq. (161).

C. ELECTRODE PROCESSES WITH ANTECEDENT PROTONATION

For irreversible polarographic waves the half-wave potential is pH-dependent, and the electrochemical step itself is preceded by the protonation of the organic molecule. An increase in organic solvent content increases the pH of acidic solutions [843, 860]. This results in a shift of the half-wave potentials for reduction waves with antecedent protonation to more negative potentials (if at the same time the pK_A value of the depolarizer does not increase significantly).

On addition of an organic solvent to an aqueous solution, the rate of protolytic interactions of bases with water usually decreases sharply. The result is a diminished rate of electrode processes limited by antecedent protonation by interaction with water. Figure 68 shows as an example the dependence of the protonation rate constant for pyridine and 2,6-lutidine by water [276] on the ethyl alcohol concentration in 0.5 M potassium chloride solution. The rate constants were calculated from the value of the slope of the dependence of the height of catalytic volume hydrogen waves on the pyridine and lutidine concentration, respectively, by the equation for limiting catalytic hydrogen current in unbuffered solutions. Figure 68 shows that an increase of ethyl alcohol concentration up to 20 vol.% decreases the rate constant (k_1) for pyridine by a factor of four, and for lutidine by a factor of ten. It was found that the logarithm of the rate constant of protonation by water in aqueous alcoholic solution at not too high alcohol concentrations changes linearly with the reciprocal value of the dielectric constant (D) for the solvent [276] (the D values were taken without considering the effect of dissolved potassium chloride).

If the ethyl alcohol concentration increases, both the basicity of the catalyst decreases (pK_A becomes smaller) and the acidity of the proton donor (water) decreases (pK_W increases). Both these factors cause a decrease of the k_1 value, which is the rate constant for the interaction of the catalyst with water. It is interesting to note that in accord with the Brönsted relationship, the change of $\log k_1$ caused by an increase of alcohol content is proportional to the sum of the absolute values of the change for pK_A of the catalyst and pK_W for the water (Fig. 69) [276]. This ratio, $\Delta \log k_1/(|\Delta pK_A| + |\Delta pK_W|)$, for the protonation of pyridine by water is close to the sum of the slopes of the Brönsted lines, which express the dependence of $\log k_1$ on pK of the acid (~0.5) and of pK_A of the catalysts (~0.58). For 2,6-lutidine the $\Delta \log k_1/(|\Delta pK_A| + |\Delta pK_W|)$ ratio is ~1.35 (see Fig. 69).

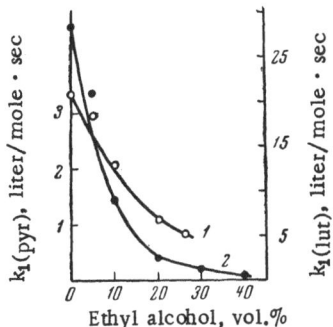

Fig. 68. The dependence of the rate constant of protonation by water for (1) pyridine, $k_{1(pyr)}$, (2) 2,6-lutidine $k_{1(lut)}$ on the ethyl alcohol concentration in aqueous alcoholic solutions.

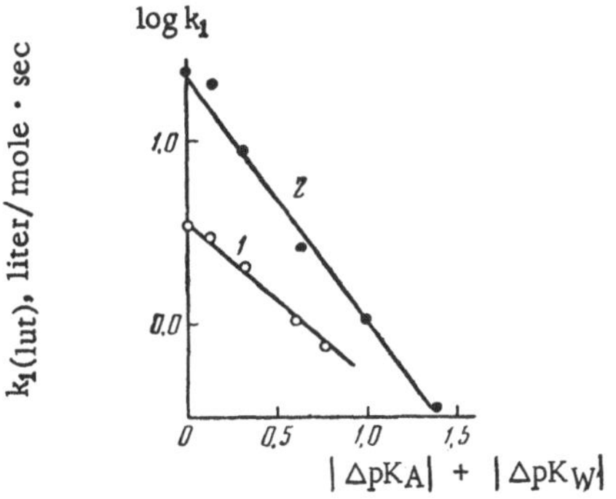

Fig. 69. The dependence of the logarithm of the rate constants of protonation by water for (1) pyridine, (2) 2,6-lutidine on total shift ΔpK_A and ΔpK_W caused by a change of ethyl alcohol content in aqueous alcoholic solutions.

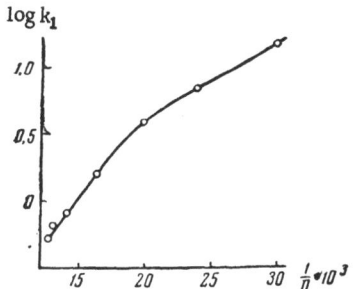

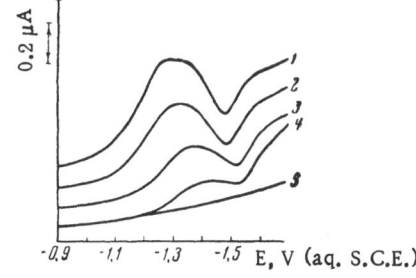

Fig. 70. The logarithm of the protonation rate constant (by water) for maleic acid dianions in aqueous dioxane solutions, in presence of 0.5 M KCl as a function of the reciprocal value of the dielectric constant of the medium.

Fig. 71. Polarographic curves for 0.2 mM 2-acetyl-5-bromothiophene solutions on a 0.1 M KCl + 0.1 M KOH background in water–methyl alcohol mixtures at 25 °C. Alcohol concentration (vol.%): (1) 1.6; (2) 20; (3) 40; (4) 60; (5) curve for the background for 60% alcohol-containing solution.

A decrease of the protonation rate constant for pyridine and 2,6-lutidine by water was observed not only on addition of ethyl alcohol to the solution, but also for dioxane [868].

It must be mentioned that introduction of an organic solvent into an aqueous solution does not always diminish the rate constant of protonation by water. Thus, in the protonation of maleic acid dianions with water, addition of ethyl alcohol or dioxane to the solution results not in a decrease but in an increase of the rate of protonation [869]. Figure 70 shows the dependence of $\log k_1$ on the reciprocal value of the dielectric constant of the solution. The explanation of this organic solvent effect is [868] that on increasing the concentration the dielectric constant of the solution decreases and this results in higher stability of the activated complex that is formed by interaction of the dianion with water $(A^{2-}H^{+}...OH^{-})$. This complex is evidently less polar than the starting maleic acid dianion. Considering the character of the effect of the dielectric constant of the medium on the stability of the activated complex (produced in the chemical reaction), it can be anticipated that for uncharged molecules (apparently they are more frequently encountered in the electrochemical reduction of organic compounds) addition of an organic solvent to the aqueous solution must decrease the rate constant of protonation by interaction with water or other uncharged proton donors.

The rate of the electrode process is particularly strongly decreased by an increase of organic solvent content for antecedent surface protonation. Here, besides the reduction of the protonation rate constant, the adsorptivity of the basic (electrochemically inactive) depolarizer form is also decreased.

Figure 71 shows [348] the reduction waves for 2-acetyl-5-bromothiophene in alkaline solution at different methanol concentrations. The electrochemical cleavage of the $C-Br$ bond is preceded by protonation (by water) of the carbonyl group. This process occurs at the electrode surface with adsorbed 2-acetyl-5-bromothiophene molecules. Therefore, addition of methyl alcohol to the solution shifts the wave to negative potentials and decreases the limiting current value, due both to a decrease of the 2-acetyl-5-bromothiophene surface concentration and to a reduced rate of protonation. Figure 72 shows the effect of methyl alcohol concentration on the half-wave potential and on the slope $bb'/(b'-b)$ for

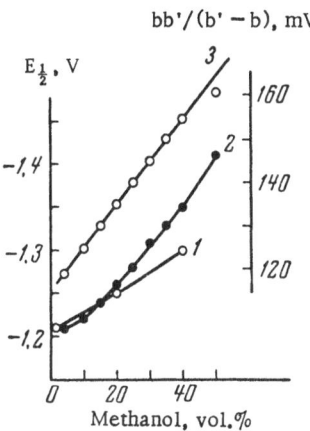

Fig. 72. Half-wave potentials (1, 2) and the slope bb'/(b' − b) (3) for the first wave of 2-acetyl-5-bromothiophene reduction as functions of methyl alcohol concentration. 2-Acetyl-5-bromothiophene concentration (mM): (1) 0.20; (2, 3) 0.26.

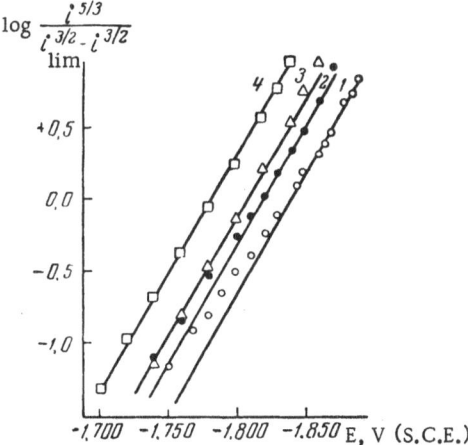

Fig. 73. Logarithmic graph of catalytic hydrogen waves in 0.5 M KCl containing 30 mM 2,6-lutidine. Ethyl alcohol concentration (in weight %): (1) 0; (2) 10; (3) 20; (4) 30.

this wave [348]. The change of the half-wave potential and of the slope is caused mainly by a change of the β^0 value (see section C in Chapter VI).

Stradyn', Reikhmanis, and Gavar [870] observed an increased rate of protonation on addition of ethyl alcohol to the solution in the reduction of 2-nitrofuran and its derivatives in neutral solutions. With increasing alcohol concentration in the solution, the four-electron reduction wave decreases to a one-electron wave (reduction into anion radicals) and at more negative potentials, a new wave appears and begins to grow, corresponding to the reduction of the unprotonated anion radicals of 2-nitrofuran [870]. A great number of similar examples can be cited (see [843, 851, 860, 674, 871-873]).

It was already mentioned that if the proportion of organic solvent in an aqueous mixture increases, the half-wave potential, corresponding to an electrode process with antecedent protonation, becomes more negative [718, 843, 860, 874-877]. Schwabe proved [843, 860], using the example of the first benzaldehyde wave, that the observed shift in half-wave potentials to negative values with increasing concentration of the organic solvent is significantly larger than it would be if it were caused by the pH change of the solution only. A larger shift of half-wave potentials than that corresponding to the pH change was also observed for the first reduction wave of p-nitrobenzene in phthalate buffer [877], where the shift of the half-wave potential is 85 mV when the apparent pH changes by one unit. Schwabe assumed that the additional wave shift is caused by the inhibition of the electrode process by solvent molecules adsorbed on the electrode. This conclusion is contradicted by a series of observations. Thus, it was shown [878] that an increase in ethyl alcohol and isopropyl content does not decrease but instead, to a certain degree, increases the reversibility of the first waves for benzaldehyde and acetophenone in acid solution (see also [717]). Furthermore, a reversible wave was observed for benzophenone in the presence of acid in dimethyl formamide medium [520] and in pure pyridine [879], and for benzil in aqueous alcohol and dimethyl formamide [880].

In the benzaldehyde and acetophenone reduction in acid medium at intermediate organic solvent concentrations, the electrochemical step is close to reversible. This is preceded by the surface protonation of the carbonyl group [524]. Therefore, if the concentration of the organic solvent increases, the half-wave potential not only shifts as a result of pH change and the decrease of the pK_A value of the protonated depolarizer* [see Eq. (125)] but also due to a decrease in its surface concentration. The appearance of this latter effect is caused by the incomplete reversibility of the electrochemical step of the process [878].

A decrease of the surface current component caused by increased concentration of the organic solvent often "improves" the shape of the wave for complex processes, which means that the wave becomes more similar to the theoretical wave. Thus, in the investigation of catalytic hydrogen waves in unbuffered solutions of pyridine and its derivatives, the observed current values are somewhat higher on the initial portion of the wave than the calculated values (see, for example, Fig. 55). This current increase is caused by a sufficient quantity of adsorbed catalyst at potentials corresponding to the initial portion of the wave, which causes a significant surface current. With increasing negative potential, the adsorptivity of the catalyst sharply decreases, and the current becomes a purely volume current. The decrease of catalyst adsorptivity caused by addition of alcohol to the solution (see section C in Chapter III) changes the catalytic wave due to pyridine and its homologs into a purely volume wave along its entire length. Figure 73 shows as an example in E vs. $\log[i^{5/3}/(i_{lim}^{3/2} - i^{3/2})]$ coordinates the graph for the wave of the catalytic current in unbuffered aqueous and aqueous alcoholic solution of 2,6-lutidine [276]. It can be seen from this graph that the semilogarithmic wave plot in these coordinates at an alcohol concentration in the solution above 10% is a straight line, with the theoretically predicted slope $(59)^{-1}$ mV^{-1}.

For mixed volume−surface kinetic waves (see Chapter VII) the fraction of the surface component of the current decreases on addition of alcohol to the polarographed solution, due to decreased adsorptivity of the electrochemically inactive depolarizer form. For this reason graphs constructed according to Eq. (128) become less steep with increasing alcohol concentration [697] (Fig. 74).

---

* The pK_A values for "onium" acids decrease with increasing alcohol concentration, as follows from the pK_A data of pyridine and 2,6-lutidine [276].

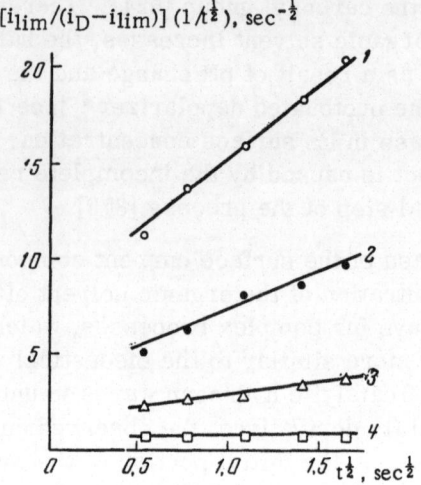

Fig. 74. The dependence of the $(1/t^{\frac{1}{2}})[i_{lim}/(i_D - i_{lim})]$
function on $t^{\frac{1}{2}}$ for the volume–surface limiting current,
limited by the rate of recombination of maleic acid monoanions in acetate buffer solution at pH 5.1 (without alcohol) at 25°C. Ethyl alcohol concentration (in weight %):
(1) 0; (2) 10; (3) 20; (40) 30.

D. PROCESSES WITH ADSORPTION OF THE ELECTROCHEMICAL REACTION PRODUCTS

The adsorptivity of organic molecules on the mercury electrode increases with increasing size (at potentials that are not too
negative). For this reason the products must have a much higher
adsorptivity in comparison with the adsorptivity of starting materials in electrode processes resulting in dimer formation. The
effect of the reaction-product adsorptivity on electrode processes
can be studied principally when the electron transfer takes place
so fast that the electrochemical step can be considered to be reversible or almost reversible.

Such electrode processes, as was shown in Chapter VIII,
take place during the first wave on polarographic curves of aromatic aldehydes and ketones in acid solutions, and also in the reduction of N-alkyl pyridinium salts. The shape of the polarographic wave for such processes is described by Eq. (133). The graph

of the wave in E vs. $\log[i^{2/3}/(i_{lim}-i)]$ coordinates must be a straight line with a reciprocal slope value of about 59 mV. Figure 75 shows the graph [878] constructed in these coordinates, for the benzaldehyde wave in solutions with different isopropyl alcohol contents.* It can be seen from this graph that in the absence of alcohol (line 1), the linear relationship with the correct theoretical slope is valid only on the lower section of the wave, i.e., at small current densities. The upper portion of the curve is displaced toward negative potentials, and this displacement increases with current intensity. The shift of the upper portion of the curve for benzaldehyde is caused [878] by the inhibiting effect caused by dimeric product accumulating on the electrode. With increasing current intensity the coverage of the surface by products increases, and thus the inhibiting effect becomes more pronounced. Addition of isopropyl alcohol to the solution, which reduces the adsorptivity of the dimeric products, shifts the wave shape toward the theoreti-

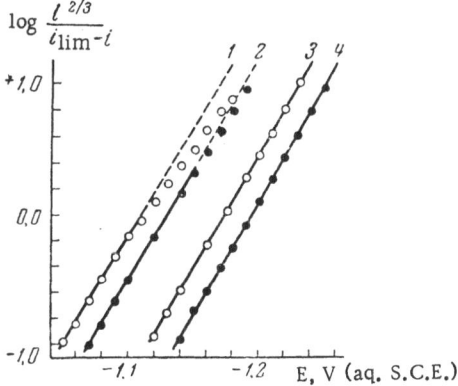

Fig. 75. Semilogarithmic plot of the first benzaldehyde wave (c = 1.5 mM) in acetate buffer solution (0.1 M CH_3COONa + 0.1 M CH_3COOH). Isopropyl alcohol concentration (vol.%) (in parentheses the apparent pH): (1) 0 (4.37); (2) 12 (4.48); (3) 32 (4.85); (4) 42 (5.09). Electrode characteristics: m = 0.91 mg/sec, t = 0.28 sec.

_____

*In [878] it is incorrectly stated that the experiments were made in water–ethyl alcohol mixtures.

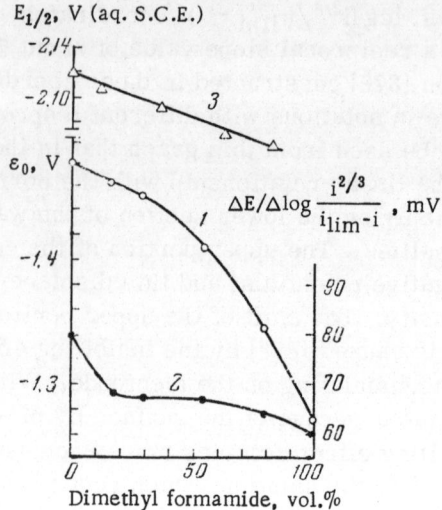

Fig. 76. The characteristic potential ε_0 (1) and the reciprocal slope value of the semilogarithmic graph (2) in E vs. $\log[i^{2/3}/(i_{lim} - i)]$ coordinates for the N-ethyl pyridinium iodide wave (c = 0.64 mM) and the half-wave potential for the rubidium reduction wave (3) as functions of dimethyl formamide concentration in 0.05 M $(C_2H_5)_4NI$.

cal values (see Fig. 75). When the alcohol concentration is about 30%, the entire wave follows Eq. (133). Similar phenomena were observed for the first wave on acetophenone polarograms [878].

In the polarography of N-ethyl pyridinium solutions in a supporting electrolyte containing tetraalkyl ammonium salt, the average value of the reciprocal slope for the semilogarithmic plot, constructed using Eq. (133), also decreases with increasing concentration of the organic solvent (see curves 2 in Figs. 76 and 77) and approaches the theoretical value ($\sim$59 mV). In this case, as in the polarography of aldehydes and ketones, the organic solvent decreases the adsorptivity of the dimer (N,N'-diethyl hydrodipyridyl) at the electrode and thus diminishes the inhibiting action for the electrode process.

It must be mentioned that ε_0 (and also the half-wave potential) for the ethyl pyridinium wave becomes less negative if the concentration of the organic solvent increases in the solution (curves 1 in Figs. 76 and 77) as happens for the reversible reduction of metal cations under conditions where complex formation does not take place (see curves 3 in Figs. 76 and 77 for the reduction wave of rubidium ions). This shift is caused by the increase of the zero-activity coefficients of ions by a decrease of the dielectric constant of the solution, which means a change of E_0.

As was already mentioned, the ε_0 (and $E_{1/2}$) for the first waves for aromatic aldehydes and ketones becomes more negative, when the organic solvent content increases (see Fig. 75). This was observed even though the electrons are transferred on a protonated, positively charged particle. This behavior of carbonyl-containing compounds is due to the surface protonation that precedes their reduction, and the rate of protonation decreases with increasing organic solvent content both due to the reduced activity of hydrogen ions and also because of decreased concentration of unprotonated particles adsorbed at the electrode.

In conclusion, it must be mentioned that only in the presence of organic solvents that diminish the adsorption effects can purely volume catalytic and kinetic waves be observed. The character of these waves can be described by relatively simple theoretical

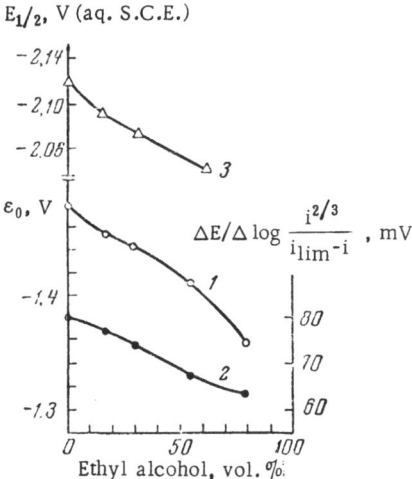

Fig. 77. The dependence of ε_0 (1) and the reciprocal slope value of the semilogarithmic graph (2) for the N-ethyl pyridinium iodide and half-wave potential of rubidium reduction (3) on the ethyl alcohol concentration in aqueous alcohol solutions in 0.05 M $(C_2H_5)_4NI$.

equations. Furthermore, only by polarography in organic solvents
can half-wave potentials for organic compounds be determined
that are not influenced by adsorption effects. These half-wave
potentials can be used to establish relationships with the structure
of organic compounds.

E. INTERACTION OF THE DEPOLARIZER WITH
THE COMPONENTS OF THE SOLUTION

In several instances when the composition of the solution
changes or the nature of the solvent is altered, the half-wave potential can be affected to a certain degree not only by the protolytic or desorption effects of the solvent, but also by specific interactions of the depolarizer with the solvent. Evidently this takes
place when hydrogen bonds are formed between the depolarizer
and solvent (see, for example, [881]), and also in other cases. Specific interactions thus appear in the polarography of p-nitrochlorobenzene in nonaqueous solvents [119]. If the solvent is changed
from methanol to acetonitrile, the half-wave potential for the p-
nitrochlorobenzene reduction shifts from $^{-}0.8$ V (aq. S.C.E.) to
-1.18 V; for dimethyl formamide it shifts to -1.41 V (in 0.1 M NaI).

In several instances the solvent may affect ion-pair formation. Thus, easier ion-pair formation [632] shifts the half-wave
potentials of p-nitrochlorobenzene to positive potentials (-1.87,
-1.52, and -1.41 V, respectively); when the radius of the cation of
the supporting electrolyte decreases in the series of iodides
$(C_2H_5)_4NI$, CsI, and NaI.

Bezuglyi and Rapota [882] investigated the dependence of
limiting currents for reduction waves of certain aromatic aldehydes (benzaldehyde, p-chlorobenzaldehyde, phenyl acetaldehyde)
on the concentration of various alcohols in aqueous alcoholic solutions. They found that in acidic media at sufficiently high alcohol
concentrations, the $i_{lim}\sqrt{\eta}$ product (where η is the viscosity of
the solution) begins to change. A particularly strong decrease of
the wave height was observed in methyl alcohol solutions and a
significantly lower decrease in solutions of ethyl alcohol but only
at low water contents. The $i_{lim}\sqrt{\eta}$ product was almost unchanged
in solutions of isopropyl and butyl alcohol. The i_{lim} value of the
investigated compounds immediately after their addition to the
acidic alcohol solution (extrapolation to zero time) corresponded

to the normal diffusion current, but the i_{lim} value decreased rapidly with time and attained a constant level in 10-20 min. Bezuglyi and Rapota explained the current decrease with time by formation of polarographically inactive semiacetals from the alde- hydes. The constant current value attained with elapsed time cor- responds to the establishment of equilibrium. The equilibrium constants for the formation of semiacetals from these alcohols at different temperatures were determined from the dependence of equilibrium i_{lim} values on the methyl and ethyl alcohol concentra- tion.

Chapter XI

The Effect of Antecedent Metal Ion Discharge on Polarographic Reduction Waves of Organic Compounds

A. INTRODUCTION

The electrode potential value at which the surface charge is zero (in absence of specifically adsorbing compounds in the solution) is determined by the nature of the electrode metal and is called the null point of the metal (E_N). The E_N values for various metals are different. Thus for mercury E_N is equal to -0.21 V (versus a normal hydrogen electrode); for copper, -0.04 V; zinc, -0.66 to -0.69 V. The most negative E_N values are for Na (-1.7 to -1.9), Cd (-0.74 to -0.9), and Tl (-0.65 to -0.86 V).

The E_N value of the electrode material significantly affects the kinetics and mechanism of electrode processes, particularly for processes with participation of organic compounds [656-658]. The reason for this is, first, that the structure of the electrical double layer is determined by the charge of the electrode surface (see section B, Chapter V), which means that the structure depends on how far the electrode potential is from the point of zero charge; second, the adsorption of organic molecules on the electrode surface is maximum, according to Frumkin (see section A, Chapter III), in the potential region close to the zero charge.

The null point of metal alloy electrodes or amalgam electrodes is between the null points of the pure metals from which the alloy is composed. It was shown in the example of amalgams that addition to the base metal (mercury) of even a small quantity of a metal that has a substantially different E_N value sharply displaces the point of zero charge [883]. Therefore, amalgam electrodes are used instead of mercury electrodes in the investigation of the kinetics of electrode processes to change the E_N value, and thus to change the structure of the electrical double layer, while the solution composition is unchanged [562, 590]. Amalgam electrodes also are used to increase the adsorptivity of organic depolarizers and to increase the rate of the electrode process during the electrolysis [884, 885].

If metal ions that are soluble in mercury are reduced on the dropping mercury electrode, the amalgam of the corresponding metal is formed in the surface layer of the mercury electrode. If the reduction of the organic compounds occurs at more negative potentials than the potential for the reduction of these ions, then the reduction of the organic compound evidently occurs on an amalgam dropping electrode.

This chapter gives preliminary qualitative data on the effect of antecedent reduction of certain metal ions on the polarographic reduction waves of organic compounds [886].

B.   ELECTRODE PROCESSES WITHOUT
ANTECEDENT PROTONATION

An increase in the ionic concentration of the metal reduced at more positive potentials than the organic compound and forming an amalgam causes a change of the half-wave potential and shape of the reduction wave for organic compounds. In several instances a certain decrease or increase of the limiting current for subsequent diffusion reduction waves also can be observed. This latter effect can be explained by the altered character and intensity of tangential motion of the surface (which almost always takes place in a dropping mercury electrode) due to changes of the surface tension on different parts of the drop caused by nonuniform amalgam formation. This effect will not be discussed here.

We will discuss the effect of amalgam formation on the reduction waves of organic compounds adsorbed at the electrode. The shape of such waves can be described by Eq. (123) and the half-wave potential by Eq. (126), which can be conveniently given in the form

$$\varphi_{1/2} = b(\log \frac{0.81 \, k'_{el} \, t^{1/2}}{D^{1/2}} + \log \beta^{0}\Gamma_{\infty} - 0.43 a\rho_{1/2}^{2}). \tag{126b}$$

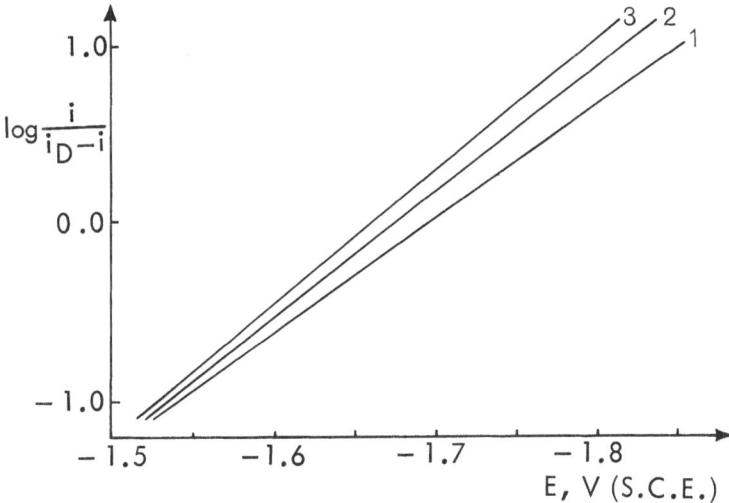

Fig. 78. The quantity $\log[i/(i_D - i)]$ as a function of E for the reduction of 0.95 mM C_2H_5I solution in 0.1 N KCl at different cadmium salt concentrations: (1) 0; (2) 0.3 mM; (3) 0.9 mM. Electrode characteristics: m = 0.97 mg/sec, t = 0.42 sec.

The potential on the polarograms, as is usual, is related to a certain reference electrode, the potential of which is independent of the solution investigated (usually the reference electrode is a saturated calomel electrode). Therefore, shift of the half-wave potential related to such an electrode is determined only by the change of the right-hand side of Eq. (126) when the depolarizer adsorptivity changes. Since $E_{1/2} = \varphi_{1/2} + E_M$, this means that the $E_{1/2}$ value is not directly affected by a change of E_M, which is considered to be zero, during the measurement of φ.

In the formation of amalgam metals whose E_N value is more negative than that of the mercury electrode, the potential of maximum adsorptivity (E_M) of organic compounds (in the absence of specific interaction of the adsorbed compound with the electrode material) also becomes more negative. If the reduction of the organic compound occurs after the reduction of the metal, then the amalgam formation shifts the E_M value to the potential region corresponding to the reduction of the organic compound. In this way, the adsorptivity of the organic compound is increased at the same potentials measured against the independent reference electrode.

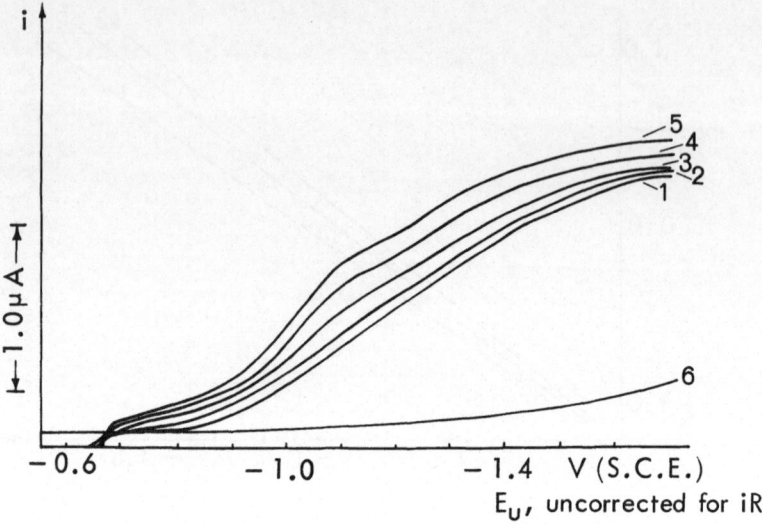

Fig. 79. Polarographic curves for 0.51 mM bromoethynyl cyclohexanol in 0.1 N KCl at different cadmium salt concentrations:
(1) 0; (2) 0.4; (3) 0.8; (4) 1.3; (5) 1.7 mM; (6) residual current.
The waves are shifted along the current coordinate to the almost-
complete compensation of Cd^{2+} reduction currents.

If there is no specific interaction between the depolarizer and elec-
trode material, the k_{el}, b, a, β^0, and Γ_∞ values characteristic of
the depolarizer behavior on the electrode should evidently remain
constant during the amalgam formation. At the same time, the absolute $\varphi_{1/2}$ value becomes smaller and, consequently, the absolute
value of the third term on the right-hand side of Eq. (126b) also
decreases. According to Frumkin, this term takes into consideration the decrease in adsorption due to change of the electrode potential away from E_M. Thus the sum of the terms in the parentheses of Eq. (126b) increases, which means that the half-wave potential becomes more positive and, as follows from the expression
for bb'/(b' − b) (see page 209), the slope of the wave decreases
(the wave becomes more flat).

The shift of E_N on formation of amalgam also changes the
ψ_1-potential value (the potential drop in the diffusion part of the
double layer), which is measured under identical conditions by the
φ_a potential from the potential of zero charge E_N ($\varphi_a = E − E_N$).

If E_N becomes more negative than in the potential region corresponding to the left branch of the electrocapillary curve, the absolute value of φ_a decreases, and therefore the absolute value of the ψ_1 potential also decreases (see section B, Chapter V). In the electroreduction of an uncharged compound, the change in the half-wave potential of the wave agrees in direction and magnitude with the change in ψ_1. Therefore, increased depolarizer adsorption caused by amalgam formation and the change of the ψ_1 potential must shift the half-wave potential to less negative potentials, and both effects are superimposed.

Experimental results entirely confirmed these conclusions. Figure 78 shows a semilogarithmic graph for the reduction of methyl iodide, which follows the deposition of cadmium ions at several cadmium ion concentrations [886]. It can be seen from the figure that with increasing cadmium ion concentration the reduction wave for methyl iodide becomes steeper and shifts to less negative potentials. It was shown by special experiments [886] that addition of methyl iodide to a solution containing cadmium salts does not change the half-wave potential of Cd^{2+}, which means that methyl iodide does not form a complex with Cd^{2+}. The half-wave potential shift of the methyl iodide wave toward less negative potentials was also observed for the antecedent deposition of Tl^+ ions, but the slope of the wave did not change as much as for Cd^{2+} ions. In the reduction of methyl iodide on thallium amalgam, the main factor determining the half-wave potential shift seems to be the change of ψ_1 potential, while for cadmium amalgam a major role is exhibited by the change of adsorptivity. Even a specific interaction of the methyl iodide molecule with the cadmium in the amalgam cannot be excluded, and such an interaction results in a significant increase of the adsorptivity for methyl iodide.

A sharp change of the shape of the wave, accompanied by a doubling of the wave, is observed for the two-electron reduction of bromoethynyl cyclohexanol with increasing cadmium content in the amalgam (Fig. 79). The lower part of the wave shifts much more than the upper part of the wave. This can be explained by assuming a specific adsorptive interaction of the cadmium in the amalgam with the original bromo derivative, resulting in formation of a mercury organic salt and possibly in the formation of a cadmium organic salt to a certain degree. It must be mentioned

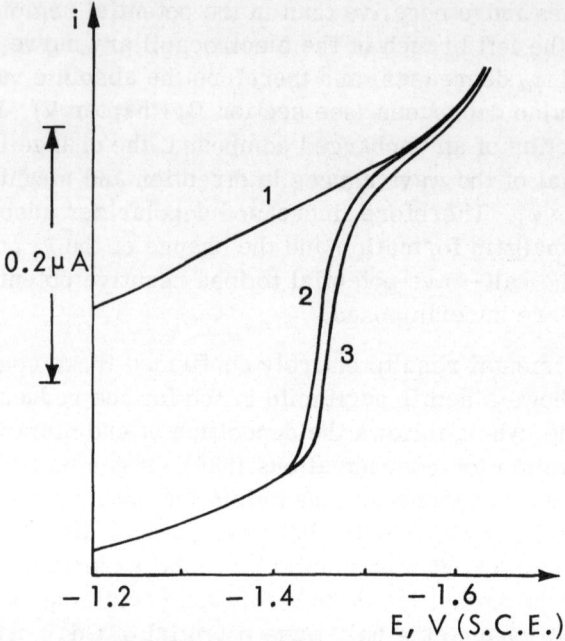

Fig. 80. Residual current curves on a dropping mercury electrode
with m = 1.10 mg/sec and t = 0.27 sec. (1) 0.1 M KNO$_3$ solution
after careful oxygen removal; (2) same solution in presence of
25 mM 2,6-lutidine; (3) 25 mM lutidine solution in the presence
of 0.43 mM TlNO$_3$ in 0.1 M KNO$_3$. Curve 3 is displaced on the
current coordinate so that its lower part coincides with curve 2.

that the resolution of two-electron reduction waves into two waves
for halide derivatives is accompanied by appearance of a maximum
on the first step. Such a maximum is observed when the conditions
of adsorption change, and when the intermediate products of the
electrochemical (or chemical) reaction (often mercury organic
salts) are stabilized by certain factors.

Increased adsorption of 2,6-lutidine on the dropping electrode after preceding reduction of thallium ions on the electrode
is indicated by a small shift of the step on capacity current curves
(charging curves) toward negative potentials that somewhat exceeds the experimental error. This step corresponds to the desorption of 2,6-lutidine (see section C, Chapter III). Figure 80
shows examples of residual current curves on the dropping electrode (charging currents) in 25 mM 2,6-lutidine in the presence
and absence of Tl$^+$ ions in the solution.

A significantly larger adsorptivity increase can be found for
the adsorption prewave (when the adsorptivity of the electrode reaction product is relatively not high) and for surface catalytic hydrogen waves in unbuffered quinine solutions upon addition of cadmium salt to the solution (Fig. 81). The amalgam formation results in a sharp increase in the height of the adsorption prewave
(due to increased adsorptivity of the electrode product) and both
catalytic hydrogen waves (Fig. 81 does not show the second catalytic wave). Special experiments proved that addition of quinine
to a cadmium salt solution does not affect the half-wave potential
of Cd^{2+}, which means that cadmium does not form complexes with
quinine. Only at quinine concentrations in the solution exceeding
1 mM does the Cd^{2+} wave disappear due to inhibition of the electrode process by a layer of adsorbed quinine (more exactly, dihydroquinine).

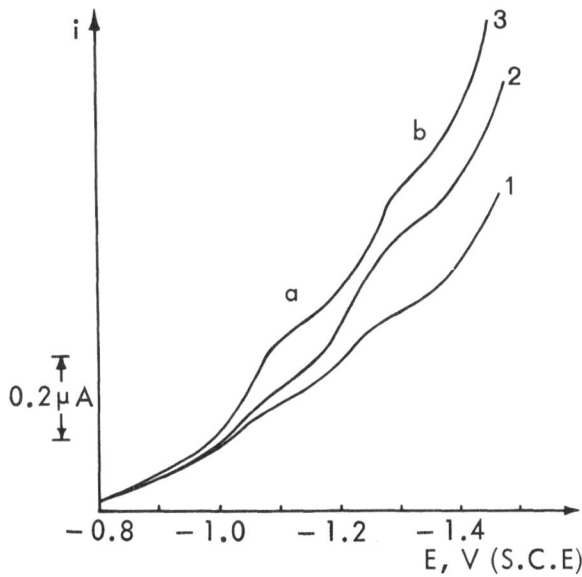

Fig. 81. Adsorption prewave (a) and the first catalytic hydrogen
wave (b) in a 0.10 mM quinine hydrochloride solution in 0.1 M
KCl at different Cd^{2+} salt concentrations: (1) 0; (2) 1.37; (3) 2.56
mM. Characteristics of the electrode: m = 2.2 mg/sec, t = 0.14
sec. The curves are shifted along the current coordinate until
their initial portions coincide.

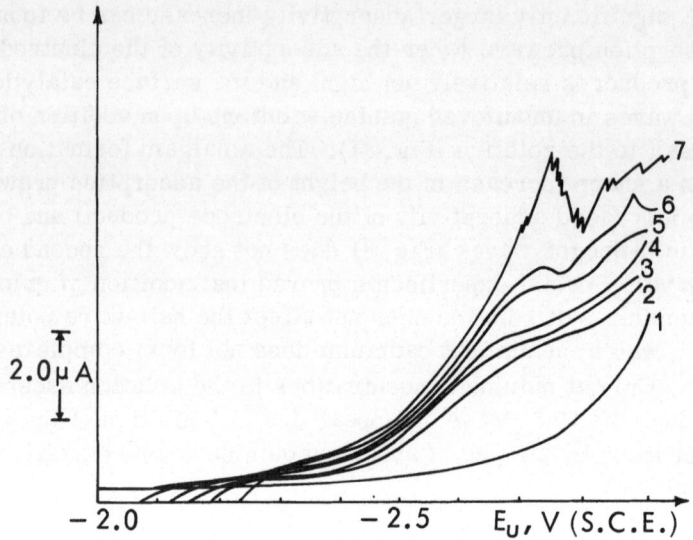

Fig. 82. Polarographic curves of a 1.5 mM dithienyl sulfide solution (catalytic hydrogen waves) in 0.05 N $(C_2H_5)_4NI$ containing
50% ethanol at different NaCl concentrations: (1) residual current;
(2) 0; (3) 0.5; (4) 1.0; (5) 1.5; (6) 3.0; (7) 4.8 mM. The curves are
shifted along the current coordinate, potentials are uncorrected for
the iR of the solution. Electrode characteristics: m = 1.4 mg/sec,
t = 0.28 sec.

An increase of catalytic hydrogen current in unbuffered dithienyl sulfide solutions on formation of sodium amalgam (see
Fig. 82) also can be explained, apparently, by the increased adsorptivity of the sulfide catalyst.

The problem is much more complicated for electrode processes with antecedent protonation reaction in buffer solutions.
The E_N shift of the electrode to negative potentials on formation
of thallium or cadmium amalgam results in a decrease of the absolute ψ_1-potential value, and consequently in a higher pH of the
solution near the electrode (see section E, Chapter V). The change
of the half-wave potential for processes with antecedent protonation, that take place at the electrode itself or in a thin reaction
layer, can be determined in the general case using Eq. (102).

It follows from Eq. (102) that for positively charged or uncharged depolarizers the half-wave potential must become more
negative, if the absolute value of the negative ψ_1 potential de-

creases. Therefore, in this case, the change of depolarizer adsorptivity and the change of the ψ_1 potential during amalgam formation shows opposite effects on half-wave potentials, which means that these effects are subtracted from each other. Depending on which of them is larger, a shift of the half-wave potential to positive or negative values is observed.

Thus, for the surface quasidiffusion reduction wave of phenyl acetaldehyde oxime in an acetate buffer solution at a pH near 5, on addition of cadmium salt to the solution, the half-wave potential becomes somewhat less negative (by about 10 and 15 mV, if the cadmium salt concentration is 0.5 and 1.0 mM, respectively) and the logarithmic graph of the wave becomes slightly steeper. In this case, apparently the effect of the ψ_1-potential change is significantly larger than the effect of the adsorptivity change. The apparent αn_a value for this wave is 0.33, z = 0 (the adsorbed oxime

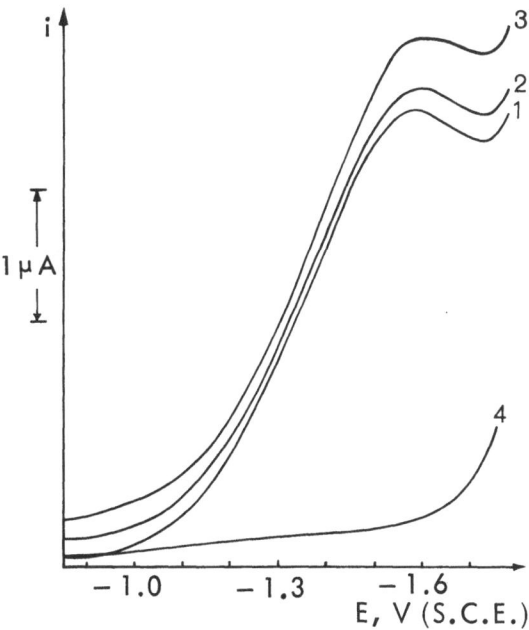

Fig. 83. The effect of cadmium salt concentration in the solution: (1) 0; (2) 0.44 mM; (3) 1.32 mM on the reduction wave for phenyl acetaldehyde oxime (1.0 mM) in an acetate buffer solution at pH 5.6. (4) Residual current. Electrode characteristics: m = 1.10 mg/sec, t = 0.27 sec.

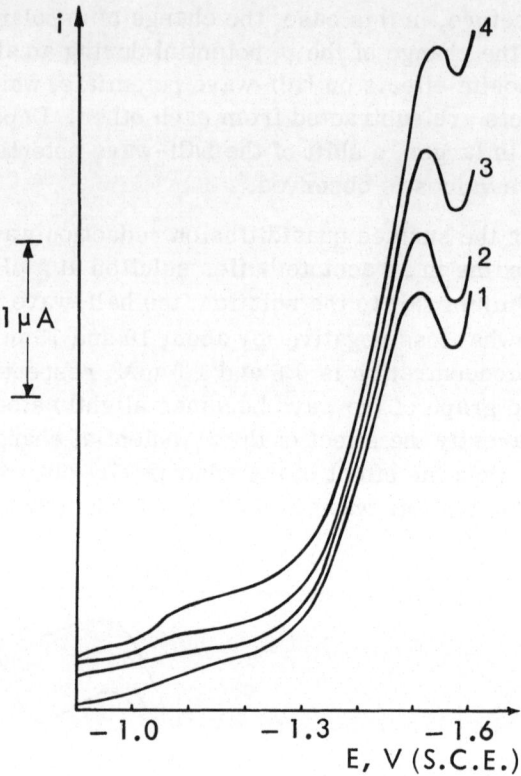

Fig. 84. Catalytic hydrogen waves, caused by quinine ($1.75 \cdot 10^{-5}$ M) in acetate buffer solution at pH 4.0 at different cadmium salt concentrations in the solution: (1) 0; (2) 0.68; (3) 1.81; (4) 3.34 mM. Electrode characteristics: m = 2.2 mg/sec, t = 0.14 sec; the curves are displaced along the current coordinate.

molecules are protonated on the electrode), and consequently, according to Eq. (102), $\Delta E_{1/2} \approx -2\Delta\psi_1$. A slight increase in the steepness of the wave evidently indicates the influence of adsorption.

The latter effect becomes more pronounced if the wave ceases to be quasidiffusion-controlled and becomes a kinetic wave. In this case, instead of the limiting current step, a decrease appears on the wave [for the phenyl acetaldehyde oxime this occurs in less acid solutions (see Fig. 83)].

Figure 83 shows that for this case an increase of cadmium salt concentration in the solution hardly changes the potentials and the slope of the lower portion of the wave, but diminishes the slope of the decrease to a certain degree and shifts the potential of the maximum to more negative potentials (part of the increase of the wave height can be due both to increased depolarizer adsorptivity and also to increased tangential movement on the electrode surface).

Similar behavior was also observed for surface catalytic hydrogen waves in buffer solutions (Fig. 84). With increasing cadmium concentration the catalytic wave increases somewhat. The maximum on the wave shifts significantly to negative potentials, while the potentials of the starting section of the wave remain practically unchanged. In Fig. 84 a sharp increase of the adsorption prewave can be noticed with increasing cadmium concentration.

Literature

1. A. I. Shatenshtein. Isotopic Exchange and Hydrogen Substitution in Organic Compounds. Moscow, Izd. Akad. Nauk SSSR (1960).
2. M. I. Kabachnik. Dokl. Akad. Nauk SSSR 83:407 (1952).
3. R. Brdička. Z. Physik. Chem., Sonderheft, Juli, 165 (1958).
4. I. M. Kolthoff and J. J. Lingane. Polarography (2 vols.). New York, John Wiley and Sons, Inc. (1952)..
5. T. A. Kryukova, C. I. Sinyakova, and T. V. Aref'eva. The Polarographic Method. Moscow, Goskhimizdat (1959).
6. J. Heyrovský. The Principles of Polarography. New York, Academic Press (1966).
7. A. N. Frumkin, V. S. Bagotskii, Z. A. Iofa, and B. N. Kabanov. The Kinetics of Electrode Processes. Moscow, Izd. MGU (1952).
8. A. N. Frumkin. Z. Physik. Chem. 164A:121 (1933).
9. P. Delahay. New Instrumental Methods in Electrochemistry. New York, John Wiley and Sons, Inc. (1954).
10. P. Delahay. J. Am. Chem. Soc. 75:1430 (1953).
11. D. Ilkovič. Coll. Czech. Chem. Comm. 6:498 (1934).
12. H. Matsuda. Bull. Chem. Soc. Japan 26:342 (1953).
13. S. G. Mairanovskii and M. B. Neiman. Izv. Akad. Nauk SSSR, Otd. Khim. Nauk, No. 3:420 (1955).
14. J. J. Lingane and B. A. Loveridge. J. Am. Chem. Soc. 66:1425 (1944); 68:395 (1946); 72:438 (1950); Anal. Chem. 21:52 (1949).
15. T. Kambara and I. Tachi. Sborník I mezinar. polarograf. sjezdu, Praha (1950), I díl. Praha, Přírodoved. vydavatelství (1951), p. 126.
16. H. Strehlow and M. Stackelberg. Z. Elektrochem. 54:51 (1950).
17. M. Stackelberg. Z. Elektrochem. 57:338 (1953).
18. T. A. Kryukova, C. I. Sinyakova, and T. V. Aref'eva. The Polarographic Method. Moscow, Goskhimizdat (1959), p. 614.
19. W. Hans, W. Henne, and E. Meurer. Z. Elektrochem. 58:836 (1954).
20. T. A. Kryukova. Zavodsk. Lab. 9:699 (1940).
21. T. A. Kryukova and B. N. Kabanov. Zh. Fiz. Khim. 15:475 (1941).
22. J. Koutecký. Českosl. Časop. Fys. 2:117 (1952).
23. A. Vlček. Chem. Listy 47:1428, 1440 (1953).
24. L. Meites and T. Meites. J. Am. Chem. Soc. 72:3686 (1950); 73:395 (1951).

25.   W. Hans and W. Jensch. Z. Elektrochem. 56 : 648 (1952).
26.   H. Strehlow, O. Madrich, and M. Stackelberg. Z. Elektrochem. 55 : 244 (1951).
27.   H. Kapulla and H. Berg. J. Electroanal. Chem. 1 : 108 (1959).
28.   J. Koutecký and M. Stackelberg. Progress in Polarography, Vol. 1. New York, Interscience Publishers, Inc. (1962).
29.   J. Heyrovský and D. Ilkovič. Coll. Czech. Chem. Comm. 7 : 198 (1935).
30.   J. Jordan, Anal. Chem. 27 : 1708 (1955).
31.   W. Stricks and I. M. Kolthoff. J. Am. Chem. Soc. 78 : 2085 (1956).
32.   Y. Okinaka and I. M. Kolthoff. J. Am. Chem. Soc. 79 : 3326 (1957).
33.   I. M. Kolthoff and Y. Okinaka. Anal. Chim. Acta 18 : 83 (1958).
34.   Y. Okinaka and I. M. Kolthoff. J. Am. Chem. Soc. 82 : 324 (1960).
35.   J. Weber and J. Koutecký. Coll. Czech. Chem. Comm. 20 : 980 (1955).
36.   S. G. Mairanovskii. Zh. Fiz. Khim. 32 : 2456 (1958).
37.   B. S. Bagotskii. Zh. Fiz. Khim. 22 : 1466 (1948).
38.   A. N. Frumkin. Acta Phys.-Chim. URSS 18 : 38 (1943).
39.   P. Kivalo, K. B. Oldham, and H. A. Laitinen. J. Am. Chem. Soc. 75 : 4148 (1953).
40.   S. G. Mairanovskii, V. A. Ponomarenko, N. V. Barashkova, and A. D. Snegova. Dokl. Akad. Nauk SSSR 134 : 387 (1960).
41.   I. M. Kolthoff and C. S. Miller. J. Am. Chem. Soc. 63 : 1405 (1941).
42.   R. Brdička and K. Wiesner. Naturwiss. 31 : 247 (1943).
43.   K. Wiesner. Z. Elektrochem. 49 : 164 (1943).
44.   R. Brdička. Advances in Polarography, Vol. 2. London, Pergamon Press (1960), p. 655.
45.   K. Wiesner. Chem. Listy 41 : 6 (1947).
46.   J. Koutecký and R. Brdička. Coll. Czech. Chem. Comm. 12 : 337 (1947).
47.   E. Budevski. Izv. Bulgar. Akad. Nauk, Otd. Fiz. Mat. Nauk, Seriya Fiz. 3:43 (1952-1954).
48.   D. Kern. J. Am. Chem. Soc. 75 : 2473 (1953).
49.   V. Hanuš. Chem. Zvesti 8 : 702 (1954).
50.   J. Koutecký. Sborník I mezinar. polarograf. sjezdu, Praha (1950), I díl. Praha, Přírodoved. vidavtelství (1951), p. 826.
51.   J. Koutecký. Coll. Czech. Chem. Comm. 18 : 183 (1953).
52.   J. Koutecký. Coll. Czech. Chem. Comm. 18 : 311 (1953).
53.   J. Koutecký. Coll. Czech. Chem. Comm. 18 : 597 (1953).
54.   J. Koutecký. Nature 174 : 233 (1954).
55.   J. Koutecký. Coll. Czech. Chem. Comm. 19 : 1045 (1954).
56.   Ibid., 19 : 1093 (1954).
57.   Ibid., 19 : 857 (1954).
58.   Ibid., 20 : 116 (1955).
59.   Ibid., 21 : 652 (1956).
60.   Ibid., 21 : 1056 (1956).
61.   Ibid., 22 : 160 (1957).
62.   J. Koutecký, R. Brdička, and V. Hanuš. Coll. Czech. Chem. Comm. 18 : 611 (1953).
63.   J. Koutecký and J. Koryta. Coll. Czech. Chem. Comm. 19 : 845 (1954).

64. J. Čížek, J. Koryta, and J. Koutecký. Coll. Czech. Chem. Comm. 24: 663 (1959).
65. J. Čížek, J. Koryta, and J. Koutecký. Coll. Czech. Chem. Comm. 24: 3844 (1959).
66. M. Neiman. Zh. Fiz. Khim. 22: 1454 (1948).
67. K. H. Henke and W. Hans. Z. Elektrochem. 59: 676 (1955).
68. J. Mecke. Z. Elektrochem. 66: 601 (1962).
69. P. Delahay. J. Am. Chem. Soc. 73: 4944 (1951).
70. P. Delahay. J. Am. Chem. Soc. 74: 3056 (1952).
71. P. Delahay and T. J. Adams. J. Am. Chem. Soc. 74: 1437 (1952).
72. P. Delahay and J. E. Strassner. J. Am. Chem. Soc. 74: 893 (1952).
73. P. Delahay and G. L. Stiehl. J. Am. Chem. Soc. 74: 3500 (1952).
74. R. Brdička and J. Koutecký. J. Am. Chem. Soc. 76: 907 (1954).
75. M. Smutek. Coll. Czech. Chem. Comm. 20: 247 (1955).
76. J. Koutecký and J. Koryta. Electrochim. Acta 3: 318 (1961).
77. R. Brdička, V. Hanuš, and J. Koutecký. Progress in Polarography, Vol. 1. New York, Interscience Publishers, Inc. (1962), p. 146.
78. J. Koutecký and J. Čížek. Coll. Czech. Chem. Comm. 21: 836 (1956).
79. J. Koutecký and J. Čížek. Coll. Czech. Chem. Comm. 21: 1063 (1956).
80. V. G. Levich. Acta Physico-chim. URSS 17: 257 (1942); 19: 117, 133 (1944).
81. J. Koutecký and V. G. Levich. Dokl. Akad. Nauk SSSR 117: 441 (1957).
82. R. Brdička and K. Wiesner. Coll. Czech. Chem. Comm. 12: 138 (1947).
83. R. Brdička. Coll. Czech. Chem. Comm. 19 (Suppl. II): 41 (1954).
84. É. S. Levin. Dokl. Akad. Nauk SSSR 151: 1375 (1963).
85. Z. Galus and R. N. Adams. J. Electroanal. Chem. 4: 248 (1962).
86. R. Brdička and C. Tropp. Biochem. Z. 289: 301 (1937).
87. R. Brdička and K. Wiesner. Coll. Czech. Chem. Comm. 12: 39 (1947).
88. R. Sellner and M. Bucher. Acta Biol. Med. Ger. 7: 427 (1961).
89. I. M. Kolthoff and E. P. Parry. J. Am. Chem. Soc. 73: 3718 (1951).
90. I. M. Kolthoff and E. P. Parry. J. Am. Chem. Soc. 73: 5315 (1951).
91. S. L. Miller. J. Am. Chem. Soc. 74: 4130 (1952).
92. Z. Pospíšil. Coll. Czech. Chem. Comm. 18: 337 (1953).
93. I. M. Kolthoff and J. Jordan. J. Am. Chem. Soc. 74: 570 (1952).
94. S. I. Sinyakova, I. V. Markova, and N. G. Galfayan. Material of the 2nd All-Union Polarographic Meeting, Kazan', 1962. Izd. Kazanskogo Universiteta (1962), p. 118.
95. K. Fülöp and L. J. Csányi. Acta Chim. Acad. Sci. Hung. 38: 193 (1963).
96. I. M. Reibel'. Material of the 1st All-Union Polarographic Meeting, Kishinev, 1959. Kishinev, Izd. Shtiintsa (1962), p. 323.
97. H. A. Laitinen and W. A. Ziegler. J. Am. Chem. Soc. 75: 3045 (1953).
98. S. I. Sinyakova. Dissertation. Moscow, Geokhimiya (1963).
99. R. Höltje and R. Geyer. Z. Anorg. Allgem. Chem. 246: 258 (1941).
100. G. P. Haight, Jr. Anal. Chem. 23: 1505 (1951).
101. G. P. Haight, Jr. and W. F. Sager. J. Am. Chem. Soc. 74: 6056 (1952).
102. G. P. Haight, Jr. J. Am. Chem. Soc. 76: 4718 (1954).

103. S. I. Sinyakova and M. I. Glinkina. Proceedings of the 4th All-Union Electrochemical Meeting, Moscow, 1956. Moscow, Izd. Akad. Nauk SSSR (1959),
p. 201.

104. G. A. Rechnitz and H. A. Laitinen. Anal. Chem. 33:1473 (1961).

105. M. G. Johnson and R. J. Robinson. Anal. Chem. 24 : 366 (1952).

106. I. M. Kolthoff and I. Hodara. J. Electroanal. Chem. 4 : 369 (1962).

107. I. M. Kolthoff and I. Hodara. J. Electroanal. Chem. 5 : 2 (1963).

108. I. M. Kolthoff and I. Hodara. J. Electroanal. Chem. 5 :165 (1963).

109. I. M. Kolthoff and I. Hodara. Bull. Res. Council Israel, Sect. A, 11A : 203
(1962).

110. F. Mánok and B. Tokés. Studia Univ., Babes-Bolyai. Chem. No. 2 : 35 (1959).

111. S. I. Sinyakova and G. G. Karanovich. Trudy Komissii po analit. khim.
2(5) : 65 (1949); J. Koryta. Coll. Czech. Chem. Comm. 20 : 667 (1955).

112. Z. R. Grabowski and A. Grabowska. Roczniki Chem. 30:1245 (1956).

113. Yu. K. Delimarskii and V. D. Skobets. Dopovidi Akad. Nauk URSR No. 7 :
928 (1962).

114. A. Blažek and J. Koryta. Coll. Czech. Chem. Comm. 18 : 326 (1953).

115. R. C. Bower and I. M. Kolthoff. J. Am. Chem. Soc. 81 :1836 (1959).

116. E. F. Orlemann and D.M.H. Kern. J. Am. Chem. Soc. 75 : 3058 (1953).

117. C. Groden. Australian J. Chem. 11 : 255 (1958).

118. B. Kastening. Naturwiss. 47 : 443 (1960).

119. L. Holleck and B. Kastening. Rev. Polarog. (Kyoto) 11 :129 (1963).

120. A. V. Gorodyskii, Yu. K. Delimarskii, and V. F. Grishehenko. Dokl. Akad.
Nauk SSSR 150 : 578 (1963).

121. L. Schwaer and K. Suchy. Coll. Czech. Chem. Comm. 7 : 25 (1935).

122. W. Hans and M. Stackelberg. Z. Elektrochem. 54 : 62 (1950); H. Schmid and
M. Stackelberg. J. Polarog. Soc. (London) 8 : 49 (1962).

123. J. J. Lingane and L. W. Niedrach. J. Am. Chem. Soc. 71 :196 (1949).

124. A. I. Alekperov. Azerb. Khim. Zh. 1 : 73 (1963).

125. S. I. Zhdanov and R. G. Pats. Elektrokhimiya 1 : 947 (1965).

126. S. I. Zhdanov. Dokl. Akad. Nauk SSSR 143 : 902 (1962).

127. L. Holleck. Z. Elektrochem. 49 : 400 (1943).

128. I. M. Kolthoff, W. E. Harris, and G. Matsuyiama. J. Am. Chem. Soc.
66 :1782 (1944).

129. A. N. Frumkin and S. I. Zhdanov. Dokl. Akad. Nauk SSSR 92 : 629 (1953).

130. S. I. Zhdanov and A. N. Frumkin. Dokl. Akad. Nauk SSSR 92 : 789 (1953).

131. I. Mašek. Coll. Czech. Chem. Comm. 18 : 583 (1953).

132. S. I. Zhdanov and A. N. Frumkin. Zh. Fiz. Khim. 29:1459 (1955).

133. Z. R. Grabovski and S. I. Zhdanov. Zh. Fiz. Khim. 31 :1162 (1957).

134. S. I. Zhdanov, V. I. Zykov, and T. V. Kalish. Proceedings of the 4th All-
Union Electrochemical Meeting, Moscow, 1956. Moscow, Izd. Akad. Nauk
SSSR (1959), p. 164.

135. S. I. Zhdanov. Zh. Fiz. Khim. 37 : 387 (1963).

136. A. N. Frumkin and S. I. Zhdanov. Dokl. Akad. Nauk SSSR 96 : 793 (1954).

137. V. Hanuš and R. Brdička. Khimiya 1 :28 (1951); Chem. Listy 44 : 291 (1950).

138. R. Brdička. Coll. Czech. Chem. Comm. 12:212 (1947).
139. E. G. Clair and K. Wiesner. Nature 165:202 (1950).
140. K. Wiesner. Anal. Chem. 27:1712 (1955).
141. J. Volke and V. Volková. Chem. Listy 49:490 (1955).
142. Y. Nagata and I. Tachi. Bull. Chem. Soc. Japan 27:290 (1954).
143. J. Volke and V. Volková. Coll. Czech. Chem. Comm. 22:1777 (1957).
144. N. H. Furman and C. E. Bricker. J. Am. Chem. Soc. 64:660 (1942).
145. A. Ryvolová and V. Hanuš. Coll. Czech. Chem. Comm. 21:853 (1956).
146. R. P.Buck. Anal. Chem. 35:1853 (1963).
147. M. Maturova, A. Nemečkova, and F. Santavý. Coll. Czech. Chem. Comm. 27:1021 (1962).
148. J. Kůta and E. Krejčí. Coll. Czech. Chem. Comm. 24:258 (1959).
149. E. Gergely and T. Iredale. J. Chem. Soc. 3226 (1953).
150. V. D. Bezuglyi, V. N. Dmitrieva, T. S. Tarasyuk, and N. A. Izmailov. Zh. Obshch. Khim. 30:2415 (1960).
151. V. D. Bezuglyi and E. Yu. Novik. Zavodsk. Lab. 27:544 (1961).
152. Ya. I. Tur'yan. Zh. Fiz. Khim. 36:2103 (1962).
153. Yu. A. Vakhrushev and Ya. I. Tur'yan. Zh. Fiz. Khim. 37:1650 (1963).
154. Ya. I. Tur'yan and Yu. A. Varkhrushev. Zh. Fiz. Khim. 37:1921 (1963).
155. P. Elving, J. Markowitz, and I. Rosenthal. J. Electrochem. Soc. 101:195 (1954).
156. P. Elving and Ching-Siang Tang. J. Am. Chem. Soc. 72:3244 (1950).
157. L. Fey. Studii Cercetari Chim. Acad. R.P.R. (Rumania) 7:69 (1956).
158. E. Budevski. Izv. Bulgar.Akad.Nauk, Ser.Fiz. 4:119 (1954-55).
159. I. M. Kolthoff and A. Liberti. J. Am. Chem. Soc. 70: 1885 (1948).
160. H. J. Gardner and W. P. Georgans. J. Chem. Soc. 1956: 4180.
161. H. Lund. Acta Chem. Scand. 13: 249 (1959).
162. W. Kemula, Z. Grabowski, and E. Bartel. Roczniki Chem. 33:1125 (1959).
163. P. O. Kane. J. Electroanal. Chem. 2:152 (1961).
164. K. Wiesner, M. Wheatley, and J. Los. J. Am. Chem. Soc. 76:4858 (1954).
165. M. Becker and H. Strehlow. Z. Elektrochem. 64 : 42 (1960).
166. M. Becker and H. Strehlow. Z. Elektrochem. 64:818 (1960).
167. A. I. Shatenshtein. The Theories of Acids and Bases. Moscow, Goskhimizdat (1949), p. 183.
168. S. G. Mairanovskii and M. B. Neiman. Dokl. Akad. Nauk SSSR 87: 93 (1951).
169. J. Kůta. Prace Konferencji Elektrochem., Warszawa (1955). Warszawa. Publ. P. W. N. (1957), p. 555.
170. J. Kůta. Coll. Czech. Chem. Comm. 20:1068 (1955).
171. J. Kůta. Coll. Czech. Chem. Comm. 22: 1411 (1957).
172. É. S. Levin. Dokl. Akad. Nauk SSSR 144:159 (1962).
173. É. S. Levin and N. A. Osinova. Zh.Obshch. Khim. 32:2084 (1962).
174. P. Rüetschi. Z. Phys. Chem. 5: 323 (1955).
175. P. Rüetschi and W. Vielstich. Z. Phys. Chem. 4:124 (1955).
176. H. W. Nürnberg. Advances in Polarography, Vol. 2. London, Pergamon Press (1960), p. 694.
177. H. W. Nürnberg, G. van Riesenbeck, and M. Stackelberg. Z. Elektrochem. 64:130 (1960).

178. H. W. Nürnberg, G. van Riesenbeck, and M. Stackelberg. Coll. Czech. Chem. Comm. 26:126 (1961).

179. D. Wolf. J. Electroanal. Chem. 5:186 (1963).

180. M. Stackelberg, D. Wolf, and H. Schmidt. Rev. Polarog. (Kyoto) 11(1-2):41 (1963).

181. H. Strehlow and M. Becker. Z. Elektrochem. 64:813 (1960).

182. H. Strehlow. Z. Elektrochem. 66:392 (1962).

183. P. Delahay and S. Oka. J. Am. Chem. Soc. 82:329 (1960).

184. R. P. Bell and A. O. McDougall. Trans. Faraday Soc. 56:1281 (1960).

185. M. B. Neiman and M. I. Gerber. Zh. Analit. Khim. 2:135 (1947).

186. R. Bieber and G. Trümpler. Helv. Chim. Acta 30:706 (1947).

187. K. Veselý and R. Brdička. Coll. Czech. Chem. Comm. 12:313 (1947).

188. M. A. Loshkarev and A. I. Chernikov. Zh. Fiz. Khim. 27:1778 (1953).

189. R. Brdička. Coll. Czech. Chem. Comm. 20:387 (1955).

190. R. Brdička. Z. Elektrochem. 59:787 (1955).

191. R. Brdička and L. Nemec. Rev. Polarog. (Kyoto) 11(1-2):5 (1963).

192. I. Smoler. Chem. Listy 47:1667 (1953); Chem. Zvesti 8:867 (1954).

193. I. Smoler. J. Electroanal. Chem. 6:465 (1963).

194. S. Ono, M. Takagi, and T. Wasa. J. Am. Chem. Soc. 75:4369 (1953).

195. S. Ono, M. Takagi, and T. Wasa. Bull. Chem. Soc. Japan 31:356 (1958).

196. S. Ono, M. Takagi, and T. Wasa. Bull. Chem. Soc. Japan 31:364 (1958).

197. J. M. Los and N. J. Gaspar. Can. J. Chem. 37:495 (1959).

198. J. M. Los and N. J. Gaspar. Z. Elektrochem. 64:41 (1960).

199. J. M. Los and N. J. Gaspar. Rev. Trav. Chim. Pays-Bas 76:112 (1960).

200. M. Takagi, S. Ono, and T. Wasa. Rev. Polarog. (Kyoto) 11(3-4):210 (1963).

201. J. Kůta. Coll. Czech. Chem. Comm. 24:2532 (1959).

202. S. Ono, M. Takagi, and T. Wasa. Coll. Czech. Chem. Comm. 26:141 (1961).

203. O. H. Müller and J. P. Baumberger. J. Am. Chem. Soc. 61:590 (1939).

204. T. Wasa, M. Takagi, and S. Ono. Bull. Chem. Soc. Japan 34:518 (1961).

205. J. Volke. Coll. Czech. Chem. Comm. 23:1486 (1958).

206. S. G. Mairanovskii, D. I. Dzhaparidze, and O. I. Sorokin. Izv. Akad. Nauk SSSR, Ser. Khim. No. 5:795 (1964).

207. A. Kirrman and J.-M. Savéant. Compt. Rend. 247:1192 (1958); J.-M. Savéant and Y. Brault. Compt. Rend. 258:3698 (1964).

208. A. Kirrman and P. Federlin. Bull. Soc. Chim. France 1958:944.

209. P. Federlin. Bull. Soc. Chim. France 1958:949.

210. Dang-Quoc-Quan. Compt. Rend. 251:2927 (1960).

211. K. Wiesner. Coll. Czech. Chem. Comm. 12:64 (1947); J. M. Los and K. Wiesner. J. Am. Chem. Soc. 75:6346 (1953).

212. J. M. Los, L. B. Simpson, and K. Wiesner. J. Am. Chem. Soc. 78:1564 (1956).

213. W. G. Overend, A. R. Peacock, and J. B. Smith. J. Chem. Soc. 1961:3487.

214. P. Zuman and H. Zinner. Chem. Ber. 95:2089 (1962).

215. J. Koryta and I. Kössler. Coll. Czech. Chem. Comm. 15:241 (1950).

216. J. J. Lingane. Chem. Rev. 29:1 (1941).

217. J. Koryta. Sborník I mezinar. polarograf. sjezdu, Praha (1950), I díl. Praha, Přírodoved. vydavatelství (1951), p. 794.

218. J. Koryta. Z. Elektrochem. 64:26 (1960).
219. J. Koryta. Coll. Czech. Chem. Comm. 24:2903 (1959).
220. J. Koryta. Coll. Czech. Chem. Comm. 24:3057 (1959).
221. J. Dandoy and L. Gierst. J. Electroanal. Chem. 2:116 (1961).
222. A. G. Stromberg. Zh. Fiz. Khim. 27:1287 (1953).
223. Ya. I. Tur'yan. Zh. Fiz. Khim. 31:2423 (1957).
224. Ya. I. Tur'yan and G. F. Serova. Dokl. Akad. Nauk SSSR 125:595 (1959); Zh. Fiz. Khim. 34:1009 (1960).
225. Ya. I. Tur'yan and G. F. Serova. Zh. Fiz. Khim. 31:1976 (1957).
226. Ya. I. Tur'yan. Dokl. Akad. Nauk SSSR 146: 848 (1962).
227. H. B. Mark, Jr. and C. N. Reilley. J. Electroanal. Chem. 4:189 (1962).
228. H. B. Mark, Jr. and C. N. Reilley. Anal. Chem. 35:195 (1963).
229. H. B. Mark, Jr. J. Electroanal. Chem. 7:276 (1964).
230. W. Kemula and Sz. Rosolowski. Roczn. Chem. 36:1417 (1962); 37:941 (1963); 38:905 (1964).
231. S. Tribalat and D. Delafosse. Analyt. Chim. Acta 19: 74 (1958).
232. P. Papoff. J. Am. Chem. Soc. 81:3254 (1959).
233. J. Koryta. Electrochim. Acta 1:26 (1959).
234. J. Koryta. Acta Chim. Acad. Sci. Hung. 9: 363 (1956).
235. M. E. Peover. Trans. Faraday Soc. 60:417 (1964).
236. P. Valenta. Coll. Czech. Chem. Comm. 25: 853 (1960).
237. G. Semerano and E. Vianello. Proceedings of the 8th Mendeleev Convention, Moscow, 1959. Izd. Akad. Nauk SSSR (1959).
238. J.-M. Savéant and E. Vianello. Electrochim. Acta 8:905 (1963).
239. T. Berzins and P. Delahay. J. Am. Chem. Soc. 75:4205 (1953).
240. O. Dračka. Coll. Czech. Chem. Comm. 25: 338 (1960).
241. A. C. Testa and W. H. Reinmuth. Anal. Chem. 32:1512 (1960).
242. H. B. Herman and A. J. Bard. Anal. Chem. 36:510 (1964).
243. O. Fischer, O. Dračka, and E. Fischerová. Coll. Czech. Chem. Comm. 26: 1505 (1961).
244. C. Furlani and G. Morpurgo. J. Electroanal. Chem. 1: 351 (1960).
245. S. W. Feldberg and C. Auerbach. Anal. Chem. 36:505 (1964).
246. O. Dračka. Coll. Czech. Chem. Comm. 28:3194 (1963).
247. A. K. N. Reddy, M. A. V. Devanathan, and J. O'M. Bockris. J. Electroanal. Chem. 6:61 (1963).
248. G. S. Alberts and I. Shain. Anal. Chem. 35:1859 (1963).
249. T. J. Katz, W. H. Reinmuth, and D. E. Smith. J. Am. Chem. Soc. 84:802 (1962).
250. B. Breyer and F. Gutman. Australian J. Sci. 8:21, 163 (1946).
251. B. Breyer and F. Gutman. Trans. Faraday Soc. 42:645, 650, 785 (1946).
252. B. Breyer and S. Hacobian. Australian J. Sci. 14:118, 153 (1952).
253. B. Breyer, J. R. Beevers, and H. H. Bauer. J. Electroanal. Chem. 2:60 (1961).
254. H. Gerischer. Z. Phys. Chem. 198:286 (1951); 201:55 (1952); 202:292, 302 (1953).
255. V. G. Levich, B. I. Khaikin, and V. A. Kir'yanov. Dokl. Akad. Nauk SSSR 139:925 (1961).
256. D. E. Smith. Anal. Chem. 35:602, 610 (1963).

257. H. L. Hung, J. R. Delmastro, and D. E. Smith. J. Electroanal. Chem. 7:1 (1964).

258. G. C. Barker, H. W. Nürnberg, and G. Boltsan. Report of the 14th TsITTsE Convention, Moscow, 1963; G. C. Barker and H. W. Nürnberg. Naturwiss. 51:191 (1964); H. W. Nürnberg and G. C. Barker. Naturwiss. 51:192 (1964).

259. I. Langmuir. J. Am. Chem. Soc. 40:1361 (1918).

260. I. Langmuir. J. Am. Chem. Soc. 39:1883 (1917).

261. A. N. Frumkin. Trudy Khim. Inst. L. Ya. Karpova, No. 4:56 (1925).

262. S. G. Mairanovskii. Elektrokhimiya 1: 164 (1965).

263. B. B. Damaskin. Dokl. Akad. Nauk SSSR 144:1073 (1962).

264. B. B. Damaskin and N. B. Grigor'ev. Dokl. Akad. Nauk SSSR 147:135 (1962).

265. L. D. Klyukina and B. B. Damaskin. Izv. Akad. Nauk SSSR, Otd. Khim. Nauk No. 6:1022 (1962).

266. R. Lerkkh and B. B. Damaskin. Zh. Fiz. Khim. 38:1154 (1964).

267. R. Lerkkh and B. B. Damaskin. Zh. Fiz. Khim. 39:211 (1965).

268. M. Volmer. Z. Phys. Chem. 115:253 (1925).

269. W. Lorenz. Z. Elektrochem. 62:193 (1958).

270. E. Helfand, H. L. Frisch, and J. L. Lebowitz. J. Chem. Phys. 34:1937 (1961); F. P. Buff and F. H. Stillinger. J. Chem. Phys. 39:1911 (1963).

271. R. Parsons. Report of the 14th TsITTsE Convention, Moscow, 1963; R. Parsons. J. Electroanal. Chem. 7:136 (1964).

272. B. B. Damaskin. J. Electroanal. Chem. 7: 155 (1964).

273. A. N. Frumkin and B. B. Damaskin. Dokl. Akad. Nauk SSSR 129:862 (1959).

274. W. Lorenz and W. Müller. Z. Phys. Chem. 25:161 (1960).

275. W. Lorenz. Z. Phys. Chem. 18(1-2):1 (1958).

276. S. G. Mairanovskii and V. P. Gul'tyai. Elektrokhimiya 1(4):460 (1965).

277. S. E. Khalafalla, A. M. Shams, E. Din, and S. A. Marei. Rec. Trav. Chim. 78:513 (1959).

278. R. Brdička. Coll. Czech. Chem. Comm. 12:522 (1947).

279. A. N. Frumkin. Trudy Khim. Inst. L. Ya. Karpova No. 5: 3 (1926).

280. G. Gouy. Ann. Chim. Phys. 8(8):294 (1908).

281. J. Butler. Proc. Roy. Soc. 122A: 399 (1929).

282. V. I. Melik-Gaikazyan. Zh. Fiz. Khim. 26:560 (1952).

283. M. A. Proskurnin and A. N. Frumkin. Trans. Faraday Soc. 31:110 (1935).

284. A. N. Frumkin and V. I. Melik-Gaikazyan. Dokl. Akad. Nauk SSSR 77:855 (1951).

285. W. Lorenz and F. Möckel. Z. Elektrochem. 60:507, 939 (1956); W. Lorenz and E. Schmalz. Z. Elektrochem. 62:301 (1958).

286. P. Delahay. Double Layer and Electrode Kinetics. New York, Interscience Publishers (1965), p. 118.

287. J. Koryta. Coll. Czech. Chem. Comm. 18:206 (1953).

288. P. Delahay and I. Trachtenberg. J. Am. Chem. Soc. 79: 2355 (1957).

289. P. Delahay and I. Trachtenberg. J. Am. Chem. Soc. 80:2094 (1958).

290. P. Delahay and C. Fike. J. Am. Chem. Soc. 80:2628 (1958).

291. W. H. Reinmuth. J. Phys. Chem. 65:473 (1961).

292. G. Gouy. Ann. Chim. Phys. 29(7):145 (1903); 9(8): 75, 116 (1906).

293. A. N. Frumkin. Electrocapillary Phenomena and Electrode Processes. Odessa (1919).

294.   L. Gierst, D. Bermane, and P. Corbusier. Ric. Sci., Vol. 29, Suppl. Contributi
       di Polarografia (1959), p. 75.
295.   L. Gierst. Transactions of the Symposium on Electrode Processes, Philadelphia,
       1959 (E. Yeager, ed.). New York, John Wiley and Sons (1961), p. 294.
296.   B. E. Conway and R. G. Barradas. Electrochim. Acta 5 :319 (1960).
297.   R. G. Barradas and B. E. Conway. Electrochim. Acta 5 : 349 (1960).
298.   E. Blomgren and J. O'M. Bockris. J. Phys. Chem. 63:1475 (1959).
299.   M. A. Gerovich. Dokl. Akad. Nauk SSSR 96:543 (1954); 105: 1278 (1955);
       M. A. Gerovich and N. S. Polyanovskaya. Nauchn. Dokl. Vysshei Shkoly,
       Khimiya i Khim. Tekhnol. No. 4: 651 (1958).
300.   M. A. Gerovich and G. F. Rybal'chenko. Zh. Fiz. Khim. 32:109 (1958).
301.   J. Zwierzykowska. Roczn. Chem. 38: 671 (1964).
302.   E. Cherneva and A. Gorodetskaya. Zh. Fiz. Khim. 13: 1117 (1939).
303.   M. A. V. Devanathan and M. J. Fernando. Trans. Faraday Soc. 58: 368 (1962).
304.   G. A. Korchinskii. Ukr. Khim. Zh. 29:1031 (1963).
305.   B. Kučera. Ann. Phys. 11(4): 529, 698 (1903).
306.   J. Řiha. Advances in Polarography, Vol. 1. London, Pergamon Press (1960),
       p. 210.
307.   B. B. Damaskin and G. A. Tedoradze. Dokl. Akad. Nauk SSSR 152:1151 (1963).
308.   H. A. Laitinen and B. Mosier. J. Am. Chem. Soc. 80: 2363 (1958).
309.   A. A. Moussa, H. M. Sammour, and D. A. Hickson. J. Phys. Chem. 62:1017
       (1958).
310.   R. S. Hansen, R. E. Minturn, and D. A. Hickson. J. Phys. Chem. 60 :1185 (1956).
311.   M. Breiter and P. Delahay. J. Am. Chem. Soc. 81: 2938 (1959).
312.   B. Breyer and S. Hacobian. Australian J. Sci. A5:500 (1952).
313.   B. Breyer and S. Hacobian. Australian J. Sci. 9: 7 (1956).
314.   A. N. Frumkin and B. B. Damaskin. J. Electroanal. Chem. 3: 36 (1962).
315.   J. Heyrovský and J. Forejt. Z. Phys. Chem. 193: 77 (1943).
316.   M. Matyáš. Chem. Listy 46: 65 (1952).
317.   R. Kalvoda. Coll. Czech. Chem. Comm. 25 : 3071 (1960).
318.   J. Heyrovský, F. Šorm, and J. Forejt. Coll. Czech. Chem. Comm. 12 : 11 (1947).
319.   J. E. B. Randles. Trans. Faraday Soc. 44: 327 (1948).
320.   A. Ševčík. Coll. Czech. Chem. Comm. 13: 314 (1948).
321.   R. Kalvoda. J. Electroanal. Chem. 1 : 314 (1960).
322.   M. Kalousek and M. Rálek. Coll. Czech. Chem. Comm. 19 : 1099 (1954).
323.   D. I. Dzhaparidze and G. A. Tedoradze. Izv. Akad. Nauk SSSR, Otd. Khim.
       Nauk No. 10:1718 (1962).
324.   W. Lorenz. Z. Elektrochem. 59: 730 (1955).
325.   W. H. Reinmuth. Anal. Chem. 33: 322 (1961).
326.   H. A. Laitinen. Anal. Chem. 33:1458 (1961).
327.   F. C. Anson. Anal. Chem. 33:1123, 1498, 1838 (1961).
328.   A. J. Bard. Anal. Chem. 35: 340 (1963).
329.   H. A. Laitinen and L. M. Chambers. Anal. Chem. 36: 5 (1964).
330.   S. V. Tatwawadi and A. J. Bard. Anal. Chem. 36: 2 (1964).
331.   R. W. Murray. J. Electroanal. Chem. 7: 242 (1964).
332.   H. Hurwitz and L. Gierst. J. Electroanal. Chem. 2: 128 (1961).

333. H. Hurwitz. J. Electroanal. Chem. 2: 142 (1961).

334. R. A. Osteryoung. Anal. Chem. 35: 1100 (1963).

335. H. B. Herman, S. V. Tatwawadi, and A. J. Bard. Anal. Chem. 35: 2210 (1963).

336. J. H. Christie, G. Lauer, and R. A. Osteryoung. J. Electroanal. Chem. 7: 60 (1964).

337. R. A. Osteryoung, G. Lauer, and F. C. Anson. Anal. Chem. 34: 1833 (1962); J. Electrochem. Soc. 110: 926 (1963).

338. P. Delahay. Rev. Polarog. (Kyoto) 11(1-2):141 (1963).

339. M. S. Shul'man. Kolloidn. Zh. 19: 384 (1957).

340. I. Rusznák, K. Fukker, and I. Králik. Acta Chim. Acad. Sci. Hung. 9(1-4): 49 (1956).

341. I. Rusznák, I. Králik, and K. Fukker. Z. Phys. Chem. (BRD) 17: 56, 61 (1958).

342. P. Zuman. Coll. Czech. Chem. Comm. 20: 883 (1955).

343. K. R. Voronova and A. G. Stromberg. Zh. Obshch. Khim. 31: 2786 (1961).

344. E. L. Colichman. J. Am. Chem. Soc. 74: 722 (1952).

345. T. A. Kryukova and A. N. Frumkin. Zh. Fiz. Khim. 23: 819 (1949).

346. V. D. Bezuglyi and E. K. Saliichuk. Dokl. Akad. Nauk SSSR 158: 1390 (1964).

347. E. M. Skobets and N. S. Kavetskii. Zavodsk. Lab. 15: 1299 (1949).

348. S. G. Mairanovskii and A. D. Filonova. Elektrokhimiya 1: 1044 (1965).

349. A. Bresle. Acta Chem. Scand. 10:947 (1956).

350. S.G. Marianovskii and É. F. Mairanovskaya. Izv. Akad. Nauk SSSR, Otd. Khim. Nauk No. 5: 922 (1961).

351. R. Brdička and E. Knobloch. Z. Elektrochem. 47: 721 (1941).

352. O. H. Müller. J. Biol. Chem. 145: 425 (1942).

353. R. Brdička. Z. Elektrochem. 48:278 (1942).

354. M. Senda, M. Senda, and I. Tachi. Rev. Polarog. (Kyoto) 10: 142 (1962).

355. As. Trifonov. Izv. Khim. Inst. Bulgar. Akad. Nauk 4: 21 (1956).

356. M. Voříšková. Coll. Czech. Chem. Comm. 12: 607 (1947).

357. B. Breyer and H. H. Bauer. Australian J. Chem. 8: 472 (1955).

358. B. Breyer and H. H. Bauer. Australian J. Chem. 8: 480 (1955).

359. B. Jámbor. Acta Chim. Acad. Sci. Hung. 4: 55 (1954).

360. B. Lovreček and T. Jukić-Markušić. Croat. Chem. Acta 28: 255 (1956).

361. P. Zuman, J. Tenygl, and M. Brezina. Coll. Czech. Chem. Comm. 19: 46 (1954).

362. P. Zuman. Coll. Czech. Chem. Comm. 19: 1140 (1954).

363. P. Zuman and F. Santavý. Coll. Czech. Chem. Comm. 18: 28 (1953).

364. H. Berg and W. Böckel. Monatsber. Deut. Akad. Wiss. Berlin 3: 32 (1961).

365. L. Starka, A. Vystrčil, and B. Stárková. Chem. Listy 51: 1440 (1959).

366. S. I. Zhdanov. Zh. Obshch. Khim. 31: 3874 (1961).

367. J. Koryta and P. Zuman. Coll. Czech. Chem. Comm. 18: 197 (1953).

368. P. Zuman, J. Koryta, and R. Kalvoda. Coll. Czech. Chem. Comm. 18: 351 (1953).

369. P. Zuman. Coll. Czech. Chem. Comm. 20: 646 (1955).

370. M. Březina and P. Zuman. Polarography in Medicine, Biochemistry, and Pharmacy. New York, Interscience Publishers, Inc. (1958), p. 352.

371. S. G. Mairanovskii. Zh. Fiz. Khim. 36: 165 (1962).

372. Y. Asahi. Bull. Chem. Soc. Japan 34:1185 (1961).
373. I. Bartek, M. Černoch, and F. Santavý. Coll. Czech. Chem. Comm. 19:605 (1954).
374. I. M. Kolthoff and Y. Okinaka. J. Am. Chem. Soc. 82:3528 (1960).
375. F. Péter, I. Rusznák, Gy. Pályi, and I. Szabados. Magy. Kém. Folyóirat 66:178 (1960).
376. F. Péter, Gy. Pályi, and I. Szabados. Magy. Kém. Folyóirat 67:428 (1961).
377. Gy. Pályi, F. Péter, and I. Szeberényi. Acta Chim. Acad. Sci. Hung. 32: 387 (1962).
378. F. Péter and I. Szabados. Magy. Kém. Folyóirat 68:145 (1962).
379. I. M. Kolthoff and C. Barnum. J. Am. Chem. Soc. 63:520 (1941).
380. S. Fiala. Chem. Listy 39:14 (1945).
381. K. Wiesner. Coll. Czech. Chem. Comm. 12:594 (1947).
382. M. A. Loshkarev and A. A. Kryukova. Zh. Fiz. Khim. 23:209 (1949).
383. M. A. Loshkarev, A. Krivtsov, and A. A. Kryukova. Zh. Fiz. Khim. 23:221 (1949).
384. J. Heyrovský. Discussions Faraday Soc. 1:212 (1947).
385. M. A. Loshkarev and A. A. Kryukova. Zh. Fiz. Khim. 23:1457 (1949).
386. M. A. Loshkarev and A. A. Kryukova. Zh. Fiz. Khim. 26:731 (1952).
387. M. A. Loshkarev and A. A. Kryukova. Zh. Fiz. Khim. 26:737 (1952).
388. A. A. Kryukova and M. A. Loshkarev. Zh. Fiz. Khim. 30:2236 (1956).
389. M. A. Loshkarev and A. A. Kryukova. Zh. Fiz. Khim. 31:452 (1957).
390. H. B. Miller and V. A. Pleskov. Dokl. Akad. Nauk SSSR 74:323 (1950).
391. A. G. Stromberg and M. S. Guterman. Zh. Fiz. Khim. 27:993 (1953).
392. A. G. Stromberg and L. S. Zagainova. Dokl. Akad. Nauk SSSR 97:107 (1954).
393. L. S. Zagainova and A. G. Stromberg. Dokl. Akad. Nauk SSSR 105:747 (1955).
394. A. G. Stromberg and L. S. Zagainova. Zh. Fiz. Khim. 31:1042 (1957).
395. M. A. Loshkarev. Dokl. Akad. Nauk SSSR 72:729 (1950).
396. S. L. Bonting and B. S. Aussen. Rev. Trav. Chim. Pays-Bas 73:455 (1954).
397. V. V. Losev. Dokl. Akad. Nauk SSSR 110:111 (1955).
398. V. V. Losev. Dokl. Akad. Nauk SSSR 111:626 (1956).
399. H. A. Laitinen and W. J. Subcasky. J. Am. Chem. Soc. 80:2623 (1958).
400. J. Říha. Electrochim. Acta 6:75 (1962).
401. G. A. Tedoradze, S. G. Mairanovskii and L. D. Klyukina. Izv. Akad. Nauk SSSR, Otd. Khim. Nauk No. 7:1352 (1961).
402. Gy. Josepovits. Acta Chim. Acad. Sci. Hung. 9:397 (1956).
403. R. W. Schmid and C. N. Reilley. J. Am. Chem. Soc. 80:2087 (1958).
404. J. Weber, J. Koutecký, and J. Koryta. Z. Elektrochem. 63:583 (1959).
405. J. Kůta and I. Smoler. Z. Elektrochem. 64:285 (1960).
406. J. Kůta, J. Weber, and J. Koutecký. Coll. Czech. Chem. Comm. 25:2376 (1960).
407. J. Koutecký and J. Weber. Coll. Czech. Chem. Comm. 25:1423 (1960); J. Weber and J. Koutecký. Coll. Czech. Chem. Comm. 25:2993 (1960).
408. J. Kůta and I. Smoler. Progress in Polarography, Vol. 1. New York, Interscience Publishers, Inc. (1962), p. 43.
409. P. Silvestroni and L. Rampazzo. J. Electroanal. Chem. 7:73 (1964).

410. L. Holleck and H. I. Exner. Sborník I mezinar. polarograf. sjezdu, Praha 1950, I díl. Praha, Přírodoved. vydavatelství (1951), p. 97.

411. B. Kastening and L. Holleck. Z. Elektrochem. 63:166 (1959).

412. L. Holleck and B. Kastening. Z. Elektrochem. 63:177 (1959).

413. B. Kastening and L. Holleck. Z. Elektrochem. 64: 823 (1960).

414. L. Holleck. Z. Naturforsch. 186: 439 (1963).

415. F. Péter, I. Szabados, and Gy. Pályi. Magy. Kém. Folyóirat 68:101 (1962).

416. L. Holleck, B. Kastening, and R. D. Williams. Z. Elektrochem. 66:396 (1962).

417. S. R. Missan, E. I. Becker, and L. Meites. J. Am. Chem. Soc. 83:58 (1961).

418. W. Müller and W. Lorenz. Z. Phys. Chem. (BRD) 27:23 (1961).

419. J. H. M. Rek. Doctoral Thesis, State Univ. Utrecht, 1963; from Electroanal. Abstr. 1, ref. N1478 (1963).

420. A. Aramata and P. Delahay. J. Phys. Chem. 68:880 (1964).

421. O. Dračka. Coll. Czech. Chem. Comm. 24:3523 (1959).

422. L. Gierst and A. L. Juliard. J. Phys. Chem. 57: 701 (1953).

423. O. Fischer. Coll. Czech. Chem. Comm. 27:1119 (1962).

424. R. C. Bowers and A. M. Wilson. J. Am. Chem. Soc. 80:2968 (1958).

425. T. Biegler and B. Breyer. Rev. Polarog. (Kyoto) 7(1): 31 (1959).

426. B. Breyer. Progress in Polarography, Vol. 2. New York, Interscience Publishers, Inc. (1962), p. 495.

427. M. Senda, M. Senda, and I. Tachi. J. Electrochem. Soc. Japan, Overseas Ed. 27:E21 (1959).

428. A. Mirri and P. Favero. Ric. Sci. 29:106 (1959).

429. M. Person and J. Tirouflet. Compt. Rend. 251: 2532 (1960).

430. J. Tirouflet and P. Fournari. Compt. Rend. 248: 1184 (1959).

431. T. A. Kryukova. Zh. Fiz. Khim. 24:437 (1950).

432. A. N. Frumkin and B. P. Bruns. Acta Physicochim. URSS 1:232 (1934).

433. T. A. Kryukova and B. N. Kabanov. Zh. Fiz. Khim. 12:1454 (1939).

434. T. A. Kryukova and B. N. Kabanov. Zh. Fiz. Khim. 15:473 (1941).

435. T. A. Kryukova and B. N. Kabanov. Zh. Obshch. Khim. 15:294 (1945).

436. T. A. Kryukova. Zh. Fiz. Khim. 20:1179 (1946).

437. A. N. Frumkin and V. G. Levich. Zh. Fiz. Khim. 21:1183 (1947).

438. E. Laviron. Bull. Soc. Chim. France No. 3:418 (1962).

439. J. Volke and J. Holubek. Coll. Czech. Chem. Comm. 27:1777 (1963).

440. A. A. Pozdeeva and S. I. Zhdanov. Proceedings of the 3rd International Polarographic Congress, Southampton, 1946.

441. M. K. Polievktov, S. G. Mairanovskii, and M. G. Gonikberg. Kinetika i kataliz, Vol. 7 (1966).

442. M. Fedoronko and H. Berg. Z. Phys. Chem. (DDR) 220:120 (1962).

443. M. Fedoronko and H. Berg. Chem. Zvesti 16:28 (1962).

444. P. Elving and A. F. Krivis. Anal. Chem. 29:1292 (1957).

445. L. Holleck, B. Kastening, and H. Vogt. Electrochim. Acta 8: 255 (1963).

446. D. J. Casimir, A. J. Harle, and L. E. Lyons. J. Chem. Soc. 5297:1961.

447. S. I. Zhdanov and A. N. Frumkin. Dokl. Akad. Nauk SSSR 122:412 (1958).

448. P. Silvestroni. Ric. Sci. 24:1695 (1954).

449. A. Vlček. Coll. Czech. Chem. Comm. 19:221 (1954).
450. J. Volke. Coll. Czech. Chem. Comm. 25 : 3397 (1960).
451. M. Person. Compt. Rend. 255 : 301 (1962).
452. K. Kačirková. Coll. Czech. Chem. Comm. 1 : 477 (1929).
453. L. Meites. J. Am. Chem. Soc. 76:5927 (1954).
454. I. M. Kolthoff, W. Stricks, and N. Tanaka. J. Am. Chem. Soc. 77:5211 (1955).
455. O. Grubner. Coll. Czech. Chem. Comm. 19:444 (1954).
456. Y. Nagata, K. Kitao, and I. Tachi. J. Electrochem. Soc. Japan 29(4):E233 (1961).
457. P. Zuman. Coll. Czech. Chem. Comm. 19:602 (1954).
458. E. Fischerová and O. Fischer. Coll. Czech. Chem. Comm. 26:2570 (1961).
459. M. Senda, M. Senda, and I. Tachi. Rev. Polarog. (Kyoto) 10:142 (1962).
460. M. Nishiyama, M. Maruyama, and H. Hamaguchi. Bull. Chem. Soc. Japan 37:616 (1964).
461. T. Biegler. J. Electroanal. Chem. 4:317 (1962).
462. Z. Bozyk. Acta Polon. Pharmac. 18:25 (1961).
463. A. N. Frumkin. Dokl. Akad. Nauk SSSR 85:373 (1952).
464. A. N. Frumkin. Usp. Khim. 24:933 (1955).
465. A. N. Frumkin. Proceedings of the 4th All-Union Electrochemical Convention, Moscow, 1956. Izd. Akad. Nauk SSSR (1959).
466. A. N. Frumkin. Transactions of a Symposium on Electrode Processes. New York, John Wiley and Sons, Inc. (1961), p. 1.
467. A. N. Frumkin. Report of the 14th TsITTsE Convention, Moscow, 1963; A. N. Frumkin. Electrochim. Acta 9:465 (1964).
468. M. Stackelberg and W. Stracke. Z. Elektrochem. 53:118 (1949).
469. W. Kemula and A. Cisak. Roczniki Chem. 29: 275 (1954).
470. P. J. Elving, I. Rosenthal, and A. I. Martin. J. Am. Chem. Soc. 77:5218 (1955).
471. S. G. Mairanovskii and L. D. Bergel'son. Zh. Fiz. Khim. 34:236 (1960).
472. L. G. Feoktistov and S. I. Zhdanov. Izv. Akad. Nauk SSSR, Otd. Khim. Nauk No. 1:45 (1963); L. G. Feoktistov, A. P. Tomilov, and M. M. Gol'din. Izv. Akad. Nauk SSSR, Otd. Khim. Nauk 1352 (1963).
473. N. T. Vagramyan and L. I. Antropov. Zh. Fiz. Khim. 25: 419 (1951).
474. W. W. Clark. Oxidation Reduction Potentials of Organic Compounds. Baltimore, Williams and Wilkins Co. (1960).
475. P. J. Elving. Pure Appl. Chem. 7:423 (1963).
476. H. J. Gardner and L. E. Lyins. Rev. Pure Appl. Chem. (Australia) 3:134 (1953).
477. E. Knobloch. Coll. Czech. Chem. Comm. 14:508 (1949).
478. M. B. Neiman, S. G. Mairanovskii, B. M. Kovarskaya, É. G. Rozantsev, and É. G. Gintsberg. Izv. Akad. Nauk SSSR, Ser. Khim. No. 8:1518 (1964).
479. N. Tanaka and R. Tamamushi. Sborník I mezinar. polarograf. sjezdu, Praha (1951), p. 486.
480. S. G. Mairanovskii. Dokl. Akad. Nauk SSSR 142:1120 (1962); S. G. Mairanovskii. J. Electroanal. Chem. 3:166 (1962).
481. Yu. P. Kitaev, G. K. Budnikov, and A. E. Arbuzov. Dokl. Akad. Nauk SSSR 127:808 (1959); Yu. P. Kitaev and G. K. Budnikov. Zh. Obshch. Khim. 33:1396 (1963).

482.　B. A. Arbuzov and E. A. Berdnikov. Izv. Akad. Nauk SSSR, Otd. Khim. Nauk No. 1:165 (1962).

483.　J. Mollin and F. Kasparek. Coll. Czech. Chem. Comm. 25:451 (1960); 26: 2438 (1961).

484.　L. Holleck and R. Schindler. Z. Elektrochem. 62:942 (1958).

485.　C. Calzolari and A. Donda. Ann. Chim. (Rome) 43:753 (1953).

486.　P. J. Elving and E. C. Olson. J. Am. Chem. Soc. 79:2697 (1957).

487.　H. Lund. Coll. Czech. Chem. Comm. 25:3313 (1960).

488.　J. Krupička and J. Gut. Coll. Czech. Chem. Comm. 27:546 (1962).

489.　A. Ryvolová. Advances in Polarography, Vol. 3. London, Pergamon Press (1960), p. 861.

490.　A. Foffani and E. Fornasari. Gazz. Chim. Ital. 83:1051 (1953).

491.　E. Fornasari and A. Foffani. Gazz. Chim. Ital. 83:1059 (1953).

492.　S. G. Mairanovskii, N. V. Barashkova, and F. D. Alashev. Zh. Fiz. Khim. 36:562 (1962).

493.　G. Hom. Monatsber. Deut. Akad. Wiss. Berlin 3:387 (1961).

494.　P. Zuman. Z. Phys. Chem. Sonderheft 243 (1958).

495.　F. Santavý. Z. Phys. Chem. Sonderheft 210 (1958).

496.　S. I. Zhdanov and M. K. Polievktov. Zh. Obshch. Khim. 31:3870 (1960).

497.　S. Zhdanov and L. Mirkin. Coll. Czech. Chem. Comm. 26:370 (1961).

498.　J. Volke, R. Kubiček, and F. Santavý. Coll. Czech. Chem. Comm. 25:871, 1510 (1960).

499.　J. Volke. Acta. Chim. Acad. Sci. Hung. 9:233 (1956).

500.　J. Holubek and J. Volke. Coll. Czech. Chem. Comm. 25:3286 (1960).

501.　J. Volke. Coll. Czech. Chem. Comm. 21:246 (1956).

502.　M. Maturová, V. Preininger, and F. Santavý. Die Polarographie in der Chemotherapie, Biochemie und Biologie (I. Jenaer Simposium, 1962). Berlin, Akademie-Verlag (1964), p. 85.

503.　J. Asahi. Ibid., p. 74.

504.　P. Zuman. Coll. Czech. Chem. Comm. 25:3245, 3252 (1960).

505.　V. G. Mairanovskii, L. E. Kholodov, and V. G. Yashunskii. Zh. Obshch. Khim. 33:347 (1963).

506.　O. Manoušek and P. Zuman. Coll. Czech. Chem. Comm. 29:1457 (1964).

507.　B. Jámbor. Acta Chim. Hung. 4:55 (1954).

508.　B. Jámbor. Die Polarographie in der Chemotherapie, Biochemie und Biologie (I. Jenaer Simposium, 1962). Berlin, Akademie-Verlag (1964), p. 49.

509.　E. Laviron. Ibid., p. 63.

510.　M. Vajda and F. Ruff. Ibid., p. 112.

511.　H. Wagner and H. Berg. J. Electroanal. Chem. 2:452 (1961).

512.　A. Nemecková, M. Maturová, M. Pergál, and F. Santavý. Coll. Czech. Chem. Comm. 26:2749 (1961).

513.　J. W. Smith and I. G. Waller. Trans. Faraday Soc. 46:290 (1950).

514.　P. Zuman and V. Horák. Coll. Czech. Chem. Comm. 27:187 (1962).

515.　E. A. Abrahamson. J. Am. Chem. Soc. 81:3919 (1959).

516.　O. Hrdý. Die Polarographie in der Chemotherapie, Biochemie und Biologie (I. Jenaer Simposium, 1962). Berlin, Akademie-Verlag (1964), p. 109.

517. J. W. Baker, W. C. Davies, and M. L. Hemming. J. Chem. Soc. 692 (1940).
518. P. J. Elving and P. G. Grodzka. Anal. Chem. 33:2 (1961).
519. A. Kirrman, J.-M. Savéant, and N. Moe. Compt. Rend. 253:1106 (1961).
520. P. H. Given and M. E. Peover. J. Chem. Soc. 385 (1960).
521. P. H. Given and M. E. Peover. Advances in Polarography, Vol. 3. London, Pergamon Press (1960), p. 948.
522. G. J. Hoijtink, J. Van Schooten, E. Boor, and W. Y. Aalbersberg. Rec. Trav. Chim. 73: 355 (1954).
523. P. J. Elving, J. C. Komyathy, R. E. Van Atta, C. S. Tang, and I. Rosenthal. Anal. Chem. 23:1218 (1951).
524. S. G. Mairanovskii and V. N. Pavlov. Zh. Fiz. Khim. 38:1804 (1964).
525. D. H. Geske and A. H. Maki, J. Am. Chem. Soc. 82:2671 (1960).
526. B. Kastening. Electrochim. Acta 9:241 (1964).
527. W. Kemula and R. Sioda. Bull. Acad. Polon. Sci., Ser. Chim. 10:513 (1962); 11:395 (1963); W. Kemula and R. Sioda. J. Electroanal. Chem. 7:233 (1964).
528. S. G. Mairanovskii, V. M. Belikov, Ts. B. Korchemnaya, and S. S. Novikov. Izv. Akad. Nauk SSSR, Otd. Khim. Nauk No. 3:522 (1962).
529. P. Zuman. Advances in Polarography, Vol. 3. London, Pergamon Press (1960), p. 812.
530. S. G. Mairanovskii. Izv. Akad. Nauk SSSR 132:1352 (1960).
531. S. G. Mairanovskii and L. I. Lishcheta. Coll. Czech. Chem. Comm. 25: 3025 (1960).
532. R. Brdička. Coll. Czech. Chem. Comm. 19(Suppl. 2):49 (1954).
533. L. Onsager. J. Chem. Phys. 2:599 (1934).
534. P. Debye. Trans. Electrochem. Soc. 82:265 (1942).
535. M. Eigen and L. de Maeyer. Z. Elektrochem. 59:986 (1955).
536. R. Bell. Quart. Rev. 13:169 (1959).
537. R. Bell. The Proton in Chemistry. Ithaca, New York, Cornell University Press (1959), p. 120.
538. H. H. Jaffe and G. O. Doak. J. Am. Chem. Soc. 77:4441 (1955).
539. L. I. Lishcheta. Dissertation. Moscow, N. D. Zelinskii Inst. Organic Chemistry, Akad. Nauk SSSR (1963).
540. S. G. Mairanovskii. Dokl. Akad. Nauk SSSR 149:1373 (1963).
541. S. G. Mairanovskii. Zh. Fiz. Khim. 33:691 (1959).
542. S. G. Mairanovskii and M. K. Polievktov. Zh. Fiz. Khim. 37:885 (1963).
543. S. A. Maron and V. K. La Mer. J. Am. Chem. Soc. 61:2018 (1939).
544. R. Bell and T. Spencer. Proc. Roy. Soc. 251A:41 (1959).
545. S. G. Mairanovskii, V. M. Belikov, Ts. B. Korchemnaya, V. A. Klimova, and S. S. Novikov. Izv. Akad. Nauk SSSR, Otd. Khim. Nauk No. 10:1787 (1960).
546. S. G. Mairanovskii and L. I. Lishcheta. Izv. Akad. Nauk SSSR, Otd. Khim. Nauk No. 10, 1749 (1961).
547. S. Glasstone, K. Laidler, and G. Eyring. The Theory of Absolute Reaction Rates. Moscow, IL (1948), pp. 402, 417, 500, 501, 533. [Originally published as The Theory of Rate Processes. New York, McGraw-Hill, 1941.]
548. V. M. Belikov, S. G. Mairanovskii, Ts. B. Korchemnaya, and S. S. Novikov. Izv. Akad. Nauk SSSR, Otd. Khim. Nauk No. 11, 2103 (1962).

549.   S. G. Mairanovskii. Collection of Papers of the International Symposium on Nitro Compounds, Warsaw (1963), p. 92.

550.   V. M. Belikov, S. G. Mairanovskii, Ts. B. Korchemnaya, and V. P. Gul'tyai. Izv. Akad. Nauk SSSR, Ser. Khim. No. 3, 439 (1964).

551.   A. N. Frumkin. Zh. Fiz. Khim. 24 : 244 (1950).

552.   S. V. Bagotskii and I. E. Yablokova. Zh. Fiz. Khim. 27 : 1663 (1953).

553.   A. N. Frumkin. J. Electroanal. Chem. 9 : 173 (1965).

554.   A. N. Frumkin and G. M. Florianovich. Dokl. Akad. Nauk SSSR 80 : 907 (1952); Zh. Fiz. Khim. 29 : 1827 (1955).

555.   T. A. Kryukova. Dokl. Akad. Nauk SSSR 65 : 517 (1949).

556.   S. I. Zhdanov, V. I. Zykov, and T. V. Kalish. Proceedings of the 4th All-Union Electrochemical Convention, Moscow, 1956. Moscow, Izd. Akad. Nauk SSSR (1959), p. 164.

557.   V. I. Zykov and S. I. Zhdanov. Zh. Fiz. Khim. 32 : 644 (1958).

558.   A. N. Frumkin and N. V. Nikolaeva-Fedorovich. Vestnik MGU No. 4, 179 (1957).

559.   N. V. Nikolaeva-Fedorovich, B. B. Damaskin, and O. A. Petrii. Coll. Czech. Chem. Comm. 25 : 2982 (1960).

560.   M. Breiter, M. Kleinerman, and P. Delahay. J. Am. Chem. Soc. 80 : 5111 (1958).

561.   W. H. Reinmuth, L. B. Rogers, and L. C. I. Hummerstedt. J. Am. Chem. Soc. 81 : 2947 (1959).

562.   P. Delahay and M. Kleinerman. J. Am. Chem. Soc. 82 : 4509 (1960).

563.   Z. Grabowski. Proceedings of the 4th All-Union Electrochemical Convention, Moscow, 1956. Moscow, Izd. Akad. Nauk SSSR (1959), p. 233.

564.   Z. Grabowski and E. Bartel. Roczniki Chem. 34 : 611 (1960).

565.   H. Matsuda. J. Phys. Chem. 64 : 336 (1960).

566.   L. Gierst and H. Hurwitz. Z. Elektrochem. 64 : 36 (1960).

567.   H. Hurwitz. Z. Elektrochem. 65 : 178 (1961).

568.   S. Rangarajan. Can. J. Chem. 41 : 983 (1963).

569.   S. Rangarajan. Can. J. Chem. 41 : 1469 (1963).

570.   B. B. Damaskin. Usp. Khim. 30 : 230 (1964).

571.   R. Parsons. Advances in Electrochemistry and Electrochemical Engineering, Vol. 1. New York, Interscience Publishers, Inc. (1961), p. 1.

572.   H. W. Nürnberg and M. Stackelberg. J. Electroanal. Chem. 4 : 1 (1962).

573.   V. G. Levich. Dokl. Akad. Nauk SSSR 67 : 309 (1949).

574.   V. G. Levich. Dokl. Akad. Nauk SSSR 124 : 869 (1959).

575.   É. S. Levin and Z. I. Fodiman. Zh. Fiz. Khim. 28 : 601 (1954).

576.   R. M. Powers and R. A. Day, Jr. J. Am. Chem. Soc. 80 : 808 (1958).

577.   P. J. Elving and J. T. Leone. J. Am. Chem. Soc. 80 : 1021 (1958).

578.   Ya. P. Stradyn' and V. V. Teraud. Izv. Akad. Nauk Latv. SSR, Ser. Khim. No. 2, 169 (1964); Ya. P. Stradyn'. Electrochim. Acta 9 : 711 (1964).

579.   V. D. Bezuglyi, L. A. Mel'nik, and V. N. Dmitrieva. Zh. Obshch. Khim. 34 : 1048 (1964).

580.   M. A. Vorsina and A. N. Frumkin. Zh. Fiz. Khim. 19 : 171 (1945).

581.   D. C. Grahame and B. A. Soderberg. J. Chem. Phys. 22 : 449 (1954).

582.   J. J. Lothe and L. B. Rogers. J. Electrochem. Soc. 101 : 258 (1954).

LITERATURE 339

583. Ya. I. Tur'yan. Dokl. Akad. Nauk SSSR 113: 631 (1957).
584. S. N. Mukherjee and A. Chakravarti. J. Indian Chem. Soc. 39:149 (1962).
585. M. Shikata and I. Tachi. Coll. Czech. Chem. Comm. 10:368 (1938).
586. N. G. Lordi and E. M. Cohen. Anal. Chim. Acta 25:281 (1961).
587. J. Slavík, L. Slavíková, V. Preininger, and F. Santavý. Coll. Czech. Chem. Comm. 21:1058 (1956); F. Santavý. Die Polarographie in der Chemotherapie Biochemie und Biologie (I. Jenaer Simposium, 1962). Berlin, Akademie-Verlag (1964), p. 1.
588. A. N. Frumkin and O. A. Petry. Dokl. Akad. Nauk SSSR 147:418 (1962).
589. A. N. Frumkin, O. A. Petry, and N. V. Nikolaeva-Fedorovich. Electrochim. Acta 8:177 (1963).
590. A. N. Frumkin, O. A. Petry, and N. V. Nikolaeva-Fedorovich. Dokl. Akad. Nauk SSSR 147:878 (1962).
591. B. Breyer, H. H. Bauer, and J. R. Beevers. Australian J. Chem. 14:479 (1961); H. H. Bauer and P. B. Goodwin. Australian J. Chem. 15:391 (1962).
592. P. Elving, I. Rosenthal, and M. Kramer. J. Am. Chem. Soc. 73:1717 (1951).
593. M. Ashworth. Coll. Czech. Chem. Comm. 13:229 (1948).
594. L. Hummelstedt and L. Rogers. J. Electrochem. Soc. 106:248 (1959).
595. V. D. Bezuglyi and E. Yu. Novik. Zh. Fiz. Khim. 34:795 (1960).
596. J. Volke and V. Volková. Coll. Czech. Chem. Comm. 20:1332 (1955).
597. V. S. Bagotskii. Dokl. Akad. Nauk SSSR 58:1387 (1947).
598. N. V. Nikolaeva, N. S. Shapiro, and A. N. Frumkin. Dokl. Akad. Nauk SSSR 86:581 (1952).
599. B. N. Rybakov and N. V. Nikolaeva-Fedorovich. Dokl. Akad. Nauk. SSSR 137:1135 (1960).
600. N. V. Nikolaeva-Fedorovich and A. N. Frumkin. Dokl. Akad. Nauk SSSR 137:1135 (1960).
601. L. Holleck. Naturwiss. 43:13 (1956).
602. D. Grahame. Chem. Rev. 41:467 (1947).
603. D. Grahame. Z. Elektrochem. 62:264 (1958).
604. B. V. Ershler. Zh. Fiz. Khim. 20:679 (1946).
605. O. A. Esin and V. M. Shikhov. Zh. Fiz. Khim. 17:236 (1943).
606. S. G. Mairanovskii and Ya. P. Stradyn'. Izv. Akad. Nauk SSSR, Otd. Khim. Nauk No. 12, 2239 (1961).
607. S. Siekierski. Roczniki Chem. 30:1083 (1956).
608. S. Siekierski. Roczniki Chem. 28:90 (1954).
609. A. N. Frumkin, B. B. Damaskin, and N. V. Nikolaeva-Fedorovich. Dokl. Akad. Nauk SSSR 115:751 (1957).
610. D. Grahame. J. Electrochem. Soc. 98:343 (1951).
611. J. Zezula. Chem. Listy 47:492 (1953).
612. P. Herasymenko and J. Slendyk. Z. Phys. Chem. 149A:123 (1930).
613. N. A. Izgaryshev and Kh. M. Ravikovich. ZhRFKhO, Ch. Khim. 62:255 (1930).
614. R. M. Vasenin and S. V. Gorbachev. Zh. Fiz. Khim. 28:2157 (1954).
615. M. Fedoreňko. Chem. Zveski 12:17 (1958).
616. J. Kůta. Coll. Czech. Chem. Comm. 22:1677 (1957).
617. E. Fischerová and O. Fischer. Chem. Zveski 14:743 (1960).
618. R. W. Schmid and C. N. Reilley. J. Am. Chem. Soc. 80:2102 (1958).

619.   V. I. Zykov and S. I. Zhdanov. Zh. Fiz. Khim. 32:791 (1958).

620.   Ly Shou-Zhun and S. I. Zhdanov. Zh. Fiz. Khim. 37:1750 (1963).

621.   H. Dehn and G. Schober. Monatsh. Chem. 93:1448 (1962).

622.   É. A. Maznichenko, B. B. Damaskin, and Z. A. Iofa. Dokl. Akad. Nauk SSSR 138:1377 (1961).

623.   B. B. Damaskin and N. V. Nikolaeva-Fedorovich. Zh. Fiz. Khim. 35:1279 (1961).

624.   L. Gierst, J. Tondeur, R. Cornelissen, and F. Lamy. Report of the 14th TsITTsE Convention, Moscow, 1963.

625.   L. G. Feoktistov and S. I. Zhdanov. Ibid.; Electrochim. Acta, Vol. 9 (1964).

626.   L. G. Feoktistov. Abstracts of the Papers of the 4th All-Union Conference on the Electrochemistry of Organic Compounds, Moscow, 1962. Moscow, Izd. Akad. Nauk SSSR (1962), p. 23.

627.   S. Sat'yanarayan and N. V. Nikolaeva-Fedorovich. Dokl. Akad. Nauk SSSR 141:1139 (1961).

628.   A. Aramata and P. Delahay. J. Phys. Chem. 66:2710 (1962).

629.   K. Asada, P. Delahay, and A. K. Sundarem. J. Am. Chem. Soc. 83:3396 (1961).

630.   A. N. Frumkin, O. A. Petry, and N. V. Nikolaeva-Fedorovich. Dokl. Akad. Nauk SSSR 137:896 (1961).

631.   M. E. Peover and J. D. Davies. J. Electroanal. Chem. 6:46 (1963).

632.   L. Holleck and D. Becher. J. Electroanal. Chem. 4:321 (1962).

633.   V. Cermack. Coll. Czech. Chem. Comm. 20:983 (1955).

634.   J. J. Tondeur, A. Dombret, and L. Gierst. J. Electroanal. Chem. 3:225 (1962).

635.   V. Volková. Advances in Polarography, Vol. 3. London, Pergamon Press (1960), p. 840.

636.   V. Volková. Nature 185(4715):743 (1960).

637.   J. Kůta and J. Weber. Report of the 14th TsITTsE Convention, Moscow, 1963; Electrochim. Acta 9:541 (1964).

638.   D. J. Pietrzyk and L. B. Rogers. Anal. Chem. 34:936 (1962).

639.   Z. Grabowski and W. Kemula. Polarographie in der Chemotherapie, Biochemie und Biologie (I. Jenaer Simposium, 1962). Berlin, Akademie-Verlag (1964), p. 377.

640.   A. Vincenz-Chodkowska and Z. R. Grabowski. Report of the 14th TsITTsE Convention, Moscow, 1963; Electrochim. Acta 9:789 (1964).

641.   V. D. Bezuglyi. Abstracts of Papers of the 8th Mendeleev Conference, Moscow, (1958), No. 13, p. 37; Abstracts of the Papers of the 1st All-Union Polarographic Meeting, Kishinev, 1959. Kishinev, Izd. "Shtiintsa" (1959).

642.   V. D. Bezuglyi. Analytical Methods for Chemical Reagents and Preparations, 3rd Edition. Moscow, Izd. IREA (1962), p. 11; Polarography in the Chemistry and Technology of Polymers. Khar'k. Gos. Univers. (1964), p. 9.

643.   M. Wien and J. Schiele. Physik. Z. 32:545 (1931).

644.   B. L. Timan. Dokl. Akad. Nauk SSSR 112:894 (1957); Zh. Fiz. Khim. 31:2143 (1957); 33:1189 (1959).

645.   G. Haijtink. Rec. Trav. Chim. 76:869, 887 (1957).

646.   N. S. Hush and J. W. Scarrott. J. Electroanal. Chem. 7:26 (1964).

647. E. Fornasari, G. Giacometti, and G. Riggati. Advances in Polarography, Vol. 3. London, Pergamon Press (1960), p. 895.

648. G. Riggati. Ibid., p. 904.

649. Ya. P. Stradyn' and S. A. Giller. Izv. Akad. Nauk Latv.SSR No.10, 121 (1958).

650. J. Page, J. Smith, and J. Waller. J. Phys. Coll. Chem. 53:545 (1948).

651. P. Stewart and W. Bonner. Anal. Chem. 22:793 (1950).

652. Ya. P. Stradyn'. The Polarography of Nitro-Compounds. Riga, Izd. Akad. Nauk Latv. SSR (1961).

653. B. Breyer, H. Bauer, and S. Hacobian. Australian J. Chem. 7:305 (1954).

654. H. Bauer and P. Elving. Australian J. Chem. 12:335 (1959).

655. B. Breyer and S. Hacobian. Australian J. Chem. 7: 225 (1954).

656. L. I. Antropov. Tr. Erevansk. politekhn. inst. im. K. Marksa 2:71, 97 (1946).

657. L. I. Antropov. Proceedings of the 2nd All-Union Conference on Theoretical and Applied Electrochemistry. Kiev (1949), p. 139.

658. L. I. Antropov. Zh. Fiz. Khim. 24:1428 (1950).

659. E. Vianello and E. Fornasari. Ric. Sci. 29:124 (1959).

660. W. H. Reinmuth. Anal. Chem. 33:322 (1961).

661. H. Matsuda and P. Delahay. Coll. Czech. Chem. Comm. 25:2977 (1960).

662. H. A. Laitinen and J. E. B. Randles. Trans. Faraday Soc. 51:54 (1955).

663. A. B. Ershler, G. A. Tedoradze, and S. G. Mairanovskii. Dokl. Akad. Nauk SSSR 145:1324 (1962).

664. L. W. Marple, L. E. I. Hummelstedt, and L. B. Rogers. J. Electrochem. Soc. 107(5):437 (1960).

665. W. Kemula and A. Cisak. Roczniki Chem. 31:337 (1957).

666. G. A. Tedoradze, A. B. Érshler, and S. G. Mairanovskii. Abstracts of the Papers of the 4th All-Union Conference on the Electrochemistry of Organic Compounds, Moscow, 1962. Izd. Akad. Nauk SSSR (1962); Izv. Akad. Nauk SSSR, Otd. Khim. Nauk No. 2, 235 (1963).

667. G. A. Tedoradze, A. B. Érshler, and S. G. Mairanovskii. Die Polarographie in der Chemotherapie, Biochemie und Biologie (I. Jenaer Simposium, 1962). Berlin, Akademie-Verlag (1964), p. 371.

668. S. G. Mairanovskii and A. N. Frumkin. Rev. Polarog. (Kyoto) 11(1-2):96 (1963).

669. S. G. Mairanovskii. Report of the 14th TsITTsE Convention, Moscow, 1963; Electrochim. Acta 9:803 (1964).

670. V. G. Levich, B. I. Khaikin, and E. D. Belokolos. Elektrokhimiya 1:1273 (1965).

671. S. G. Mairanovskii, N. V. Barashkova, and Yu. B. Vol'kenshtein. Izv. Akad. Nauk SSSR, Ser. Khim. No. 9, 1539 (1965).

672. J. Holubek and J. Volke. Coll. Czech. Chem. Comm. 27:680 (1962).

673. A. Ryvolová. Coll. Czech. Chem. Comm. 25: 420 (1959).

674. J. M. Kolthoff and D. J. Lehmicke. J. Am. Chem. Soc. 70:1879 (1948).

675. H. Berg, E. Bauer, and D. Tresselt. Advances in Polarography, Vol. 1. London, Pergamon Press (1960), p. 382.

676. J.-M. Savéant. Compt. Rend. 257:448 (1963).

677. J.-M. Savéant. Compt. Rend. 258:585 (1964).

678. L. Holleck and O. Lehmann. Monatsh. Chem. 92: 33 (1961).

679. L. Holleck and O. Lehmann. Ber. Bunsenges. Phys. Chem. 67:609 (1963).

680. Zh. Z. Brainina. Dokl. Akad. Nauk SSSR 130 :797 (1960).

681. A. G. Stromberg and Kh. Z. Brainina. Zh. Fiz. Khim. 35:2016 (1961).

682. O. Manoušek and P. Zuman. Coll. Czech. Chem. Comm. 29:1718 (1964).

683. Ya. P. Stradyn', S. A. Giller, and Yu. K. Yur'ev. Dokl. Akad. Nauk SSSR
      129: 816 (1959).

684. Ya.P.Stradyn'. Polarography of Nitro Compounds. Riga, Izd. Akad. Nauk
      Latv. SSR (1961), pp. 87, 91.

685. M. Suzuki and P. Elving. Coll. Czech. Chem. Comm. 25 : 3202 (1960).

686. P. J. Elving, J. M. Markowitz, and I. Rosenthal. J. Electrochem. Soc. 101:195
      (1954).

687. P. Federlin, F. Meziou, and S. Piekarski. Compt. Rend. 255:1349, 1394
      (1962).

688. A. Kirrman and P. Federlin. Bull. Soc. Chim. France 1958: 944.

689. P. J. Elving and J. M. Markowitz. J. Org. Chem. 26:18 (1960).

690. C. E. Bennett and P. J. Elving. Coll. Czech. Chem. Comm. 25 : 3213 (1960).

691. F. L. Lambert and K. Kobayshi. Chem. Age India 30: 949 (1958).

692. L. G. Feoktistov and S. I. Zhdanov. Izv. Akad. Nauk SSSR, Otd. Khim. Nauk
      No. 12, 2127 (1962).

693. L. G. Feoktistov. Dissertation. Moscow, Inst. Elektrokhimii Akad. Nauk SSSR
      (1963).

694. S. G. Mairanovskii. Abstracts of Papers Presented at the 5th All-Union Con-
      ference of Organic Electrochemistry, Moscow, 1965. Moscow, Izd. Akad.
      Nauk SSSR (1965).

695. S. G. Mairanovskii, N. V. Barashkova, and Yu. B. Vol'kenshtein. Elek-
      trokhimiya 1 : 72 (1965).

696. S. G. Mairanovskii and L. I. Lishcheta. Izv. Akad. Nauk SSSR, Otd. Khim.
      Nauk 1984 (1962).

697. S. G. Mairanovskii, E. D. Belokolos, B. I. Khaikin, V. P. Pul'tyan, and
      L. I. Lishcheta. Elektrokhimiya Vol. 2 (1966).

698. D. M. H. Kern. J. Am. Chem. Soc. 76:1011 (1954).

699. D. M. H. Kern. J. Am. Chem. Soc. 76:4234 (1954).

700. P. Kivalo. Acta Chem. Scand. 9:221 (1955).

701. J. Koryta and Z. Zábransky. Coll. Czech. Chem. Comm. 25 : 3153 (1960).

702. J. Biernat and J. Koryta. Coll. Czech. Chem. Comm. 25:38 (1960).

703. D. M. H. Kern. J. Am. Chem. Soc. 81:1563 (1959).

704. J. Koutecký and V. Hanuš. Coll. Czech. Chem. Comm. 20:124 (1955).

705. J.-M. Savéant and E. Vianello. Compt. Rend. 256:2597 (1963).

706. B. Emmert. Ber. 42:1507 (1909).

707. P. Tompkins. According to: S. Wawzonek. Anal. Chem. 21:64 (1949).

708. S. G. Mairanovskii. Dokl. Akad. Nauk SSSR 110: 593 (1956).

709. S. G. Mairanovskii. Proceedings of the Fourth All-Union Electrochemical Con-
      vention, Moscow, 1956. Moscow, Izd. Akad. Nauk SSSR (1959), p. 223.

710. M. Fournier. Compt. Rend. 232:1673 (1951).

711. G. Buchanan and R. Werner. Australian J. Chem. 7:239, 312 (1954).

712. I. Tachi. Bull. Chem. Soc. Japan 28:25 (1955).

713. E. Colichman and P. O'Donovan. J. Am. Chem. Soc. 76: 3588 (1954).
714. S. G. Mairanovskii. Izv. Akad. Nauk SSSR, Otd. Khim. Nauk, No. 12: 2140 (1961).
715. W. Kemula, Z. Grabowski, and M. Kalinowski. Naturwiss. 47 : 514 (1960).
716. R. Pasternak. Helv. Chim. Acta 31: 753 (1948).
717. V. N. Pavlov, Ya. M. Zolotovitskii, S. G. Mairanovskii, and G. A. Tedoradze. Elektrokhimiya 1: 427 (1965).
718. M. Suzuki and P. J. Elving. J. Phys. Chem. 65: 391 (1961).
719. D. Stočesova. Coll. Czech. Chem. Comm. 14: 615 (1949).
720. M. E. Vol'pin, S. I. Zhdanov, and D. N. Kursanov. Dokl. Akad. Nauk SSSR 112: 264 (1958).
721. S. I. Zhdanov and A. N. Frumkin. Dokl. Akad. Nauk SSSR 122: 412 (1958).
722. P. Zuman, J. Chodkowski, and F. Sanatavý. Coll. Czech. Chem. Comm. 26: 380 (1961).
723. V. F. Lavrushin, V. D. Bezuglyi, and G. G. Belous. Zh. Obshch. Khim. 33: 1711 (1963).
724. Ya. P. Stradyn'. Report of the Fourteenth TsITTsE Convention, Moscow, 1963.
725. T. I. Ivcher, E. N. Zil'berman, and E. M. Perepletchikova. Zh. Fiz. Khim. 39: 749 (1965).
726. R. Brdička. Arkiv Kemi, Mineral, Geol. 26B(19): 1 (1948).
727. F. Herles and A. Vančura. Bull. Intern. Acad. Sci. Bohême 33 : 119 (1932).
728. J. Heyrovský and J. Babička. Coll. Czech. Chem. Comm. 2: 370 (1930); Chem. News 141 : 369 (1930).
729. R. Brdička. Coll. Czech. Chem. Comm. 5: 112, 148 (1933).
730. J. Pech. Coll. Czech. Chem. Comm. 6: 190 (1934).
731. I. D. Ivanov. Polarography of Proteins. Moscow, Izd. Akad. Nauk SSSR (1961).
732. R. Brdička. Research 1 : 25 (1947).
733. H. F. W. Kirkpatrick. Quart. J. Pharm. Pharmac. 18: 245 (1945); 18: 338 (1945); 19: 8 (1946); 19: 127, 526 (1946); 20: 87 (1947).
734. A. G. Stromberg. Zh. Fiz. Khim. 20: 409 (1946).
735. E. Knobloch. Coll. Czech. Chem. Comm. 12: 407 (1947).
736. E. Knobloch. Coll. Czech. Chem. Comm. 25 : 3330 (1960).
737. J. Kŭta and I. Drabek. Coll. Czech. Chem. Comm. 20: 902 (1955).
738. M. Stackelberg and H. Fassbender. Z. Elektrochem. 62: 834 (1958).
739. M. Stackelberg, W. Hans, and W. Jensch. Z. Elektrochem. 62: 839 (1958).
740. M. Stackelberg and H. W. Nürnberg. Rev. Polarog. (Kyoto) 6: 27 (1958).
741. H. W. Nürnberg. Advances in Polarography, Vol. 2. London, Pergamon Press (1960), p. 694.
742. W. Lamprecht, S. Gudbjarnason, and H. Katzelmeier. Z. Physiol. Chem. 322: 52 (1960).
743. R. Brdička. Bull. Intern. Acad. Sci. Bohême 37: 122 (1936).
744. R. Brdička. Coll. Czech. Chem. Comm. 11 : 614 (1939).
745. J. Klumpar. Coll. Czech. Chem. Comm. 13: 11 (1948).
746. K. Kornfeld, M. Rink, G. van Riesenbeck, and M. Stackelberg. J. Electroanal. Chem. 6: 54 (1963).
747. S. G. Mairanovskii. Izv. Akad. Nauk SSSR, Otd. Khim. Nauk No. 4: 615 (1953).

748. S. G. Mairanovskii. Izv. Akad. Nauk SSSR, Otd. Khim. Nauk No. 5: 805 (1953).
749. S. G. Mairanovskii. Usp. Khim. 33: 75 (1964); J. Electroanal. Chem. 6: 77 (1963).
750. S. G. Mairanovskii, J. Koutecký, and V. Hanuš. Zh. Fiz. Khim. 37: 18 (1963).
751. S. G. Mairanovskii. Dokl. Akad. Nauk SSSR 114: 1272 (1957).
752. A. N. Frumkin and E. P. Andreeva. Dokl. Akad. Nauk SSSR 90: 417 (1953).
753. M. Finkelstein, R. C. Petersen, and S. D. Ross. J. Am. Chem. Soc. 81: 2361 (1959).
754. S. D. Ross, M. Finkelstein, and R. C. Petersen. J. Am. Chem. Soc. 82: 1582 (1960).
755. L. Robert, R. Goldstein, and J. Polonovski. Mikrochim. Acta No. 2: 318 (1959).
756. E. L. Colichman. Anal. Chem. 26: 1204 (1954).
757. E. L. Colichman and D. L. Love. J. Org. Chem. 18: 40 (1953).
758. S. G. Mairanovskii, N. V. Barashkova, and F. D. Alashev. Zh. Fiz. Khim. 35: 435 (1961).
759. V. Hanuš, S. G. Mairanovskii, and J. Koutecký. Zh. Fiz. Khim. 36: 2010 (1962).
760. J. Koutecký, V. Hanuš, and S. G. Mairanovskii. Zh. Fiz. Khim. 34: 651 (1960).
761. G. A. Tedoradze and S. G. Mairanovskii. Izv. Akad. Nauk SSSR, Otd. Khim. Nauk No. 3: 577 (1963).
762. Ya. M. Zolotovitskii and G. A. Tedoradze. Izv. Akad. Nauk SSSR, Ser. Khim. No. 12: 2133 (1964).
763. B. I. Khaikin, Ya. M. Zolotovitskii, and G. A. Tedoradze. Elektrokhimiya 1: 23 (1965).
764. Ya. M. Zolotovitskii, G. A. Tedoradze, and B. I. Khaikin. Elektrokhimiya 1: 130 (1965).
765. M. Kalousek. Coll. Czech. Chem. Comm. 13: 105 (1948).
766. A. A. Vlček. Coll. Czech. Chem. Comm. 20: 413 (1955).
767. C. J. Nyman, J. L. Ragle, and P. F. Linde. Anal. Chem. 32: 352 (1960).
768. B. C. Southworth, R. Osteryoung, K. D. Fleisher, and F. C. Nachod. Anal. Chem. 33: 208 (1961).
769. D. C. Grahame. J. Am. Chem. Soc. 63: 1207 (1941).
770. S. G. Mairanovskii. Izv. Akad. Nauk SSSR, Otd. Khim. Nauk No. 5, 784 (1962).
771. L. D. Klyukina and S. G. Mairanovskii. Izv. Akad. Nauk SSSR, Otd. Khim. Nauk No. 7: 1183 (1963).
772. M. K. Polievktov and S. G. Mairanovskii. Izv. Akad. Nauk SSSR, Ser. Khim. No. 3: 413 (1965).
773. S. G. Mairanovskii. Dokl. Akad. Nauk SSSR 142: 1327 (1962).
774. S. G. Mairanovskii and L. I. Lishcheta. Izv. Akad. Nauk SSSR, Otd. Khim Nauk No. 2: 227 (1962).
775. S. G. Mairanovskii, J. Koutecký, and V. Hanuš. Zh. Fiz. Khim. 36: 2621 (1962).
776. V. G. Levich, B. I. Khaikin, and S. G. Mairanovskii. Dokl. Akad. Nauk SSSR 145: 605 (1962).
777. S. G. Mairanovskii. Dokl. Akad. Nauk SSSR 120: 1294 (1958).
778. S. G. Mairanovskii. Dokl. Akad. Nauk SSSR 133: 162 (1960).
779. V. Vojíř. Coll. Czech. Chem. Comm. 18: 629 (1953).

780. V. Vojíř. Coll. Czech. Chem. Comm. 26:289 (1961).
781. H. Sunahara, D. N. Ward, and A. C. Griffin. J. Am. Chem. Soc. 82:6017 (1960).
782. M. Shinagawa and H. Nezu. Bull. Chem. Soc. Japan 33:272 (1960).
783. M. Shinagawa, H. Imaj, and H. Nezu. Bull. Chem. Soc. Japan 34:445 (1961).
784. A. N. Frumkin, D. I. Dzhaparidze, and G. A. Tedoradze. Dokl. Akad. Nauk SSSR 152:164 (1963).
785. M. K. Polievktov and S. G. Mairanovskii. Abstracts of Papers Presented at the Fourth Conference on Organic Electrochemistry, Moscow, 1962. Moscow, Izd. Akad. Nauk SSSR (1962).
786. S. G. Mairanovskii, L. D. Klyukina, and A. N. Frumkin. Dokl. Akad. Nauk SSSR 141:147 (1961).
787. S. I. Zhdanov and P. Zuman. Coll. Czech. Chem. Comm. 29:960 (1964).
788. S. G. Mairanovskii. Zh. Fiz. Khim. 37:451 (1963).
789. E. Pungor and Gy. Farsang. Acta Chim. Acad. Sci. Hung. 25:293 (1960); Ann. Univ. Sci. Budapest Lorando Eotveos Nominatae, Sect. Chim. 2:299 (1960).
790. E. Pungor and Gy. Farsang. Acta Chim. Acad. Sci. Hung. 27:175 (1961); J. Electroanal. Chem. 2:291 (1961).
791. V. Trkal. Coll. Czech. Chem. Comm. 21:945 (1956).
792. H. Sunahara, D. N. Ward, and A. C. Griffin. J. Am. Chem. Soc. 82:6023 (1960).
793. H. Sunahara. Rev. Polarog. (Kyoto) 9:165 (1961).
794. H. Sunahara. Rev. Polarog. (Kyoto) 9:222 (1961).
795. H. Sunahara. Rev. Polarog. (Kyoto) 9:233 (1961).
796. A. Basiński, M. Kuik, and J. Ceynowa. Roczniki Chem. 36:1889 (1962).
797. M. Březina. Advances in Polarography, Vol. 3. London, Pergamon Press (1960), p. 933.
798. V. Kalous. Ibid., p. 924.
799. M. Shinagawa, H. Nezu, H. Sunahara, F. Nakashima, H. Okashita, and T. Yamada. Ibid., p. 1142.
800. R. Brdička. Nature 139:330, 1020 (1937).
801. M. Březina and V. Gultjaj. Coll. Czech. Chem. Comm. 28:181 (1963).
802. V. Chmelař, M. Březina, and V. Kalous. Coll. Czech. Chem. Comm. 28:197 (1963).
803. B. Alexandrov, M. Březina, and V. Kalous. Coll. Czech. Chem. Comm. 28:210 (1963).
804. S. G. Mairanovskii and M. B. Neiman. Dokl. Akad. Nauk SSSR 87:805 (1952).
805. K. Schwabe and H. J. Bär. Rev. Polarog. (Kyoto) 11:117 (1963).
806. M. Březina. Coll. Czech. Chem. Comm. 24:4031 (1959).
807. V. Chmelár and J. Nosek. Coll. Czech. Chem. Comm. 24:3084 (1959).
808. J. Říha. Electrochim. Acta 6:75 (1962).
809. I. D. Ivanov. Dokl. Akad. Nauk SSSR 138:952 (1961).
810. M. Ito. J. Electrochem. Soc. Japan 27:78 (1959).
811. S. G. Mairanovskii and É. F. Mairanovskaya. Izv. Akad. Nauk SSSR, Ser. Khim. No. 5:947 (1964).
812. H. Berg. Physikalische Chemie biogener Makromoleküle (II Jenaer Simposium, 1963). Berlin. Akademie-Verlag (1964), p. 213.

813.   K. Marha. Coll. Czech. Chem. Comm. 22:153 (1957).

814.   A. Kočent. Naturwiss. 45:628 (1958).

815.   G. Gorin, J. E. Spessard, G. A. Wessler, and J. P. Oliver. J. Am. Chem. Soc. 81:3193 (1959).

816.   I. D. Ivanov and E. E. Rakhleeva. Polarographie in der Chemotherapie, Biochemie und Biologie (I Jenaer Simposium, 1962). Berlin, Akademie-Verlag (1964), p. 207.

817.   T. I. Shevchenko and V. I. Gorodyskii. Polarography in Medicine and Biology. Kiev, Goskhimidizdat USSR (1964).

818.   S. G. Mairanovskii and É. F. Mairanovskaya. Izv. Akad. Nauk SSSR, Otd. Khim. Nauk No. 5:937 (1963).

819.   S. G. Mairanovskii. Zh. Fiz. Khim. 36:389 (1962).

820.   V. Kalous and Z. Pavliček. Clin. Chim. Acta 8:170 (1963).

821.   S. G. Mairanovskii. Elektrokhimiya 1(10):1263 (1965).

822.   M. Strier and J. Cavagnol. J. Am. Chem. Soc. 80:1565 (1958).

823.   H. C. Brown and R. R. Holmes. J. Am. Chem. Soc. 77:1727 (1955); H. C. Brown, D. Gintis, and L. Domash. J. Am. Chem. Soc. 78:5387 (1956).

824.   V. A. Pal'm. Usp. Khim. 30:1069 (1961).

825.   P. Zuman. Coll. Czech. Chem. Comm. 25:3225 (1960).

826.   P. Zuman and M. Kuik. Naturwiss. 45:541 (1958).

827.   D. B. Stevančevič. Bull. Inst. Nucl. Sci. 9(167):57 (1959).

828.   V. F. Toropova and G. L. Elizarova. Material of the Second Polarographic Meeting. Kazan', Izd. Kazanskogo Univers. (1962), p. 134.

829.   V. F. Toropova and G. L. Elizarova. Ibid., p. 133.

820.   V. F. Toropova and G. L. Elizarova. Zh. Analit. Khim. 27:282 (1962).

831.   E. Pungo and E. Rokosinyi-Hollós. Acta Chim. Acad. Sci. Hung. 22:69 (1948).

832.   P. Zuman. Chem. Listy 52:1349 (1958).

833.   H. Matschiner and K. Issleib. Z. Anorg. Allgem. Chem. 340 (1965).

834.   N. Ya. Khlopin. Zh. Obshch. Khim. 18:364, 1019 (1948); N. Ya. Khlopin, N. A. Rafalovich, and G. P. Aksenova. Zh. Analit. Khim. 3:16 (1948).

835.   N. Ya. Khlopin, A. N. Rafalovich, and G. P. Aksenova. Zh. Obshch. Khim. 18:1009 (1948).

836.   As. Trifonov and N. Elenkova. Izv. Khim. Inst. Bulgarsk. Akad. Nauk 4:35 (1956).

837.   E. Knobloch. Advances in Polarography, Vol. 3. London, Pergamon Press (1960), p. 875.

838.   S. I. Zhdanov and M. K. Polievktov. Zh. Obshch. Khim. 31:3870 (1961).

839.   S. I. Zhdanov and A. A. Pozdeeva. Elektrokhimiya, Vol. 2 (1966).

840.   Z. Ostrowski and H. Fischer. Electrochim. Acta 8:37 (1963).

841.   S. G. Mairanovskii. Proceedings of the Third International Polarographic Conference in Southampton, 1964.

842.   K. Schwabe. Advances in Polarography, Vol. 3. London, Pergamon Press (1960), p. 911.

843.   K. Schwabe. Progress in Polarography, Vol. 1. New York, Interscience Publishers, Inc. (1962), p. 333.

844.   A. L. Markham and Ya. I. Tur'yan. Zh. Obshch. Khim. 22:1715 (1952).

845.   Ya. I. Tur'yan and M. A. Fishman. Zh. Obshch. Khim. 29:781 (1955).

846. Ya. I. Tur'yan and Yu. S. Milyavskii. Zh. Neorg. Khim. 5:2242 (1960).
847. I. Tachi and R. Takahashi. Coll. Czech. Chem. Comm. 25:3111 (1960).
848. R. Takahashi. Rev. Polarogr. (Kyoto) 9:116 (1961); 11:190 (1963).
849. A. M. Zan'ko and F. A. Manusova. Zh. Obshch. Khim. 10:1171 (1940).
850. I. Zlolowski and I. M. Kolthoff. J. Am. Chem. Soc. 64:1297 (1942); 66:1431 (1944).
851. N. A. Izmailov and V. D. Bezuglyi. Trudy Komissii po Analit. Khim. Akad. Nauk SSSR 4(7):29 (1952).
852. J. M. Hale and R. Parsons. Advances in Polarography, Vol. 3. London, Pergamon Press (1960), p. 829.
853. V. A. Pleskov. Usp. Khim. 16:254 (1947); Zh. Fiz. Khim. 22:.351 (1948).
854. H. Strehlow. Z. Elektrochem. 56:827 (1952).
855. H. Koepp, H. Wendt, and H. Strehlow. Z. Elektrochem. 64:483 (1960).
856. I. V. Nelson and R. T. Iwamoto. Anal. Chem. 33:1795 (1961).
857. I. M. Kolthoff. J. Polarog. Soc. 10(2):22 (1964).
858. A. I. Brodskii. Physical Chemistry, Vol. 2. ONTI (1948).
859. A. Cisak. Roczniki Chem. 36:1895 (1962).
860. K. Schwabe. Z. Phys. Chim. (Leipzig), Sonderheft, 293 (1958).
861. K. Schwabe and H. Jehring. Z. Anal. Chem. 173:36 (1960).
862. S. G. Mairanovskii and A. D. Filonova. Elektrokhimiya, Vol. 3 (1967).
863. N. A. Izmailov. The Electrochemistry of Solutions. Izd. Kharkovsk. Univers. Kharkov (1959).
864. K. Schwabe and S. Ziegenbalg. Z. Elektrochem. 62:172 (1958); K. Schwabe and M. Kunz. Z. Elektrochem. 64:1188 (1960).
865. A. N. Frumkin. Z. Phys. Chem. 103:43 (1923).
866. R. A. Maizlish, I. P. Tverdovskoi, and A. N. Frumkin. Zh. Fiz. Khim. 28:87 (1954).
867. S. G. Mairanovskii and A. D. Filonova. Elektrokhimiya, Vol. 3 (1967).
868. V. P. Gul'tyai and S. G. Mairanovskii. Elektrokhimiya, Vol. 2 (1966).
869. V. P. Gul'tyai and S. G. Mairanovskii. Elektrokhimiya 1:1295 (1965).
870. Ya. P. Stradyn', G. O. Reikhmanis, and R. A. Gavar. Elektrokhimiya 1:955 (1965).
871. E. Coufalik and F. Santavý. Coll. Czech. Chem. Comm. 19:457 (1954).
872. H. Wagner and H. Berg. J. Electroanal. Chem. 1:61 (1959/1960).
873. G. Horn and P. Zuman. Coll. Czech. Chem. Comm. 25:3401 (1960).
874. S. Vavra and N. P. Rudenko. Vestn. MGU, Khimiya No. 3:51 (1962).
875. H. J. Gardner. Chem. Ind. (London) 29:819 (1951).
876. O. D. Shreve and E. C. Markham. J. Am. Chem. Soc. 71:2993 (1949).
877. L. E. I. Hummelstedt and L. B. Rogers. J. Electrochem. Soc. 106:248 (1959).
878. S. G. Mairanovskii and V. N. Pavlov. Polarography and Kinetics of Chemical Reactions. Moscow, Izd. IREA (1964), p. 10.
879. C. Cisak and P. J. Elving. Rev. Polarog. (Kyoto) 11(1-2):21 (1963).
880. R. H. Philip, Jr., R. L. Flurry, and R. A. Day, Jr. J. Electrochem. Soc. 111:328 (1964).
881. M. E. Peover. J. Chem. Soc. 4540 (1962).
882. V. D. Bezuglyi and R. M. Rapota. Zh. Fiz. Khim. 38:2182 (1964).
883. A. N. Frumkin and F. J. Servis. Zh. Fiz. Khim. 1:52 (1930).

884.   L. I. Antropov. Report of the Fourth All-Union Electrochemical Meeting, Moscow, 1956. Izd. Akad. Nauk SSSR, Moscow (1959).

885.   M. G. Smirnova, V. A. Smirnov, and L. I. Antropov. Trudy Novocherkasskogo politekh. inst. 79:43 (1959).

886.   S. G. Mairanovskii. Elektrokhimiya, Vol. 3 (1967).

Index